Proton Exchange Membrane Fuel Cells

Scrivener Publishing
100 Cummings Center, Suite 541J
Beverly, MA 01915-6106

Publishers at Scrivener
Martin Scrivener (martin@scrivenerpublishing.com)
Phillip Carmical (pcarmical@scrivenerpublishing.com)

Proton Exchange Membrane Fuel Cells

Electrochemical Methods and Computational Fluid Dynamics

Edited by
Inamuddin
Omid Moradi
and
Mohd Imran Ahamed

This edition first published 2023 by John Wiley & Sons, Inc., 111 River Street, Hoboken, NJ 07030, USA and Scrivener Publishing LLC, 100 Cummings Center, Suite 541J, Beverly, MA 01915, USA

For more information about Scrivener publications please visit www.scrivenerpublishing.com.

Wiley Global Headquarters
111 River Street, Hoboken, NJ 07030, USA

For details of our global editorial offices, customer services, and more information about Wiley products visit us at www.wiley.com.

Library of Congress Cataloging-in-Publication Data

ISBN 9781119829331

Front cover images supplied by Wikimedia Commons
Cover design by Russell Richardson

Set in size of 11pt and Minion Pro by Manila Typesetting Company, Makati, Philippines

Printed in the USA

10 9 8 7 6 5 4 3 2 1

Contents

Preface

Proton exchange membrane fuel cells (PEMFCs) are among the most awaited near-future stationary clean energy devices just next to rechargeable batteries. Despite the appreciable improvement in their cost and durability, which are the two major commercialization barriers, our ambition has not yet reached at par mainly due to the use of expensive metal-catalyst, less durable membrane, and poor insight into the ongoing phenomena inside the PEMFCs. Efforts are being made to optimize the use of precious metals of platinum as catalyst layers or find alternatives that can be durable for more than 5,000 h. The computational models are also being developed and studied to get an insight into the shortcomings and get a remedy. The announcement by various companies including Toyota for PEMFC-based cars by 2025 has accelerated the current research on PEMFCs. A breakthrough is urgently needed. The membranes, catalysts, polymer electrolytes, and, especially, the understanding of diffusion layers need thorough revision and improvement to achieve the target.

Proton Exchange Membrane Fuel Cells: Electrochemical Methods and Computational Fluid Dynamics is intended to present experimental and computational techniques that are used to understand the phenomena inside the PEMFCs. The latest potential materials for electrode and membrane application, which are currently under active research, are thoroughly covered. The PEMFC electrolyte needs to be solid as the portable application of hydrogen is too risky. The literature on polymer electrolytes has developed recently. Some chapters are devoted to understanding these electrolytes, their shortcomings, and potential improvements. Further, the successful computational fluid dynamics models are discussed to help the reader understand diffusion phenomena that cannot be understood analytically. This book should be useful for engineers, environmentalists, governmental policy planners, non-governmental organizations, faculty, researchers, students from academics, and laboratories that are linked directly or indirectly to PEMFCs, for most awaited near-future stationary clean energy devices. Based on thematic topics, the book edition contains the following 13 chapters:

Chapter 1 is intended to represent an overview of stationary and portable applications of PEMFCs. Hydrogen, alcohol, and microbial cells are presented for existing applications. Besides, the applications of micro fuel cells are included, the challenges and promising applications of each cell are also offered.
Chapter 2 discusses the use of graphene oxide as a proton exchange membrane material. The application of modified graphene oxide membranes in fuel cells to enhance proton conductivity is reported in detail. Methods to synthesize graphene oxide membranes and techniques to characterize are discussed in light of research experiments.
Chapter 3 discusses the recent trend of graphene and its nanocomposites in the modifications of membrane-based fuel cells. The understanding of graphene features, prospects of graphene nanocomposites as membranes for fuel cell technology, and their influence on the performances of PEMFCs are also presented.
Chapter 4 discusses the carbon nanotube–based membranes for PEMFCs. In addition, the role of carbon nanotube toward PEMFCs is discussed. Various reported research papers, are reviewed and their mechanism and efficiency are also discussed.
Chapter 5 presents diverse nanoadditives and polymer matrices utilized in nanocomposite proton exchange membrane preparation for fuel cells. Thermal stability, chemical stability, and water uptake capacity of membranes are discussed in detail. The major focus is given to the proton conductivity of nanocomposite membranes with a change in percentage composition of nanoadditives.
Chapter 6 discusses various research progress and achievements of inorganic-organic composite membranes for PEMFCs. The major focus has elucidated the characteristics, advantages, design principles, and the relationship between various inorganic fillers and the properties of composite proton exchange membranes. The future development direction has also prospected.
Chapter 7 overviews the present research studies being performed on thermoset composites for bipolar plate applications. The influence of different conductive fillers and parameters of manufacturing methods on the properties of thermoset-based composite bipolar plates is discussed in detail. This chapter also includes theories of electrical conductivity and the characterization of polymer composite-based bipolar plates.

Chapter 8 discusses various design strategies of metal-organic frameworks for the PEMFCs and their performance in different matrices. This chapter mainly focuses on the advantages, limitations, and future applicability of the existing designs of metal-organic frameworks described in the literature.

Chapter 9 focuses on research in proton exchange membrane material based on fluorinated polymers. The preparation process, synthetic protocols, and properties of the various fluorinated membrane materials to achieve the effective compatibility of fuel cells are discussed. Special attention is given regarding the synthesis of fluorinated poly(aryl ether sulfone) membranes cross-linked sulfonated oligomer as the potential membrane in the construction of PEMFCs.

Chapter 10 discusses various categories of PEMFC materials, and advantages, disadvantages, and performance of each proton exchange membrane preparation method are considered in detail. The definition of a fuel cell, classifications, historical background, and its applications are explained. The critical parameters to characterize them also are discussed.

Chapter 11 details the properties of the Nafion membrane, which is the most widely used and researched proton exchange membrane. Throughout this chapter, the more relevant structural and transport models are exhaustively discussed and connected with the operation improvement of fuel cells. In addition, relevant results from different studies and authors are summarized.

Chapter 12 discusses the basics of fuel cells and their types. In addition, their working principle and application are also discussed in detail. The main focus is given to communicate the PEMFCs and the type of solid polymer electrolyte membranes along with their applications.

Chapter 13 focuses on CFD simulation and mathematical modeling of transport phenomena in PEMFCs. The effects of flow field configuration and operating parameters on PEMFCs performance are discussed. Different software and codes like ANSYS Fluent, COMSOL, lattice Boltzmann, and OpenFOAM to simulate PEMFCs are reviewed.

Highlights:

- Introduces the readers and professionals with the details of the basics, working principles, and applications of fuel cells
- Explores the features and applications of graphene oxide as proton exchange membrane material
- Focuses on progress, applications, mechanisms, and efficacy of different types of PEMFCs

- Highlights of metal-organic frameworks for the PEMFCs and their performance
- CFD simulation and mathematical modeling of transport phenomena in PEMFCs

Inamuddin
Omid Moradi
Mohd Imran Ahamed

1

Stationary and Portable Applications of Proton Exchange Membrane Fuel Cells

Shahram Mehdipour-Ataei* **and Maryam Mohammadi**

Faculty of Polymer Science, Iran Polymer and Petrochemical Institute, Tehran, Iran

Abstract
Proton exchange membrane fuel cells (PEMFCs) have been extensively evaluated for transportation applications due to the advantages such as lightweight, fast start-up, and zero emission. Some commercial products are also now being used worldwide. Moreover, because of the increasing advancement of technology and the integration of human life with new electronic technologies as well as the Internet, there is a growing trend for alternative or auxiliary sources of power for battery systems and portable devices. In addition, the requisite sources of power in areas that are remote and suffer from energy shortages are the other challenge. PEMFCs are the future vision for powering stationary and portable resources from massive power plants to cell phones. This chapter presents a variety of stationary and portable applications of PEMFCs, including hydrogen, alcohol, microbial, and micro fuel cells. Each section presents applications, achievements, and challenges. Finally, the prospects for the development of these technologies as reliable and applicable sources in the real world are presented.

Keywords: PEM fuel cell, applications, portable, stationary, hydrogen fuel cell, methanol fuel cell, microbial fuel cell, micro fuel cell

1.1 Introduction

Because of the zero or very low emission, proton exchange membrane fuel cells (PEMFCs) are promising in transportation. The first commercial application of these technologies in transportation may be urban buses.

**Corresponding author*: s.mehdipour@ippi.ac.ir

Inamuddin, Omid Moradi and Mohd Imran Ahamed (eds.) Proton Exchange Membrane Fuel Cells: Electrochemical Methods and Computational Fluid Dynamics, (1–16) © 2023 Scrivener Publishing LLC

The Scania hybrid bus is an example of this technology. In addition, fuel cells are applicable in any energy-driving device. The power of less than 1 W to several megawatts can be supplied by this technology due to the modularity, static nature, and variety. These features make a fuel cell a substitute for conventional heat engines used for transportation and power generation. Fuel cells are also an integral part of future technologies for energy conversion and storage, along with electrolyzers, batteries, flow systems, and renewable energy technologies. Lack of global market, high capital, high cost of components, and durability are the limiting factors of the mass market. However, Toyota, Hyundai, Honda, and others have commercialized their own products. Thus, the widespread usage of this technology has been made possible bypassing fossil fuel–powered to fuel cell–powered vehicles. Moreover, the use of fuel cells in US space programs continues, and PEMFCs are also considered for this purpose [1–11].

Fuel cell applications can be considered into three groups: portable, stationary, and transportation.

Low-temperature fuel cells are fit for portable and emergency power due to the short heating time. Portable fuel cells perform in the power range of 5–500 watts. Some examples of portable applications of PEMFCs in real-world include portable power generators for light personal usage in camping, continuous power systems, portable power sources as a replacement for batteries in laptops, computers, cell phones, radios, cameras, military electronics, boats, scooters, toys, kits, home lighting, emergency lights, and chargers.

Fuel cells can also regareded for stationary power generation including in the residential, commercial, and industrial sectors. In addition, by using fuel cells that operate in the range of medium to high temperature, the use of excess generated heat increases the overall efficiency and offers useful power for heating domestic water and space. A static power range of 1–50 MW can be supplied by PEMFCs. In telecommunication applications as an example of small-scale stationary power, the power range is 1–100 kW. Some applications of PEMFCs for stationary power supply include emergency backup (EPS) or uninterruptible power supply (UPS) for telecommunication networks, airports, hospitals, and training centers; remote or local power supply for small villages, buildings, and military camps; micro fuel cells, combined heat and power generation (CHP), and power regulation systems, in which surplus electricity is stored to hydrogen by electrolysis of water and converted into electricity when needed.

Transportation applications of PEMFCs include diverse types of trucks, buses, automobiles, motorcycles, bicycles, golf vehicles, service vehicles, boats, submarines, aircraft, and locomotives [3, 11–18].

Fuel cell applications can be also classified on the basis of a special need or removing a problem. High reliable power (computer equipment, communication facilities, and call as well as data processing centers), emission reduction or elimination (vehicles, industrial facilities, airports, and areas with severe emission standards for greenhouses), limited access to the electricity grid (rural or remote areas), and the availability of biogas (waste treatment plants and conversion of waste gases into electricity and heat with slight environmental impact by fuel cells) are in this classification [19].

The leading countries in the development of fuel cells include United States, Germany, Japan, Canada, and South Korea [18].

1.2 Proton Exchange Membrane Fuel Cells

PEMFCs are the most common types of fuel cell technology that are the focus of studies. The high power density, fast start-up, low manufacturing cost, long lifetime, flexibility, and widespread use in portable devices, transportation, and stationary applications are the superior characteristics of these types of cells compared to other types of fuel cells. About 90% of research studies and developments in fuel cells are in the field of PEMFCs; low operating temperatures and, therefore, reduced heat loss, small size, and lightweight make them suitable for automotive and transportation applications. They are a good choice for powering buses and commercial hydrogen vehicles as well. Polymer electrolyte membrane fuel cells have also been developed as a suitable replacement for existing batteries.

One of the most well-known research centers of PEMFCs is Los Alamos National Laboratory (LANL), which has released valuable achievements. In addition, renewable energy laboratories all over the world are representing their new successes every day [11, 17, 20–24].

1.2.1 Stationary Applications

In the early 1990s, according to the attained results from the performance and cost of PEMFCs for transportation applications, these types of cells were considered stationary, albeit with limited heat output.

Polymer electrolyte membrane fuel cells are capable of producing power in the range of a few watts to hundreds of kilowatts. Thus, these cells are applicable in almost any application that requires local power, including backup, remote, and uninterruptible power supply. The stationary application of PEMFCs includes decentralized power generation at the scale of 50–250 kW or less than 10 kW. However, it is required to focus on the small power range of 1–5 kW in the UPS or auxiliary power unit systems to be used for medium or large appliances for stationary applications [14, 24–27].

The distributed power of PEMFCs is usable for stationary applications in a variety of locations. Some of the applications include the main power source for areas where there is no access by the grid, supplementary power supply that operates in parallel with the power grid, supplementary power supply in renewable energy systems like photovoltaic and wind turbines, and emergency generators to remove power grid faults. Initial ages of stationary PEMFCs were designed for the residential power supply to use the generated heat for domestic, which significantly increases efficiency. A 250-kW stationary unit developed by Ballard Generation that runs on natural gas; other types that work with propane, hydrogen, or anaerobic digestion gas have also been designed and fully established. Besides, prototype units have been successfully achieved in the United States, Japan, and Europe [11, 13, 14, 16, 24, 25, 28, 29].

The backup power market for banks, hospitals, and telecommunications, in which there is a need for reliable power sources to prevent unexpected power downfall that causes very high cost, has attracted a lot of attention. Nonetheless, the high cost of PEMFCs is still an obstacle limits the global usage of stationary applications. Nonetheless, several commercial units such as CHP GenSys™ Blue Plug Power, Ballard FCgen™ 1020 ACS, and Ballard FCgen 10 -1030V3 fuel cell systems have been established in several places [11, 12, 14, 15].

The CHP GenSys™ fuel cell system has been installed in New York State for domestic. Ballard Power System is also available for use on telecommunication tower sites in India and Denmark to provide backup power. Besides, a model project was carried out in 2008 in Japan for the installation of ENE FARM class residential fuel cell. Small stationary units with a power of less than 10 kW were also installed for domestic usage, uninterruptible power supply, and backup power in commercial and remote locations [11, 16, 28].

Among the numerous successful projects, the Ballard Generation system is the largest plant to date. The output power of this plant is 250 kW. This PEMFC system is powered by natural gas and can be efficiently used as a backup source for emergency facilities such as hospitals. Moreover, these units produce large quantities of excess heat and hot water. These products are commonly used by the surrounding society and thus improve the efficiency [7].

The infrared effects of fuel cells like in submarines are very attractive for military applications. Many prototypes of this type have been successfully tested recently [7, 30].

The production of ammonia and direct regenerated iron are other stationary applications of fuel cell technology. By direct regenerated iron fuel cells, global production has grown significantly, which supply the present demand of hydrogen in refineries [7].

1.2.2 Portable Applications

Because of the limited capacity of batteries and the growing demand for energy for modern portable devices including laptops, mobile phones, and military radios, PEMFC systems have the potential to be supplemented or replaced with batteries. Thus, PEMFCs will serve as a vision for portable power supply. Portable systems working by PEMFC get into two main applications based on the generated power. The cells with below 100 W are applicable as substitute for battery, and the ones with more than 1 kW are suitable for portable generators [7, 11, 14, 31].

In hydrogen PEMFCs, providing safe hydrogen is a challenge. To this end, hydrogen is used as a liquid fuel for portable applications. On the other hand, compressed hydrogen is a practical answer for transportation, but it is not fit for portable devices due to the low volume energy density and storage space. In addition, carrying hydrogen is not possible. The replacement of liquid or solid fuels is a suggested solution; however, it requires a fuel refining process, which complicates systems. Direct methanol fuel cell (DMFC) is promising for portable applications. Nevertheless, the practical application of DMFCs is prevented by the low performance and high methanol crossover. Direct borohydride fuel cell is another potential system for portable applications. However, the high cost of borohydride and the purification of by-products while performing are the obstacles. Most importantly, the rapid development and success of lithium-ion (Li-ion) battery technology in powering laptops, cell phones, etc., is the remaining challenge that must be considered [26, 27, 31].

A number of large companies including Sony, Toshiba, Motorola, LG, Samsung, SFC Smart Fuel Cell, Samsung DSI, Neah power systems, MTI micro, Horizon, Jadoo power systems, Viaspace/Direct Methanol Fuel Cell Corporation, and CMR fuel cells have their own research and development units in portable PEMFCs [11].

1.2.3 Hydrogen PEMFCs

The best option for high-energy long-term power sources is hydrogen PEMFC. In such applications, hydrogen in the form of a metal or chemical hydride is stored in a tank and oxygen is supplied from the air.

An advanced hydrogen portable power cell based on a 50-W PEMFC was developed in the United States. The system was intended for commercial and military applications, which was successfully used by the Marine Corps. Moreover, Allis-Chalmers developed golf carts with hydrogen fuel cells. Testing fuel cell stacks to generate power for welders and forklifts is also another activity of Allis-Chalmers [20, 26, 27, 30, 32].

However, as mentioned earlier, the challenge of hydrogen fuel for PEMFCs for mobile and portable applications is to provide safe hydrogen.

1.2.4 Alcohol PEMFCs

1.2.4.1 Direct Methanol Fuel Cell

Polymer electrolyte membrane fuel cells with methanol fuel were considered by projects in the United States, Japan, and Europe, especially in LANL. This type of fuel cell is a promising prospect for portable applications. Any small-scale device, which works by rechargeable batteries, can work by DMFC as well. However, DMFCs are not suitable candidates for high power output applications [25].

The rapid growth of mobile communications and Internet services has led to growing demand for portable devices with high power, long operation, small size, and lightweight. Fast CPUs, high-resolution displays, and wireless connections are examples that increase the demand for power supply.

The present power supply systems for portable devices are often rechargeable lithium batteries and nickel batteries. Lithium batteries, in particular, have a major market in portable devices such as laptops and cell phones. Nonetheless, because of the low energy density and short operating time, Li-ion batteries or any other rechargeable batteries cannot meet the high power and long-lasting needs for portable devices. The DMFCs are future technology as portable power supplies. In other words, the higher energy density and grid independence of DMFCs compared to Li-ion battery systems have turned these systems into devices to complement or replace these batteries. However, many problems, including high price and methanol crossover, must be addressed for DMFCs to economically complement or replace Li-ion batteries. A research study compared the total cost of a

DMFC to Li-ion batteries for a 20-W laptop computer with a lifespan of 3,000 h in Korea. The total cost was 140 and 362 $ for the battery and DMFC, respectively. The high fuel consumption arisen from the loss of methanol by crossover is the reason of high price of DMFC. Therefore, the necessity of competition of DMFCs with Li-ion batteries includes reducing methanol crossover and price to 10^{-9} times and less than 0.5 $/kg, respectively. Under these circumstances, if the cost of DMFC production drops to 6.30 $/watt, then the DMFC will be reasonable compared to the Li-ion battery. On top of that, the regulations recently prohibited the use of highly flammable methanol on airline flights that must be cared for [14, 26, 27].

Lawrence Livermore National Laboratory, Battelle, Casio, and Ultracell are some manufacturers of DMFCs. In addition, many companies of fuel cell technology such as Antig, DMFC Corp., DTI energy, Energy Visions Inc., INI Power, MTI Micro Fuel Cells, Neah Power, Plug Power, and Smart Fuel Cell as well as communications and electricity companies like Fujitsu, IBM, LG, Motorola, NTT, Sanyo, Samsung, Sony, and Toshiba have introduced a variety of DMFC prototypes as laptop power supplies. These diverse products of companies compete with each other in weight, dimensions, power output, and concentration of methanol. The total volume of the system is one of the most important commercialization factors of DMFC prototypes for laptops, which determines power density, energy density, and the concentration of methanol solution. To gain an effective integration of DMFC system with laptop computers for optimizing power output level and operating time, Toshiba and NEC have embedded the power supply on the below back of the laptop computer. In reverse, the laptop lens was selected as the location of power supply in the prototypes by Sanyo and Samsung Electronics. Despite the successes in introducing DMFC systems for laptops, commercialization has been delayed due to the methanol ban on airplanes and high production cost, which is at least 10 times higher than Li-ion batteries [26, 27, 33].

DMFC is the smallest fuel cell, which can be developed due to the possibility of direct injection of fuel, air-breathing, and low operating temperature. Therefore, this type of fuel cell is actually a good technology for portable generators. However, DMFC currently cannot meet the power generation needs of many real portable applications [14].

1.2.4.2 Direct Dimethyl Ether Fuel Cell

Significant efforts have been made to find alternative fuels for PEMFCs to power portable energy systems. Because there are safety and technical barriers about using hydrogen and methanol. Hydrogen PEMFCs have a high power density due to the simplicity of the hydrogen molecule and its ease of oxidation. However, hydrogen suffers from problems related to low storage density, production, and infrastructure. Becaue of the more simplicity of DMFC compared to hydrogen cell, it better fits for portable applications. Performance loss of DMFC compared to hydrogen PEMFC is accepted against the easy storage and density of methanol. However, fuel storage density and toxicity of methanol are DMFC drawbacks for portable applications. Methanol is toxic when ingested or inhaled excessively, disperses rapidly in groundwater, has a colorless flame, and is more corrosive than gasoline. In this way, direct dimethyl ether polymer electrolyte fuel cell (DMEFC) has been proposed for portable applications. Dimethyl ether is a potential fuel for direct oxidation cells due to the combination of the benefits of hydrogen (pump-free feed) and methanol (high energy density storage). DMEFC stacks are currently larger than DMFC equivalents; however, the absence of a pump and lower toxicity compared to methanol significantly reduces size, weight, and complexity of the system [34].

1.2.5 Microbial Fuel Cells

Bioelectrochemical fuel cell systems with polymer membranes are another type of fuel cell in which electrochemically active microorganisms catalyze reactions. Although much research is still needed, the emerging microbial electrochemical field has a good potential to compete with modern technologies. The main drawback of microbial fuel cells (MFCs) is the lack of sufficient energy production. The expensive electrode, membrane, and cathode catalyst are also other limitations.

In addition to electricity generation, microbial fuel cells have many under investigation applications in wastewater treatment, water desalination, removal and recovery of metals, production of hydrogen in electrolysis cells, remote biosensors, elimination of pollutants, production of chemicals and fuel, and recovery of solar energy.

The application of MFCs as energy sources of spacecraft has been reported that is a validation of the applicability of these devices for practical usage. The maximum power density of MFCs is 2–3 W/m^2 under a suitable buffer and the temperature of about 30°C [12, 14, 15, 35–39].

1.2.5.1 Electricity Generation

The primary application of MFC technology is bioelectricity generation. The reports of cell phone charging by generated power from a MFC stack is a validation to this application. Regarding the low power output and the expensive materials of MFCs in comparison with the low-cost of fossil fuels, the competition of MFCs with existing technologies is not possible. Given that low cost membranes or membrane-less systems is developed, the commercialization would be accessible. Some hybrid MFC-based technologies have emerged with a bright prospect for practical applications and scaleup. These applications are discussed in the following sections [19, 35, 37, 40].

1.2.5.2 Microbial Desalination Cells

The basic principle of desalination cells is to use the potential at the anode and cathode to perform on-site desalination. Microbial desalination cells have a third chamber in which the anion exchange membrane and cation exchange membrane are located between the anode and cathode chambers. Removing the membrane by using MFCs with one chamber reduces costs and internal ohmic resistance. However, preventing short circuits and the need to keep the electrodes close to each other is a challenge in membrane-less MFCs. Other separators with higher porosity ranges such as ceramics, soil, and sand are used to decrease the distances of electrodes and improve power density. In this case, the growth of the biofilm on the separator, which may affect the performance of the MFC, must be concerned. The design should be optimized in such a way that high power density along with inexpensive constituents and simplicity of the system is brought. Ieropoulos and coworkers proposed an applicable reactor setup, which met the requirement of low-cost materials [29, 38, 39].

1.2.5.3 Removal of Metals From Industrial Waste

MFCs can be used to remove metals from industrial waste. Basic metals such as copper, nickel, iron, zinc, cobalt, and lead are used in large quantities. Generally, four mechanisms including the recovery of metals by aerobic cathodes and biocathodes in the MFC or by an external energy source have been reported. Conversion of Cu^{2+} to metallic copper at the MFC cathode, which is coupled with microbial oxidation of organic matter and power generation, is an example. In addition, the removal and recovery of uranium from contaminated sediments with poised electrodes as an electron donor is used for microorganisms [39].

1.2.5.4 Wastewater Treatment

Wastewater treatment concurrent with power generation is one of the most basic applications of MFC. As stated, the MFC technique can also be used for water desalination, but wastewater treatment is considered as the most practical application of this system. The conventional activated sludge reactors can be replaced with MFCs as bioreactor units for electricity generation, biomass production, and chemical oxygen demand (COD) removal. MFCs as biological devices can treat wastewater to inorganic materials in natural environments. The recovery of useful products such as electricity and hydrogen is one of the advantages of wastewater treatment MFCs over other primary remediation processes. Various types of industrial and domestic wastewater, including agricultural, distillation plants, food, and dairy, have been studied as substrates. Removal of bioelectrochemical organics and removal and recovery of bioelectrochemical nutrients including nitrogen and phosphorus are the processes involved in this application. Although the treatment of high toxic effluents by MFCs in not attained completely, these technologies can still lessen the COD of the effluent sufficiently to see discharge regulations prior to release into the environment. In addition, wastewater, which is rich in carbohydrates, proteins, lipids, minerals, fatty acids, etc., is treated by MFCs.

Despite all these advances, because of the complex nature of wastewater, a fully optimized configuration of MFCs as a wastewater treatment or energy recovery device has not yet been achieved. Thus, further research is needed before organizing large plants. In recent years, the first reports of practical applications of this technology in the real world are emerging [35, 37–39].

In the cases in which power recovery is not the priority, series or parallel stack geometries, as well as single cells, have been reported for practical applications. However, few prototypes have been developed for scaleup worldwide. Tube MFCs by the University of Queensland and iMETland project at the University of Alcalá de Henares in Spain are some of the efforts for this application. However, commercial wastewater treatment reactors do not exist that arise from the high cost of electrodes and separators [38].

The application of MFCs in wastewater treatment systems has also been investigated in the Live Building Challenge. Although a variety of decentralized wastewater treatment technologies are available, MFCs will be regarded as a replacement to these technologies for application in Live Building Challenge. In addition, MFCs are suitable for extensive use in places without water and electricity infrastructures [41].

1.2.5.5 Microbial Solar Cells and Fuel Cells

Harvesting solar energy and synthesizing organics are conducted by microbial solar cells using photosynthetic. Microbial carbon adsorbent cells are similar to these types of cells. Because the cathode chamber is sprayed with CO_2, it is possible that photosynthetic microorganisms degrade this greenhouse gas to organic matters. The organic matter can be converted to ethanol, biofertilizer, hydrogen, and amino acids. Sedimentary microbial fuel cells in which *in situ* generation of energy form sustainable sources is done has a dissimilar concept of MFC [38].

1.2.5.6 Biosensors

In addition to energy harvesting, MFCs can be an ideal option for biosensors to detect contaminants in remote areas. Biosensors require low-cost energy sources to operate and minimum maintenance operations. Rivers, lakes, seas, sediments, and anywhere that periodic replacement of the battery is difficult are the options for this application of MFCs. These systems based on MFCs can operate for up to five years without any maintenance. Moreover, MFCs do not need converters and operate by highly sensitive biocatalysts that have fast reaction to environmental fluctuations [35, 37, 38].

The Benthic Unattended Generator can be regarded as the first practical implementation of the MFC for the supply of oceanographic instruments using organic matter in water sediments [36].

1.2.5.7 Biohydrogen Production

Biohydrogen production is another useful application of MFCs. In this type of MFC, the production of hydrogen is made by bacterial fermentation of the substrates, which results in protons and electrons in the attendance of electrodes [37].

The conventional two-chamber MFCs can be converted into microbial electrolysis cells (MEC) to produce hydrogen. A MEC the same as a MFC consists of two chambers that are separated by an ion exchange membrane: the anode and the cathode. The produced hydrogen by the MEC can be stored and used to generate electricity. Therefore, MFCs can be combined with the MEC for electricity generation to meet electrical demand [35].

1.2.6 Micro Fuel Cells

Micro fuel cells are of intense attention due to the need for an UPS in portable electronics. The rapid growth in demand for power systems has led to that batteries cannot meet the need due to long charging times and low power capacity. These two problems can be removed by using portable micro PEMFCs. These devices can be made in small dimensions without loss of efficiency. Several methods proposed by researchers for making portable micro fuel cells. The typical power for portable electronics is in the range of 5–50 W; however, the focus is on values less than 5 W for micro power applications. [11, 18, 20, 42, 43].

The goal of using micro fuel cell is to maximize power generation for miniature energy-harvesting devices. These devices are suitable for applications that require small energy sources. Three types of fuels can feed this type of fuel cell: pure hydrogen; pure hydrocarbons like methanol, ethanol, formic acid, and ethylene glycol; and hydrogen in modified hydrocarbons. DMFCs are one of the best candidates for micro fuel cell applications. These types of cells are suitable power sources for personal digital assistance, laptops, cell phones, and hybrid battery chargers with a maximum power of 1–50 W. Pure hydrogen feed is challenging due to the lack of suitable materials for small-scale storage. Safety issue related to hydrogen carrying in wireless electronics is the other problem [33, 38, 42].

Micro biofuel cells are also of the micro fuel cells which can be implemented in implantable medical devices and micro-sensors such as pacemakers and glucose sensors, artificial valve power supplies, and the operation of robots [33].

Some of the major companies in the field of small PEMFCs are Toshiba FCP, Plug Power, P21, Matsushita, IdaTech, Hydrogenics, Eneos Celltech, Ebara Ballard, ClearEdge, and Altergy [11].

1.3 Conclusion and Future Perspective

Because of the modularity, PEMFCs can be applied to any device that needs a power source. These applications can be classified into three categories: transportation, stationary, and portable.

According to the research studies, it can be said that hydrogen PEMFCs are preferred for transportation and stationary applications due to the need for high power output. On the other hand, safety requirements for carrying hydrogen in portable devices are the remaining challenge. Methanol PEMFCs serve as a vision for portable applications. However, the low power of these cells and the toxicity of methanol are some of the limitations. Another problem is the ban on carrying methanol fuel on aircraft. Thus, other alcohols including dimethyl ether and ethanol have been studied as fuel. However, there is still a need for research and troubleshooting. Microbial fuel cells with polymer membranes are preferred in combined/hybrid applications due to their low power output. In other words, in this type of cell, electricity generation is the second priority along with other applications including wastewater treatment, desalination, production of hydrogen, and acting as biosensors.

Micro fuel cells as a subcategory of all types of cells are interesting technologies for a variety of applications, especially for portable. A variety of hydrogen, methanol, and microbial cells can be scaled down for this purpose. So far, numerous prototypes of PEMFC systems have been tested and introduced. Some of the products, especially hydrogen PEMFCs for transportations have reached the commercial level. Nevertheless, there is still a need to reduce costs, improve performance, and durability to reach the global market and widespread usage. A report on the market of PEMFCs and the forecast in the time range of 2016–2026 is available by Mordor Intelligence [44].

References

1. Baker, R.W., Membrane technology, in: *Kirk-Othmer Encyclopedia of Chemical Technology*, 2000.
2. Barbir, F., *PEM fuel cells (Theory and Practice)*, pp. 27–51, Academic press, Cambridge, Massachusetts, 2012.
3. Carrette, L., Friedrich, K.A., Stimming, U., Fuel cells: Principles, types, fuels, and applications. *Chem. Phys. Chem.*, 1, 162–193, 2000.
4. Folkesson, A., Andersson, C., Alvfors, P., Alaküla, M., Overgaard, L., Real life testing of a hybrid PEM fuel cell bus. *J. Power Sources*, 118, 349–357, 2003.
5. Gencoglu, M.T. and Ural, Z., Design of a PEM fuel cell system for residential application. *I. J. Hydrog. Energy*, 34, 5242–5248, 2009.
6. Lokurlu, A., Grube, T., Höhlein, B., Stolten, D., Fuel cells for mobile and stationary applications—Cost analysis for combined heat and power stations on the basis of fuel cells. *I. J. Hydrog. Energy*, 28, 703–711, 2003.
7. Olabi, A., Wilberforce, T., Abdelkareem, M.A., Fuel cell application in the automotive industry and future perspective. *Energy*, 214, 118955, 2021.
8. Pistoia, G., Kalogirou, S., Storvick, T., *Renewable energy focus handbook*, Oxford: Elsevier, London, 2009.
9. Rosli, R., Sulong, A., Daud, W., Zulkifley, M., Husaini, T., Rosli, M., Majlan, E., Haque, M., A review of high-temperature proton exchange membrane fuel cell (HT-PEMFC) system. *I. J. Hydrog. Energy*, 42, 9293–9314 2017.
10. Sood, R., Cavaliere, S., Jones, D.J., Rozière, J., Electrospun nanofibre composite polymer electrolyte fuel cell and electrolysis membranes. *Nano Energy*, 26, 729–745, 2016.
11. Wang, Y., Chen, K.S., Mishler, J., Cho, S.C., Adroher, X.C., A review of polymer electrolyte membrane fuel cells: Technology, applications, and needs on fundamental research. *Appl. Energy*, 88, 981–1007, 2011.

12. Behera, P.R., Dash, R., Ali, S., Mohapatra, K.K., A review on fuel cell and its applications. *Int. J. Eng. Res. Technol.*, 562–565, 2014.
13. Elmer, T., Worall, M., Wu, S., Riffat, S.B., Fuel cell technology for domestic built environment applications: State of-the-art review. *Renew. Sustain. Energy Rev.*, 42, 913–931, 2015.
14. Giorgi, L. and Leccese, F., Fuel cells: Technologies and applications. *Fuel Cells*, 6, 1–20, 2013.
15. Kamarudin, S.K., Achmad, F., Daud, W.R.W., Overview on the application of direct methanol fuel cell (DMFC) for portable electronic devices. *Int. J. Hydrog. Energy*, 34, 6902–6916, 2009.
16. Olabi, A., Wilberforce, T., Sayed, E.T., Elsaid, K., Abdelkareem, M.A., Prospects of fuel cell combined heat and power systems. *Energies*, 13, 4104, 2020.
17. Oszcipok, M., Zedda, M., Hesselmann, J., Huppmann, M., Wodrich, M., Junghardt, M., Hebling, C., Portable proton exchange membrane fuel-cell systems for outdoor applications. *J. Power Sources*, 157, 666–673, 2006.
18. Sharaf, O.Z. and Orhan, M.F., An overview of fuel cell technology: Fundamentals and applications. *Renew. Sustain. Energy Revi.*, 32, 810–853, 2014.
19. Xu, L., Zhao, Y., Doherty, L., Hu, Y., Hao, X., The integrated processes for wastewater treatment based on the principle of microbial fuel cells: A review. *Crit. Rev. Environ. Sci. Technol.*, 46, 60–91, 2016.
20. Chu, D., Jiang, R., Gardner, K., Jacobs, R., Schmidt, J., Quakenbush, T., Stephens, J., Polymer electrolyte membrane fuel cells for communication applications. *J. Power sources*, 96, 174–178, 2001.
21. Cleghorn, S., Ren, X., Springer, T., Wilson, M., Zawodzinski, C., Zawodzinski, T., Gottesfeld, S., PEM fuel cells for transportation and stationary power generation applications. *Int. J. Hydrog. Energy*, 22, 1137–1144, 1997.
22. Kamcev, J. and Freeman, B.D., Charged polymer membranes for environmental/energy applications. *Annu. Rev. Chem. Biomolecul. Eng.*, 7, 111–133, 2016.
23. Shin, D.W., Guiver, M.D., Lee, Y.M., Hydrocarbon-based polymer electrolyte membranes: importance of morphology on ion transport and membrane stability. *Chem. Rev.*, 117, 4759–4805, 2017.
24. Wilson, M.S., Zawodzinski, C., Gottesfeld, S., Landgrebe, A.R., Stationary power applications for polymer electrolyte fuel cells, in: *Proceedings of 11th Annual Battery Conference on Applications and Advances*, IEEE, 107–112, 1996.
25. Acres, G.J., Recent advances in fuel cell technology and its applications. *J. Power Sources*, 100, 60–66, 2001.
26. Wee, J.-H., A feasibility study on direct methanol fuel cells for laptop computers based on a cost comparison with lithium-ion batteries. *J. Power Sources*, 173, 424–436, 2007.

27. Wee, J.-H., Applications of proton exchange membrane fuel cell systems. *Renew. Sustain. Energy Rev.*, 11, 1720–1738, 2007.
28. Okada, O. and Yokoyama, K., Development of polymer electrolyte fuel cell cogeneration systems for residential applications. *Fuel Cells*, 1, 72–77, 2001.
29. Rezk, H., Sayed, E.T., Al-Dhaifallah, M., Obaid, M., Abou Hashema, M., Abdelkareem, M.A., Olabi, A., Fuel cell as an effective energy storage in reverse osmosis desalination plant powered by photovoltaic system. *Energy*, 175, 423–433, 2019.
30. Li, G.C., Jian, Q.F., Sun, S.Y., Application Prospects of Proton Exchange Membrane Fuel Cells to Military Affairs [J]. *Acta Armamentarii*, 4, 2007.
31. Han, H.S., Kim, Y.H., Kim, S.Y., Karng, S.W., Hyun, J.M., Development of proton exchange membrane fuel cell system for portable refrigerator. *ECS Trans.*, 42, 149, 2012.
32. Kocha, S.S., *Polymer electrolyte membrane (PEM) fuel cells, automotive applications*, pp. 473–518, Springer, New York, NY, 2013.
33. Kundu, A., Jang, J., Gil, J., Jung, C., Lee, H., Kim, S.-H., Ku, B., Oh, Y., Micro-fuel cells—Current development and applications. *J. Power Sources*, 170, 67–78, 2007.
34. Mench, M., Chance, H., Wang, C., Direct dimethyl ether polymer electrolyte fuel cells for portable applications. *J. Electrochem. Soc*, 151, A144, 2003.
35. Kumar, R., Singh, L., Zularisam, A., *Microbial fuel cells: types and applications*, pp. 367–384, Springer, 2017.
36. Lefebvre, O., Uzabiaga, A., Chang, I.S., Kim, B.-H., Ng, H.Y., Microbial fuel cells for energy self-sufficient domestic wastewater treatment—A review and discussion from energetic consideration. *Appl. Microbiol. Biotechnol.*, 89, 259–270, 2011.
37. Nandy, A. and Kundu, P.P., Configurations of Microbial Fuel Cells, in: *Progress and Recent Trends in Microbial Fuel Cells*, 1st, Dutta, K. and Kundu, P.P. (Eds.), 2018.
38. Roy, S., Marzorati, S., Schievano, A., Pant, D., Elias, S., Abraham, M., Microbial fuel cells, in: *Encyclopedia of sustainable technologies*, 2017.
39. Sanz, S.T., *Merging microbial electrochemical systems with conventional reactor designs for treating wastewater*, Universidad de Alcalá, Madrid, Spain, 2016.
40. Stambouli, A.B. and Traversa, E., Fuel cells, an alternative to standard sources of energy. *Renew. Sustain. Energy Rev.*, 6, 295–304, 2002.
41. Lee, C.-Y., Liu, H., Han, S.-K., Application of microbial fuel cells to wastewater treatment systems used in the living building challenge. *J. Environ. Health Sci.*, 39, 474–481, 2013.
42. Heinzel, A., Hebling, C., Müller, M., Zedda, M., Müller, C., Fuel cells for low power applications. *J. Power Sources*, 105, 250–255, 2002.
43. Kakaç, S., Pramuanjaroenkij, A., Vasiliev, L., *Mini-micro fuel cells: Fundamentals and applications*, Springer Science & Business Media, Netherlands, 2008.

44. Polymer electrolyte membrane fuel cells (PEMFCS) market- growth, trends, covid-19 impact, and forecasts (2021-2026). https://www.mordorintelligence.com/industry-reports/global-polymer-electrolyte-membrane-pem-fuel-cells-market-industry.

2

Graphene-Based Membranes for Proton Exchange Membrane Fuel Cells

Beenish Saba

Food Agricultural and Biological Engineering, The Ohio State University, 590 Woody Hayes Drive, Columbus, Ohio, United States

Abstract

Fuel cell technology has been invented as an alternative energy resource. However, extensive research is needed for the development of its components. Proton exchange membrane (PEM) is one of the key components of this technology. Graphene oxide and polymer composites have been blended to develop composite PEM membranes. The composite membranes have superior thermal, electrical, and conductive properties to their counterparts. Their functionality at high temperatures > 120°C makes them an ideal choice for use in transportation, portable energy storage, and household devices. There are several solvent-based synthesis methods to develop GO composite membranes that are in use. However, further investigation to improve synthesis methods is needed. GO composite membranes have higher electrical, ionic, and proton conductivity while lower membrane crossover (40%–70% for methanol) and internal resistance as compared to Nafion 117. The characteristics of composite membranes can be investigated using spectroscopy techniques. The discussion will provide features of GO, applicability in fuel cells, synthesis methods, and its compatibility with different polymers to form composite membranes.

***Keywords*:** Proton exchange membrane, fuel cells, composite membranes, graphene oxide, direct methanol fuel cells, nanomaterials

Email: beenishsabaosu@gmail.com

Inamuddin, Omid Moradi and Mohd Imran Ahamed (eds.) Proton Exchange Membrane Fuel Cells: Electrochemical Methods and Computational Fluid Dynamics, (17–32)

2.1 Introduction

Proton exchange membranes (PEMs) have proven their applications in environmentally sustainable energy generation devices like fuel cells for portable, transportation, and other purposes [1]. Fuel cells are among the promising energy generation and storage electrochemical systems. A typical fuel cell is mostly the anodic chamber and the cathodic chamber separated by PEM. Chemical energy is converted into electrical energy, and it is the most environmentally friendly emission free system [2]. The fuel cell process involves water, heat, electricity, and no harmful toxic gasses [3]. Currently, the fuel cells are facing membrane crossover, high cost, and safety issues at high temperatures. Therefore, membrane optimization such as material selection, synthesis process, and structural and morphological studies can reduce these issues.

Proton membrane efficiency and performance mainly depend on the material properties. Carbon is one of the best materials because of its outstanding environmental safety, thermal stability, electrical conductivity, and its performance in different chemical natures of media. These abovementioned properties facilitate its applicability in fuel cells [4]. Graphene has been recently used in electrochemical energy devices extensively. The attributes of graphene since its discovery and isolation in 2004 have captured the attention of many researchers [5]. Proton exchange membrane (PEMs) are the main concern in fuel cells as the other parts of fuel cell are designed according to the properties of PEM [6].

Nafion is the most common membrane used in proton exchange membrane fuel cells (PEMFCs), but it is expensive and recent studies have suggested that, if Nafion is modified with GO composites, then its conductivity can be improved significantly [7–9]. In addition, Nafion membranes are not stable at temperatures higher than 80°C its degradation at higher temperatures raised safety concerns in vehicles [10]. New polymeric low cost membranes are solutions to Nafion problems. Smitha *et al.* [10] proposed that there are 15 alternatives of Nafion membrane, and functionalization of composite membranes can increase proton conductivity, thermal stability as compared to Nafion. GO nanosheets have been functionalized (FGO) to get desired properties in PEM. FGO can optimize the structure of the membrane and have better water retention capacity, which is essential for improving proton conductivity [11, 12]. The sections below comprehensively discuss the features of GO membranes and their utilization in electrochemical fuel cells (EFCs).

2.2 Membranes

In a microbial fuel cell (MFCs), the anode and the cathode chamber must be separated by a membrane, which facilitates transfer of protons between the two chambers and prevent the transport of oxygen [13]. Membrane acts as a barrier between the aerated cathode and anaerobic anode chamber [14]. In most cases, the membranes are made of polymers [15] but, regardless of the material, they should have following characteristics:

1. Low internal resistance
2. High proton conductivity
3. Long term stability and low biofouling
4. Chemical and electrochemical stability
5. Low in cost

These key features make the membrane a proper choice for MFC [16, 17]. Ion exchange membranes (IEMs) have modifications using functional groups in the structure that allows transfer of cations or anions. IEMs can transfer counter ions while blocking the co-ions [18]. Cation exchange membranes also called PEMs have negative functional groups in their structure and they allow to transfer of cations and protons [19]. Hasani-Sadrabadi *et al.* [20] modified sulfonated groups into poly(ether ether ketone) by increasing the negative charge on membrane and then higher proton conductivity was achieved. In addition to PEMs, anion exchange membranes, mosaic IEMs, bipolar IEMs, and amphoteric IEMs are other examples of membranes used in MFCs [21, 22].

Quality and conductivity of membrane depend on its functional characteristics, and it is characterized for the functionality, thermal stability, structure analysis, and properties by following spectroscopy techniques: 1) Raman spectroscopy, 2) atomic force spectroscopy, 3) transmission electron microscopy, 4) scanning electron microscopy (SEM), 5) X-ray photoelectron spectroscopy; whereas other properties include anti-biofouling, water uptake, ion exchange capacity, and proton conductivity [23]. Composite membrane synthesis has recently gained popularity. They are lower in cost and possess good in ionic conductivity attributes.

2.3 Graphene: A Proton Exchange Membrane

In membrane energy applications, addition of inorganic filler like carbon nanotubes and graphene oxide (GO) has been in high demand. Graphene has contributed to optimizing of fuel cell performance when used in membranes. Graphene was not initially suitable for use in membranes as it was highly conductive electrically, besides mechanical strength and proton exchange qualities but GO is an oxidized form of graphene, and it has insulating properties besides mechanical strength and proton conduction. Its carbon to oxygen ratio is 2:1 [24]. GO incorporation in PEM significantly increases proton exchange capacity [2].

Graphene is a soft material of amphiphilic nature, and, if incorporated in membranes, then it enhances the surface area. It interacts with intermolecular hydrogen bonds due to the presence of epoxy, hydroxylic acid, and carboxylic acid functional groups [25, 26]. GO is a modified material, and it has oxygen as the functional group, which helps in the transfer of proton through channels and hold the water. Properties of GO include mechanical strength, gas impermeability, high surface area, electrical insulation, and hydrophilicity [27, 28]. GO is an excellent composite material with Nafion polymer. Ansari *et al.* [7] modified GO with Nafion polymer and observed enhanced ionic conductivity, whereas Kumar *et al.* (2012) [29] used it as PEM and maximum power density of 415 mW/cm^{-2} was observed at 30°C. Kim *et al.* [8] added GO coupled with phosphotungstic acid in Nafion polymer and power density jumped to 841 mW/cm^{-2}. GO has been modified with Nafion chains to obtain nanohybrids [30], and it has improved interfacial compatibility and proton conductivity to 1.6 times. Zarrin *et al.* [1] improved GO-Nafion composite membrane with functionalized groups, and its IEC was increased from 0.91 to 0.96 meq/g.

Nafion-GO composite was further mixed with other polymers to increase proton conductivity at high temperatures. Mishra *et al.* [31] mixed Nafion-GO with sulfonated polyether ether Keton (SPEEK), and a high proton conductivity 322.2 mS/m with 621.2 mW/cm^{-2} power density at 90°C was observed. Yang *et al.* [32] developed GO-polybenzimidazole (PBI) mixed composite high-temperature membranes. These continuous changes in the development of GO composite membranes have made them highly conductive and temperature-tolerant and have a significant contributions in improving the functionalization flexibility of GO.

2.4 Synthesis of GO Composite Membranes

GO composite membranes have been developed by a variety of researchers using different methods. The solution casting method is one of the commonly used methods. Cao *et al.* [33] synthesized poly(ethylene oxide) GO PEO/GO composite membrane using this method. At first, GO solution was prepared using sulfuric acid and potassium permanganate [24]. Distilled water was used as a solvent and 5 g of PEO was dissolved including 2 ml of GO solution. The mixture of PEO/GO was poured on glass slides and dried at room temperature for 48 and 12 h before characterization. There are different solution casting methods that have been developed depending on the type of solvents. Yang *et al.* [34] used isopropyle alcohol-water mixture; Wang *et al.* [26] used N,N-dimethylacetamide (DMAc) solvent; Sharma *et al.* [35] used deionized water; Lee *et al.* [36] used deionized water and isopropyl alcohol mixture. Liu *et al.* [37] used acetic acid and water mixture; He *et al.* [38] used dimethyl formamide; and Lim *et al.* [39] used dimethyl sulfoxide. Another commonly used method is Hummer method. It includes mixing of graphite flakes with H_2SO_4 and $NaNO_3$ with H_2O_2 and $KMnO_4$. The GO composite membranes have quickest electron conductivity and largest surface area. Formation of composite membranes enhances the thermal and mechanical stability of the GO membrane. Composite matrix helps to reduce methanol permeability and improves proton conductivity. However, more research is in progress to reveal more features of GO capabilities.

2.5 Graphene Oxide in Fuel Cells

High ionic conductivity and low membrane crossover make it a good choice for fuel cells. Different electrochemical and biochemical fuel cells

Table 2.1 Graphene-based membranes in fuel cells.

No.		Membrane	Contribution	Reference
1	Microbial fuel cell	Graphene oxide, poly(vinyl alcohol), and silicotungstic acid membrane	1.9 W/m^3 power density, inexpensive replacement of Nafion 117	[40]
2	Direct methanol fuel cell	Graphene and poly(vinyl alcohol)	55% reduction in methanol crossover, 148% increase in power density, and 73% improvement in tensile strength	[41]

(Continued)

Table 2.1 Graphene-based membranes in fuel cells. (*Continued*)

No.		Membrane	Contribution	Reference
3	Hydrogen fuel cell	Functionalized GO-nafion composite	Four times higher proton, ionic conductivity, and higher water uptake at 120°C	[1]
4	Alkaline fuel cell	PBI/ionic liquid functionalized/ GO	Enhanced thermal stability and high ionic conductivity 10^{-2} Scm^{-1}	[42]

have used GO composite membrane as a replacement to expensive Nafion membrane. Few studies are summarized in Table 2.1.

2.5.1 Electrochemical Fuel Cells

2.5.1.1 Hydrogen Oxide Polymer Electrolyte Membrane Fuel Cells

PEM fuel cells convert the chemical energy of hydrogen directly into electrical energy-producing water as a byproduct. PEMFCs have reduced energy use, pollution emissions, and fossil fuel dependence [43]. Hydrogen PEMFCs (Figure 2.1) consist of two electrodes, and a PEM between them.

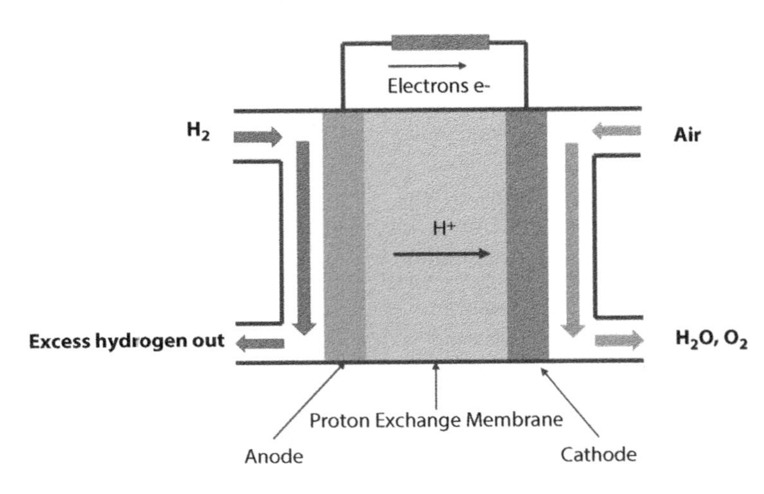

Figure 2.1 Hydrogen oxide fuel cell (adopted from Xu *et al.* [45]).

The hydrogen at the anode releases its electrons and oxidizes into positive ion (H^+), whereas oxygen takes electrons from the cathode and reduces to negative ion (O_2^- or OH^-), and it combines to form water (H_2O) while electron flows in the external circuit as electric current [44]. A variety of composite PEM membranes in PEMFCs have been tested for efficiency improvement. GO is one of the conductive filler materials, and it has excellent compatibility and advantages with Nafion polymer. Xu *et al.* [45] used PBI/GO and PBI/SGO (sulfonated GO) composite membrane with loadings of H_3PO_4 in PEMFCs and obtained a maximum power density of 600 mW/cm^{-2} at 175°C with PBI/SGO. The stability of fuel cells with high power density production at high temperatures is a significant achievement. Zarrin *et al.* [1] used 3-mercaptopropyl trimethoxysilane (MPTMS) as a precursor of sulfonic group and developed functionalized GO/Nafion membranes for PEMFCs. The composite membrane performs four times higher as compared to the Nafion membrane at 120°C, 25% humidity, and 420 mW/cm^{-2} power density. The composite membrane was stable and had shown high performance at low humidity and high temperatures. The use of GO composite membranes has increased power densities of hydrogen fuel cells performance, and its stability at higher temperatures made it useful for commercial utilization in fuel cell technology.

2.5.1.2 Direct Methanol Fuel Cells

Unique 2D structure, the flexibility of functionalization as compared to other materials, and stability at high temperatures make GO PEM an ideal membrane material to develop composite membranes for direct methanol fuel cells (DMFCs). Hydrogen fuel cells and DMFCs are similar in design and operation except for the anolyte. In DMFCs (Figure 2.2), hydrogen gas is replaced with methanol in liquid form. The methanol directly interacts with the electrode and releases its electrons. Liquid handling is much easier as compared to gas handling, and it has wide applications in portable power supplies [46]. Nafion as PEM has been used extensively in studies, but methanol crossover was the main limitation. Composite PEM of Nafion with GO has solved this issue and improved proton conductivity while short-circuiting the cell chemically. Lin *et al.* [47] used laminated GO Nafion 115 membrane in DMFCs with high methanol concentration and its methanol permeability was 70% less as compared to Nafion 115. The results are promising in terms of power density (55 mW/cm^{-2}), ionic conductivity (0.99 meq/g), and proton conductivity (2.35×10^{-2} S/cm) with a high methanol concentration (8 M). Choi *et al.* [3] synthesize GO/Nafion 112 membranes for DMFCs. The results revealed the composite membranes

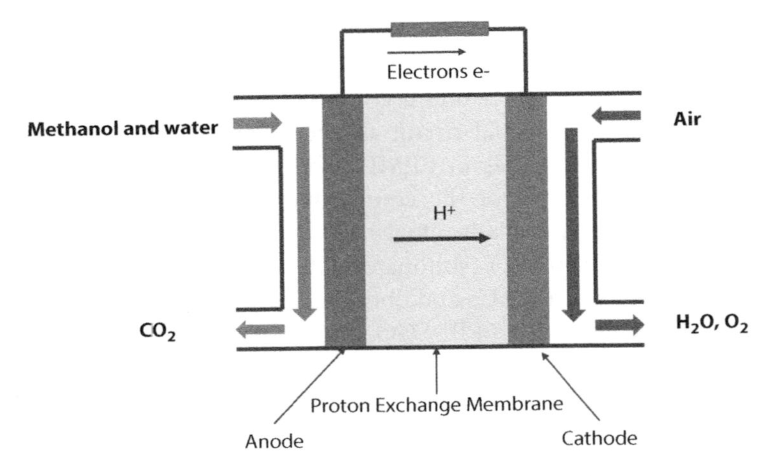

Figure 2.2 Direct methanol fuel cell (adopted from Lin *et al.* [47]).

have 98% higher ionic conductivity and 40% less methanol crossover as compared to Nafion 112 membrane. The power density observed was 114 mW/cm^{-2} at 70°C with methanol concentration 5 M. The facile fabrication of the composite membranes for DMFCs has increased power output and lower methanol crossover. Understanding the correlation between material properties and structure has paved the way to develop high ionic conductivity membranes with improved tensile strength and Young modulus. These properties will increase the longevity of the membranes.

2.5.2 Bioelectrochemical Fuel Cells

MFCs are one of the power-generating bioelectrochemical devices in which anaerobic microbial consortia are used in the anode chamber. During the metabolism of microorganisms, they use organic carbon and release electrons at the anode. The catholyte could be water, potassium ferricyanide, and algal consortium [13]. The dissolved oxygen in the catholyte takes electrons and protons to form water molecules. Electricity is generated by the flow of electrons in the external circuit [16]. The anode and the cathode chambers are separated with PEMs (Figure 2.3). In the recent decade, GO membranes have been used in MFCs as well, due to their excellent electrical, biochemical, and thermal properties [48]. The composite membranes of GO have shown good biocompatibility, and improved H^+ proton transfer was also observed by blocking other cations (K^+, Li^+, and Na^+). Shabani *et al.* [49] developed thiolated GO (TGO) and polyetehrsulfone (SPES) hybrid with GO. Composite membrane generated double the

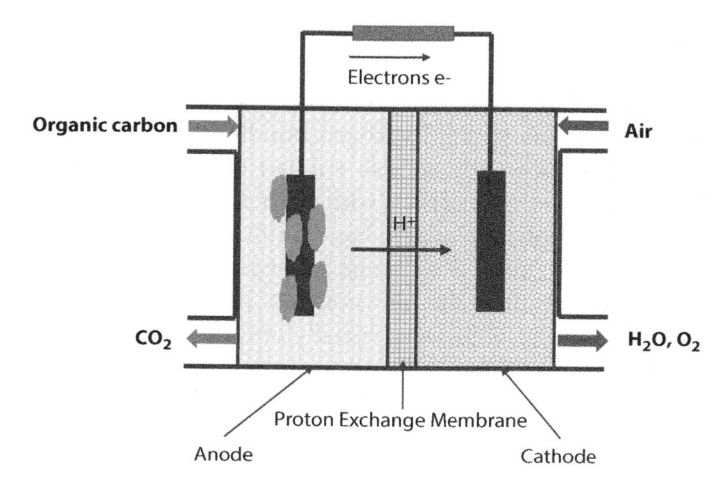

Figure 2.3 Two chambered microbial fuel cell (adopted from Shabani *et al.* [49]).

power density (66.4 mW/m^2) as compared to Nafion 117. Other parameters observed include higher columbic efficiency and chemical oxygen demand COD removal. GO-based composite membranes can be used in wastewater treating MFCs. Mondal *et al.* [9] studied the increase in sulfonated content of GO composite membrane have improved ionic and proton conductivity effect. Membranes with 3% and 5% SGO in SPBI exhibit superior (0.912 and 1.001 meq/g) ionic conductivity and (0.016 and 0.018 S cm$^{-1)}$ proton conductivity with a power density of 472.46 mW/m^2. Sulfonated composite membranes have high tensile modulus and ion exchange capacity. Spectroscopy and X-ray diffraction (XRD) analysis have confirmed these properties. The comparative efficiency in terms of power density, ionic conductivity, and a low-cost alternative of Nafion 117 membrane GO composite membranes is a promising choice for MFCs.

2.6 Characterization Techniques of GO Composite Membranes

After the development of composite membranes, a comprehensive investigation to characterize the properties of membrane is essential. Mostly used instruments are SEM, XRD analysis, Fourier transform infrared spectroscopy (FTIR), field-emission scanning electron microscope (FESEM), X-ray photoelectron spectroscopy, and energy-dispersive spectroscopy (EDS). Morphological characteristics of membranes are revealed in SEM.

Jamil *et al.* [50] observed reduced GO particles are uniformly distribute on carbon paper with the help of SEM. FESEM image confirmed by Suhaimin *et al.* [51] that GO is homogeneously distributed in the polymer matrix, which confirms strong hydrogen bonding between oxygen functional groups. The authors have also studied interaction between polar matrix, GO, and the sulfonic groups with the help of XRD, and new hydrogen bonding in the polymer matrix has significantly change the intensity of peaks, reflecting structural differences between GO, SPEEK, and GO/SPEEK. Characterization of the new composite membranes confirms the effectivity of the synthesis process visually and confirms the bonding of new functional groups that are added to enhance performance.

GO composite membranes are synthesized with the addition of sulfonated functional groups and nanomaterials to enhance proton conductivity and increase performance of PEM fuel cells. Han *et al.* [52] added sulfonated polytriazole by chemical oxidation, and azide functional groups were added (N_3-GO). FTIR spectra of GO and N_3-GO were compared to confirm the success of the reaction. The presence of stretching vibration peak of azide group at 2121 cm^{-1} was the confirmation of the azide functional group. The modification has enhanced the proton conductivity from 50% to 90% and power density from 1.21 to 1.58 W/cm^{-2}. EDS spectra are correlated to nanoparticles deposition on the composite GO and EDS elemental mapping gives atomic percentages of deposited nanoparticles. Both acid and basic functional groups can improve the proton conductivity and mechanical stability by increasing hydrophilicity [53–55]. Characterization of the modified membranes is a first step of confirmation that the reaction is successful before using the membrane in the fuel cells.

2.7 Conclusion

GO is an excellent blending material to develop polymer composite membranes with higher electrical, mechanical, and thermal properties. Its application in fuel cell technology as a membrane material has improved the performance and applicability of fuel cells. Chemical modification of GO such as sulfonation and carboxylation has further improved its functionality. However, it needs extensive research in which functional group at what temperature gives its best performance in what kind of fuel cell. Morphological characterization of the membrane using a variety of spectroscopy techniques can identify the successful reaction of the composite synthesis. The use of GO composite membranes as PEM has improved water retention, and it is very important for enhancing the conductivity of protons. The problem of membrane crossover which is the main reason for low power density and high internal resistance has also been solved. Composite GO PEMs research has entered a very crucial stage of its life cycle, where its application in various fuel cells and energy devices is in progress. Future research in this field will open new avenues of research and will uncover the new features of GO.

References

1. Zarrin, H., Higgins, D., Jun, Y., Chen, Z., Fowler, M., Functionalized graphene oxide nanocomposite membrane for low humidity and high temperature proton exchange membrane fuel cells. *J. Phys. Chem. C.*, 115, 20774, 2011.
2. Farooqui, U.R., Ahmad, A.L., Hamid, N.A., Graphene oxide: A promising membrane material for fuel cells. *Renew. Sust. Enegy. Rev.*, 82, 714, 2018.
3. Choi, B.G., Huh, Y.S., Park, Y.C., Jung, D.H., Hong, W.H., Park, H., Enhanced transport properties in polymer electrolyte composite membranes with graphene oxide sheets. *Carbon*, 50, 5395, 2012.
4. Hou, J., Shao, Y., Ellis, M.W., Moore, R.B., Yi., B., Graphene-based electrochemical energy conversion and storage: fuel cells, supercapacitors and lithium ion batteries. *Phys. Chem. Chem. Phys.*, 13, 15384, 2011.
5. Hwang, Y., Lee, J.K., Lee, C.H., Jung, Y.M., Heong, S.I., Lee, C.G. *et al.*, Stability and thermal conductivity characteristics of nanofluids. *Thermochim. Acta*, 455, 70, 007.
6. Park, J., OH, H., Ha, T., Lee, Y., Min., K., A review of the gas diffusion layer in proton exchange membrane fuel cells: durability and degradation. *Appl. Energy*, 155, 866, 2015.
7. Ansari, S., Kelarakis, A., Estevez, L., Giannelis., E.P., Oriented arrays of graphene in a polymer matrix by in situ reduction of graphite oxide nanosheets. *Small*, 6, 205, 2010.
8. Kim, Y., Ketpang, K., Jaritphun, S., Park, J.S., Shanmugam, S., A polyoxometalate coupled graphene oxide–Nafion composite membrane for fuel cells operating at low relative humidity. *J. Mater. Chem. A.*, 3, 8148, 2015.
9. Mandal, S., Papiya, F., Ash, S.N., Kundu, P.P., Composite membrane of sulphonated polybenzimidazole and sulphonated graphene oxide for potential application in microbial fuel cell. *J. Environ. Chem. Eng.*, 9, 104945, 2021.
10. Smitha, B., Sridhar, S., Khan, A.A., Solid polymer electrolyte membranes for fuel cell applications—a review. *J. Membr. Sci.*, 259, 10, 2005.
11. Enotiadis, A., Angjeli, K., Baldino, N., Nicotera, I., Gournis, D., Graphene-based Nafion nanocomposite membranes: enhanced proton transport and water retention by novel organo-functionalized graphene oxide nanosheets. *Small*, 8, 3338, 2012.

12. Pandey, R.P., Thakur, A.K., Shahi, V.K., Sulfonated polyimide/acid-functionalized graphene oxide composite polymer electrolyte membranes with improved proton conductivity and water-retention properties. *ACS. Appl. Mater. Interfaces*, 6, 16993, 2014.
13. Saba, B., Christy, A.D., Yu., Z., Co., A.C., Tuovinen, O.H., Islam., R., Characterization and performance of anodic mixed culture biofilms in submersed microbial fuel cells. *Bioelectrochemistry.*, 113, 79, 2017a.
14. Sadhasivam, T., Dhanabalan, K., Roh, S.H., Kim, T.H., Park, K.W., Jung, S., Kurkuri, M.D., Jung, H.Y., A comprehensive review on unitized regenerative fuel cells: crucial challenges and developments. *Int. J. Hydrogen Energy*, 1, 42, 4415–4433, 2017.
15. Koók, L., Quéméner, E.D.L., Bakonyi, P., Zitka, J., Trably, E., Tóth, G., Pavlovec, L., Pientka, Z., Bernet, N., Bélafi-Bakó, K., Nemestóthy, N., Behavior of two-chamber microbial electrochemical systems started-up with different ion-exchange membrane separators. *Bioresour. Technol.*, 278, 279, 2019.
16. Saba, B. and Christy, A.D., Simultaneous power generation and desalination of microbial desalination cells using Nanochloropsis salina (marine algae) vs potassium ferricyanide as catholytes. *Environ. Eng. Sci.*, 34, 185–196, 2017b.
17. Yang, E., Chae, K.J., Choi, M.J., He, Z., Kim, I.S., Critical review of bioelectrochemical systems integrated with membrane-based technologies for desalination, energy self-sufficiency, and high-efficiency water and wastewater treatment. *Desalination.*, 452, 40, 2019.
18. Luo, T., Abdu, S., Wessling, M., Selectivity of ion exchange membranes: A review. *J. Membr. Sci.*, 555, 492, 2018.
19. Harnisch, F. and Schröder, U., Selectivity versus Mobility: Separation of Anode and Cathode in Microbial Bioelectrochemical Systems. *Chem. Sus. Chem.*, 2, 921, 2009.
20. Hasani-Sadrabadi, M.M., Dashtimoghadam, E., Sarikhani, K., Majedi, F.S., Khanbabaei, G., Electrochemical investigation of sulfonated poly(ether ether ketone)/clay nanocomposite membranes for moderate temperature fuel cell applications. *J. Power Sources*, 195, 2450, 2010.
21. Daud, S.M., Kim, B.H., Ghasemi, M., Daud, W.R.W., Separators used in microbial electrochemical technologies: Current status and future prospects. *Bioresour. Technol.*, 195, 170, 2015.
22. Strathmann, H., *Ion-exchange membrane separation processes*, Elsevier, Amsterdãn, 2004.
23. Shabani, M., Younesi, H., Pontie, M., Rahimpour, A., Rahimnejad, M., Zinatizadeh, A.A., A critical review on recent proton exchange membranes applied in microbial fuel cells for renewable energy recovery. *J. Clean. Prod.*, 264, 121446, 2020.
24. Dreyer, D.R., Park, S., Bielawski, C., Ruoff, R.S., The chemistry of graphene oxide. *Chem. Soc Rev.*, 39, 228, 2010.

25. Xue, C., Zou, J., Sun, Z., Wang, F., Han, K., Zhu, H., Graphite oxide/functionalized graphene oxide and polybenzimidazole composite membranes for high temperature proton exchange membrane fuel cells. *Int. J. Hydrog. Energy*, 39, 7931, 2014.
26. Wang, L., Kang, J., Nam, J.D., Suhr, J., Prasad, A.K., Advani, S.G., Composite membrane based on graphene oxide sheets and Nafion for polymer electrolyte membrane fuel cells. *ECS Electrochem. Lett.*, 4, F1, 2014.
27. Ravikumar, Scott, K., Freestanding sulfonated graphene oxide paper: a new polymer electrolyte for polymer electrolyte fuel cells. *Chem. Commun.*, 48, 5584, 2012.
28. Sun, Y. and Shi, G., Graphene/polymer composites for energy applications. *J. Polym. Sci. Part. B. Polym. Phys.*, 51, 231, 2013.
29. Kumar, R., Xu, C., Scott, K., Graphite oxide/Nafion composite membranes for polymer electrolyte fuel cells. *RSC Adv.*, 2, 8777, 2012.
30. Peng, K.J., Lai, J.Y., Liu, Y.L., Nanohybrids of graphene oxide chemically-bonded with Nafion: preparation and application for proton exchange membrane fuel cells. *J. Membr. Sci.*, 514, 86, 2016.
31. Mishra, A.K., Kim, N.H., Jung, D., Lee, J.H., Enhanced mechanical properties and proton conductivity of Nafion-SPEEK-GO composite membranes for fuel cell applications. *J. Membr. Sci.*, 458, 128, 2014.
32. Yang, J., Liu, C., Gao, L., Wang, J., Xu, Y., He, R., Novel composite membranes of triazole modified graphene oxide and polybenzimidazole for high temperature polymer electrolyte membrane fuel cell applications. *RSC Adv.*, 5, 101049, 2015.
33. Cao, Y.C., Xu, C., Wu, X., Wang, X., Xing, L., Scott, K., A poly (ethylene oxide)/graphene oxide electrolyte membrane for low temperature polymer fuel cells. *J. Power Sources.*, 196, 8377, 2011.
34. Yang, H.N., Lee, W.H., Choi, B.S., Kim, W.J., Preparation of Nafion/Pt-containing TiO2/graphene oxide composite membranes for self-humidifying proton exchange membrane fuel cell. *J. Membr. Sci.*, 504, 20, 2016.
35. Sharma, P.P. and Kulshrestha., V., Synthesis of highly stable and high water retentive functionalized biopolymer-graphene oxide modified cation exchange membranes. *RSC Adv.*, 5, 56498, 2015.
36. Lee, D.C., Yang, H.N., Park, S.H., Kim, W.J., Nafion/graphene oxide composite membranes for low humidifying polymer electrolyte membrane fuel cell. *J. Membr. Sci.*, 452, 20, 2014.
37. Liu, Y., Wang, J., Zhang, H., Ma, C., Liu, J., Cao, S. *et al.*, Enhancement of proton conductivity of chitosan membrane enabled by sulfonated graphene oxide under both hydrated and anhydrous conditions. *J. Power Sources*, 269, 898, 2014.
38. He, Y., Wang, J., Zhang, H., Zhang, T., Zhang, B., Cao, S. *et al.*, Polydopamine-modified graphene oxide nanocomposite membrane for proton exchange membrane fuel cell under anhydrous conditions. *J. Mater. Chem. A.*, 2, 9548, 2014.

39. Lim, Y., Lee, S., Jang, H., Hossain, M.A., Gwak, G., Ju, H. *et al.*, Sulfonated poly(ether sulfone) electrolytes structured with mesonaphthobifluorene graphene moiety for PEMFC. *Int. J. Hydrog Energy.*, 1532-1538
40. Khilari, S., Pandit, S., Ghangrekar, M.M., Pradhan, D., Das, D., Graphene oxide-impregnated PVA–STA composite polymer electrolyte membrane separator for power generation in a single-chambered microbial fuel cell. *Ind. Eng. Chem. Res.*, 52, 11597, 2013.
41. Ye, Y.S., Cheng, M.Y., Xie, X.L., Rick, J., Huang, Y.J., Chang, F.C. *et al.*, Alkali doped polyvinyl alcohol/graphene electrolyte for direct methanol alkaline fuel cells. *J. Power Sources*, 239, 424, 2013.
42. Wang, C., Lin, B., Qiao, G., Wang, L., Zhu, L., Chu, F. *et al.*, Polybenzimidazole/ionic liquid functionalized graphene oxide nanocomposite membrane for alkaline anion exchange membrane fuel cells. *Mater. Lett.*, 173, 219, 2016.
43. Wang, Y., Chen, K.S., Mishler, J., Cho, S.C., Adroher, X.C., A review of polymer electrolyte membrane fuel cells: Technology, applications, and needs on fundamental research. *Appl. Energy.*, 88, 981, 2011.
44. Specchia, S., Francia, C., Spinelli, P., Polymer electrolyte membrane fuel cells, in: *Electrochemical technologies for energy storage and conversion*, R.S. Liu, L. Zhang, X. Sun, H. Liu, J. Zhnag (Eds.), pp. 601–670, Wiley-VCH Verlag GmbH & Co. KGaA, Germany, 2012.
45. Xu, C., Cao, Y., Kumar, R., Wu, X., Wang, X., Scott, K.A., polybenzimidazole/sulfonated graphite oxide composite membrane for high temperature polymer electrolyte membrane fuel cells. *J. Mater. Chem.*, 21, 11359, 2011.
46. Breeze, P., Direct methanol fuel cell, in: *Fuel Cells*, P. Breeze (Ed.), pp. 75–82, Academic Press United kingdom, Cambridge Massachusetts, United States, 2017.
47. Lin, C.W. and Lu, Y.S., Highly ordered graphene oxide paper laminated with a Nafion membrane for direct methanol fuel cells. *J. Power Sources.*, 237, 187, 2013.
48. Pandey, R.P., Shukla, G., Manohar, M., Shahi, V., Graphene oxide based nanohybrid proton exchange membranes for fuel cell applications: An overview. *Adv. Colloid Interface Sci.*, 240, 15, 2017.
49. Shabani, M., Younesi, H., Pontie, M., Rahimpour, A., Rahimnejad, M., Guo, H., Szymczyk, A., Enhancement of microbial fuel cell efficiency by incorporation of graphene oxide and functionalized graphene oxide in sulphonated polyethersulfone membrane. *Renewable Energy.*, 179, 788, 2021.
50. Jamil, M.F., Bicer, E., Kaplan, B.Y., Gursel, S.A., One step fabrication of new generation graphene based electrodes for polymer electrolyte membrane fuel cells by a novel electrophoretic deposition. *Int. J. Hydrog. Energy*, 46, 5653, 2021.
51. Suhaimin, N.S., Jaafar, J., Aziz, M., Ismail, A.F., Othman, M.H.D., Rahman, M.A., Aziz, F., Yusof, N., Nanocomposite membrane by incorporating

graphene oxide in sulfonated polyether ether ketone for direct methanol fuel cell. *Mater. Today Proc.*, 46, 2084, 2021.

52. Han, J., Lee, H., Kim, J., Kim, S., Kim, H., Kim, E., Sung, Y.E., Kim, K., Lee, J.C., Sulfonated poly(arylene ether sulfone) composite membrane having sulfonated polytriazole grafted graphene oxide for high performance proton exchange membrane fuel cells. *J. Membr. Sci.*, 612, 118428, 2020.
53. Li, J., Wu, H., Cao, L., He, X., Shi, B., Li, Y., Xu, M., Jiang, Z., Enhanced proton conductivity of sulfonated polysulfone membranes under low humidity via the incorporation of multifunctional graphene oxide. *ACS Appl. Nano Mater.*, 2, 4734, 2019.
54. Huang, Y., Cheng, T., Zhang, X., Zhang, W., Liu, X., Novel composite proton exchange membrane with long-range proton transfer channels constructed by synergistic effect between acid and base functionalized graphene oxide. *Polymer*, 149, 305, 2018.
55. Yin, Y., Wang, H., Cao, L., Li, Z., Li, Z., Gang, M., Wang, C., Wu, H., Jiang, Z., Zhang, P., Sulfonated poly (ether-ether-ketone)-based hybrid membranes containing graphene oxide with acid-base pairs for direct methanol fuel cells. *Electrochim. Acta*, 203, 178, 2016.

3

Graphene Nanocomposites as Promising Membranes for Proton Exchange Membrane Fuel Cells

Ranjit Debnath and Mitali Saha*

Department of Chemistry, National Institute of Technology Agartala, Tripura, India

Abstract

The extraordinary physical and chemical properties of graphene and its derivatives have stimulated tremendous efforts in the direction of fuel cell applications. The fuel cell technology is facing a major problem of efficient separation of protons from hydrogen. In this context, the selection of suitable material for the components of fuel cell faces the challenges in electrochemical performance, durability, and efficiency. The imperfect structure of graphene is speed up the selectivity of graphene membranes in a much better way as compared to conventional membranes, which offers a new and simpler mechanism for researchers and engineers to design the fuel cell. In the past few years, researchers and scientists have done great efforts to explore the potentiality of the graphene derivatives in the fuel cells. This chapter has highlighted the role of graphene in the modifications and applications of membrane-based fuel cells. This review has demonstrated the graphene materials as active components of membrane-based fuel cells, which will elaborate the understanding of graphene features in a better way, its compatibility with various materials, and future progress and prospects of graphene derivative–based membranes in fuel cells.

Keywords: Graphene, membrane, nanocomposites, proton conductivity, PEMFCs

**Corresponding author*: mitalichem71@gmail.com

Inamuddin, Omid Moradi and Mohd Imran Ahamed (eds.) Proton Exchange Membrane Fuel Cells: Electrochemical Methods and Computational Fluid Dynamics, (33–50) © 2023 Scrivener Publishing LLC

3.1 Introduction

Fuel cell generates electricity through chemical reaction, and it comprises two electrodes, namely, anode and cathode, where the reactions take place, producing electricity. An electrolyte carries the electrically charged particles, whereas the catalyst enhances the speed of the reactions at the anodes and cathodes. The classification of fuel cells determines the type of catalysts required for electro-chemical reactions, operating temperature range, fuel, etc., which ultimately decides the suitability of fuel cells for different applications. For the last few years, several fuel cells were developed and are still developing having potential applications, advantages, and limitations. In order to develop fuel cells with greater efficiency, researchers and scientists have designed many kinds of fuel cells with different size and varying technologies. However, they are facing the major problem of choice of the electrolyte. The materials and design of the electrodes depend mainly on the electrolyte, and, therefore, each cell has advantages and drawbacks, but none are less expensive and efficient enough, which could replace the conventional means of generating the power like coal-fired, hydroelectric, and nuclear power plants. Some cells require pure hydrogen, which demands

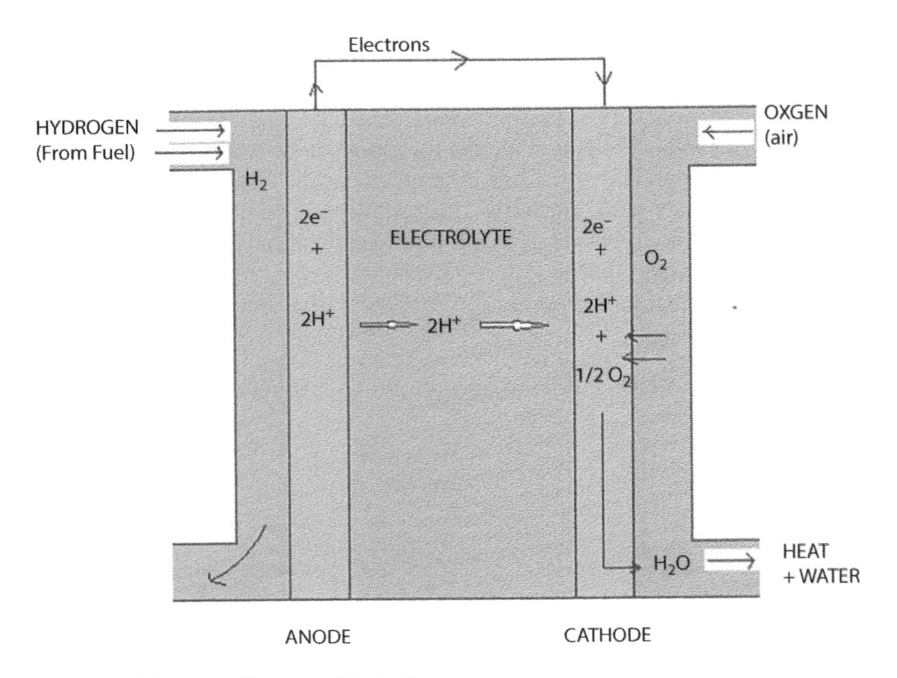

Figure 3.1 Schematic diagram of fuel cell.

extra equipment to purify the fuel. Few cells though tolerate impurities but require higher temperatures to run efficiently. In recent years, alkali, phosphoric acid (PA), and molten carbonate are used as liquid electrolytes, whereas solid oxide and proton exchange membrane (PEM) are considered as solids electrolytes. The schematic diagram of fuel cell is shown in Figure 3.1.

3.2 Recent Kinds of Fuel Cells

Alkaline fuel cells (AFCs), which operate on compressed hydrogen and oxygen in presence of potassium hydroxide as electrolyte, are the first fuel cell developed to produce electrical energy on U.S. space program and water on-board spacecraft. Their high performance is due to the rate of electro-chemical reactions, showing efficiency of about 70% with temperature operating at 150°C to 200°C. However, a key challenge to AFCs is their susceptibility to poisoning by even small amount of CO_2. A trace amount of CO_2, if comes in contact, can drastically affect the performance and durability of the cell due to formation of carbonate. This system also suffer from corrosion, wettability, conductivity, and durability of membrane, power density, anode electrocatalysis, water management, and handling higher operating temperatures and different pressures. PA fuel cells (PAFCs) are treated as most matured fuel cell to be used commercially as well as "first generation" of modern fuel cells, using electrolyte of liquid PA. They are mostly used for stationary power generation, and few are also used as power generator in large vehicles. They are also tolerant toward impurities of fossil fuels and are 85% more efficient during co-generation of heat and electricity, slightly higher as compared to combustion-based power plants. However, they are large and heavy, requiring more quantity of expensive platinum catalyst, which makes it very expensive. Molten carbonate fuel cells, also known as MCFCs, are considered as high-temperature fuel cells, currently being used as coal-based and natural gas–based power plants for electrical, industrial, and military applications. It is composed of salt mixture of molten stage carbonate having porous matrix of lithium aluminium oxide or metals as catalysts, as they operate at high temperatures of 650°C, thereby reducing the costs. MCFCs also offer high cost reductions over PAFCs due to improved efficiency up to 65% and the overall efficiencies can reach to 85%, when the waste heat is captured. However, the main disadvantage of this fuel cell technology is its durability resulting in breakdown and corrosion at high temperature and in presence of corrosive electrolyte, thereby decreasing cell life. Researchers and scientists are currently involved in changing the design of MCFCs and corrosion-resistant materials for components that can efficiently double the cell life without decreasing their performances. Solid oxide fuel cells (SOFCs) can perform at unusual high temperatures, as high as 1,000°C using a less porous compound of calcium or zirconium metal oxides as electrolyte. They are sulfur-resistant and are not poisoned by carbon monoxide, so efficiently use biogas, gases obtained from coal and natural gas. Their operation at higher temperature does not require precious metal catalyst, which reduces its cost. However, high-temperature is the main technical challenge of this technique, which results in a start-up at slow rate, requiring necessary thermal shielding to retain heat, and hence affects its durability, making it suitable for utility applications but not for transportation. Generally, they are 60% efficient to convert fuel into electricity, but the overall efficiencies could reach to 85%.

3.2.1 Proton Exchange Membrane Fuel Cells

PEM fuel cells exhibit large power density with low weight and volume in comparable to other cells, requiring only water to operate as well as oxygen and hydrogen only from the air. These cells operate at significant low temperatures (80°C), which allows them to start quickly and less wear of components, thus better durability. The presence of solid, flexible electrolyte does not lead to any leak or crack and are suitable for homes and cars. They are used some for some stationary applications and mainly for transportation applications like cars, buses, and heavy-duty trucks. However, the requirement of catalyst, like platinum for separating the hydrogen's electrons and protons, made it extremely sensitive to poisoning of carbon monoxide, and requires another reactor in addition for reducing carbon monoxide in presence of a hydrocarbon fuel. The use of reactor and the use of platinum catalyst on both sides of the membrane are the main factors for the cost of PEM cells.

Initially, in 2011, Chen *et al.* developed a nanocomposite membrane by using graphene oxide (GO) nanosheets with 3-mercaptopropyl trimethoxysilane, which showed four times better performance than Nafion at 120°C with 25% humidity [1]. Asmatulu *et al.* presented a nanocomposite membrane using nanoflakes of graphene and Nafion solutions using different amounts (1–4 wt%) and, after increasing graphene content, showed better electronic, ionic conductivities, and thermal and dielectric properties due to extraordinary high surface area–to-volume ratio and exceptionally good physical properties of graphene [2]. In 2012, Gournis and co-workers created nanocomposite membranes based on GO containing some hydrophilic functional group such as –OH, $-SO_3H$, and $-NH_2$ with

Nafion having better water retention properties than pure Nafion at temperature 140°C [3]. Tang *et al.* prepared rolled GO sheets and used it as nanocomposite membrane material with Nafion in 2013, which performed greater proton transport properties and opened new degrees of freedom for the development of new carbononaceous materials/polymer composites having extremely better properties [4]. The schematic diagram of GO is presented in Figure 3.2.

In 2014, Zhu and co-workers prepared a high temperature composite membrane using GO, 3,3′-diaminobenzidine, 5-tert-butyl isophthalic acid (GO/BuIPBI), and isocyanate modified graphite oxide/BuIPBI (iGO/BuIPBI) with different content of PA to provide proton conductivity. The proton conductivities were found to be 0.016 and 0.027 S/cm, respectively, in the case of GO/BuIPBI and iGO/BuIPBI using high acid loading at 140°C [5]. Yakun He *et al.* prepared composite-based membrane using GO modified polydopamine sheets having primary and secondary amino groups in sulfonated poly(ether ether ketone) (SPEEK) matrix, which resulted in better conductivity of proton in the composite membranes with enhancement of cell performances below 120°C under hydrous atmosphere, showing 38% increase in higher power density and 47% increase in highest current density [6]. Sharif *et al.* developed and compared between two composite membranes having pristine and FGO in both fluorinated and non-flourinated conditions [7]. Javanbakht *et al.* developed a cross-linked nanocomposite membrane using arylated sulfur-containing GO (SGO)

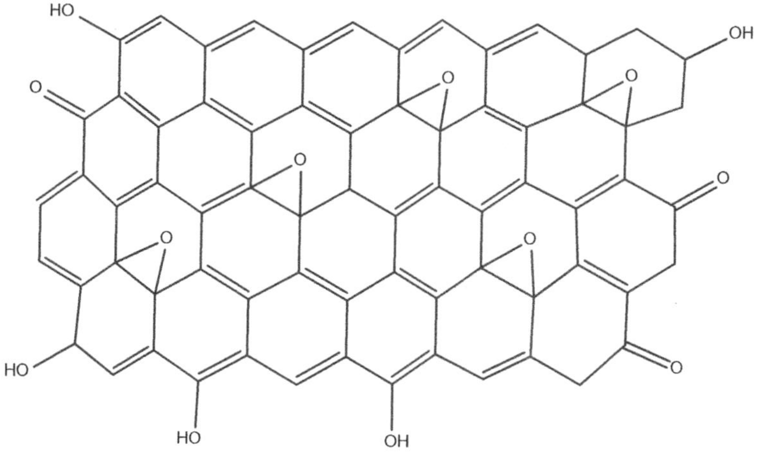

Figure 3.2 Schematic diagram of graphene oxide.

and poly(vinyl alcohol) (PVA) and the fabricated membrane showed better thermal stability, mechanical stability, and proton conductivity with power density having maximum value of 16.15 mW cm^{-2} at 30°C [8]. Ho *et al.* studied the performances of membrane based on GO and Nafion at different conditions, and results showed that current density and power density of the control fuel cell per square centimeter was 0.244678 W/cm^2 compared to the fuel cell with GO having power density per square centimeter as 0.299278 W/cm^2, an increase of 22.1% [9].

In 2015, Lü *et al.* developed membrane by combining sulfur-containing poly(ether sulfone) (SPES) matrix with GO functionalized sulfonated polymer (SPB-FGO). The results showed that proton conductivity, mechanical property, thermal stability, and oxidative stability of the nanocomposite membranes were improved than the pristine membrane of SPES [10]. Lee *et al.* developed membrane using GO, poly(2, 5-benzimidazole)–grafted GO (ABPBI-GO) and sulfonated poly(arylene ether sulfone) (SPAES), resulting in improved mechanical strength, dimensional and structural stability, and higher conductivity of proton than pristine SPAES [11]. Zhongyi Jiang *et al.* presented an approach for the construction of membrane using GO and poly(phosphonic acid), and the membranes obtained showed better conductivities of proton up to 32 mS cm^{-1} at relative humidity (RH) of 51% [12]. Ronghuan He *et al.* prepared composite membrane using polybenzimidazole combined with triazole functionalized GO (FGO) at high temperature and exhibited better proton conductivity and tensile strength than pure membrane when doped with PA [13].

In 2016, Jingtao Wang *et al.* constructed nanocomposite membranes using SPEEK and FGOs, saturated with imidazole, where the nanocomposite membrane contains 7.5% p-styrenesulfonic acid FGO, attaining the conductivity of 21.9 mS cm^{-1} at 150°C, 30 times higher than SPEEK membrane (0.69 mS cm^{-1}) [14]. Wu and co-workers fabricated a nanohybrid membrane based on Nafion doped phosphonic acid–FGO (PGO), which generated extra proton-conducting sites and improved the water retention and adsoption capacity of membranes of nanohybrids. The prepared membrane having 2 wt% by weight of PGO exhibited a proton conductivity of 0.0441 Scm^{-1} at 80°C and 40% RH and 0.277 Scm^{-1} at 100°C with 100% RH, which are 6.6 and 1.2 times higher than pristine Nafion membrane [15]. Liu *et al.* developed a nanohybrid membranes using GO and Nafion and when compared to pure Nafion membrane, and addition of GO-Nafion resulted about 35–40% increase in the performance of the fuel cell [16]. Stefan Freunberger *et al.* developed rolled GO–based membrane electrode assembly, which was ~75 times lighter, high power, and energy per weight, making it important material for portable device [17]. Fu and co-workers constructed composite membrane using inorganic filler of quaternized GO (QGO) and cross-linking agents with SPES where the water uptake and swelling ratio of SPES-10-QGO membrane possessed the least values of all the composite membranes, making it a valuable candidate for PEMFCs device [18].

GO is extraordinary potential candidate for PEMFCs as it is impermeable to O_2 and H_2 fuels but permits H^+ shuttling. Thotiyl *et al.* studied anisotropy in the conduction of proton in GO modified cell membranes by selective tuning the geometric and structural arrangement of functional groups and reported that *cis* isomer governed the amplification of proton shuttling through planes, resulting in the overall increment in the performance of fuel cell [19]. Yoo *et al.* developed a ternary hybrid membrane consisting of SPEEK, polyvinyledene fluoride-*co*-hexa fluoropropylene, and GO with 1, 3, 5, or 7 wt%. At 90°C, SPEEK showed peak proton conductivity value of 68 mS/cm (1.7 times more) with respect to the value of 122 mS/cm for hybrid membrane [20]. Kim and co-workers developed composite of GO, Nafion, and Pt nanoparticles containing TiO_2 and demonstrated that water uptake capacity was increased by maximum 15.3% with increasing GO content because of hydrophilic nature of GO with respect to Nafion and proton conductivity raised with GO but higher amount of GO reduces the performances of 2D sheets of GO due to the blocking effect [21].

In 2017, Xu *et al.* prepared composite using polybenzimidazole with radiation grafting GO (PBI/RGO) by solution-casting method and improved their proton conductivity by doping with PA. The resultant membrane of PBI/RGO-3/PA showed proton conductivity of 28.0 mS cm^{-1} at 170°C with the increase of 72.0%, compared to PBI/PA membrane at zero humidity [22]. Wang *et al.* prepared composite membranes mixing matrixes of two polymers (acidic natured SPEEK and basic natured chitosan) with four different types of nanosheets of GO nanosheets bearing imidazole brushes, PA, acid-base, or base-acid copolymer brushes. The strong electrostatic attractions form networks interconnected with each other, which increased the thermal and mechanical stabilities with large free volume and 6.7 times increase of proton conductivity [23].

The selection of materials faced many challenges in electrochemical efficiency, performances, and durability of fuel cell. However, excellent electronic, chemical, and mechanical properties of graphene derivatives made them as ideal materials for cell applications. Earlier researches suggest that graphene could be an ideal candidate for fuel cells owing to its corrosion resistance and high conductivity. Hence, in the past few years, significant works have been reported to ensure the potentiality of graphene derivatives in the fuel cells. Jiang *et al.* studied the performance of Nafion membrane at low humidity by adding multi-functional GO, which exceptionally increased the proton conductivity with 1 wt% filler under low humidity and the composite presented the highest conductivity of proton of 2.98×10^{-2} S cm^{-1} at 40% RH and at 80°C, 10 times more than pure Nafion and 135.5% enhancement of peak power density at 60°C, highlighting their potential application in PEMFCs [24]. Thotiyl *et al.* modified and developed nanoporous matrix based on GO by tuning the polarity of dopant molecule which increased the power and current densities by ~3 times as compared to pure GO membranes [25]. Zhang and co-workers prepared nanocomposite membranes using ABPBI-GO and SPEEK, which showed power densities of 831.06 and 72.25 mW cm^{-2} at low temperatures of 80°C and 120°C under 95% and 0% values of RH, respectively, whereas PA-doped polymer membrane gave 655.63 and 44.58 mW cm^{-2} only under same atmospheric conditions [26]. Vicente Compan *et al.* prepared composite membranes, containing a GO layer present between the SPEEK–polyvinyl alcohol matrix (SPEEK/PVA-GO) and the second one consisted layers of GO upon the SPEEK–polyvinyl butyral nanofibers, and it was observed that proton conductivity of SPEEK/PVA-GO membranes enhances with the rise of temperature yielding the value of 1×10^{-3} S cm^{-1} at 30°C to 8.3×10^{-3} S cm^{-1} at 130°C [27]. Shanmugam *et al.* developed a composite membrane with phosphotungstic acid (PW) and GO on sulfur-containing poly(arylene ether ketone) (SPAEK), showing better performance of fuel cell with power density value of 772 mW cm^{-2} for the composite, SPAEK/PW-mGO. However, pristine SPAEK composite membrane exhibited a power density of 10 mW cm^{-2} operated at 80°C under the RH value of 25% [28].

In 2018, Yoo *et al.* applied sulfonated (poly-arylene ether ketone) (SPAEK) and SGO to develop nano-hybrid membranes using various weight ratios of 0.5, 1, or 1.5 wt%, whereas 1.5 wt% SGO/SPAEK composite membrane showed proton conductivity of 124 mS/cm having higher amount of water absorption of 19% [29]. Ribes-Greus *et al.* prepared membranes using two different proportions of GO with PVA and SSA as cross-linking agent, and the dielectric permittivity of these membranes were significantly increased due to an interfacial polarization effect [30]. Liu *et al.* studied the performances of GO-based nanohybrid Nafion nanofiber membrane using a facile electro spinning technique which revealed that morphology, structure, proton conductivity, water uptake, mechanical properties, and swelling properties significantly improved [31]. Lyu *et al.* prepared membrane from SPAES and GO (SPAES-GO-x); SPAES-GO-3% membrane represented conductivity of 0.183 S cm^{-1} at RH value of 100% (120°C). The swelling ratio of SPAES-GO-2% membrane was found to reduce by 55.7% at 90°C in comparison to pure SPAES membrane [32]. Qingyin Wu *et al.* prepared composites using reduced GO (rGO), vanadium-substituted Dawson-type heteropoly acid (H8P2W16V2O62•20H2O), and SPEEK, and the membrane showed excellent proton conductivity of 7.90 $\times 10^{-2}$ S cm^{-1} at 50°C [33].

Zhu *et al.* prepared composite membrane by grafting GO into highly sulfur functionalized poly(ether ether ketone), and blending with Nafion, which revealed cell performance of 182 mW cm^{-2} at 25°C and 213 mW cm^{-2} at 60°C [34]. Xu *et al.* prepared three-layer membrane comprising GO-reinforced polybenzimidazole/porous polybenzimidazole/radiation, grafting GO-reinforced polybenzimidazole (PBI-RGO/PPBI/PBI-RGO), and further doping with PA, and results demonstrated significant proton conductivity and tensile strength than that of PA doped PBI membrane [35]. Zhu *et al.* prepared composite membranes based on highly SPEEK and sulfonated polybenzimidazole with sulfonated GO (s-GO), showing significant proton conductivity value and electrical performance, with good mechanical properties and proton conductivity value of 0.217 S cm^{-1} at 80°C. The maximum power density value of resultant PEMFC cell was reached 171 mW cm^{-2} at room temperature [36]. Yoo *et al.* prepared high temperature nanocomposite membranes from iron oxide nanoparticles and SGO with Nafion at low humidity, where the proton conductivity of the resultant (3 wt%) membrane at 120°C was found to be 11.62 mS cm^{-1} under humidity of 20%, 4.74 times more than pure membrane of Nafion [37]. Cao *et al.* developed a low cost composite based on sulfonated graphene cross-linked with SPEEK, and the membrane possessed good proton conductivity and prevention of methanol permeability [38].

In 2019, Lee *et al.* used GO membranes as electrolytes with controlled ratios of carbon/oxygen and less carbon/oxygen ratio showed 2.9-fold better power density with enhanced electrochemical properties compared to pristine GO [39]. Lee *et al.* prepared a composite GO and sulfonated poly(arylene thioether sulfone)–grafted GO as fillers with SPAES, which showed higher proton conductivity than the pristine SPAES [40]. In 2020, Lavanya and co-workers developed polymer nanocomposite membranes based on GO modified sulfonated poly(vinylidene fluoride-co-hexafluoropropylene) by electrospinning technique, showing good thermal stability and proton conductivity [41]. Devrim *et al.* proposed a high temperature membrane containing polybenzimidazole and GO (PBI/GO) and the data showed maximum 546- and 468-W power from PBI/GO and PBI membrane–based HT-PEMFC stacks, respectively [42]. Che *et al.* used LBL self-assembly technique to develop multilayered membranes following ordered deposition of components using GO as polyanions, poly(diallyldimethylammonium chloride) (PDDA) as polycations, and polyurethane (PU) immersed into PA solution. A proton conductivity value of 1.83×10^{-1} S/cm was seen for (PU/GO/PDDA/GO) membranes at 150°C [43].

In the last couple of years, considerable attention has been paid to graphene-based fuel cells. However, the main hindrance for the adoption of PEMFCs on commercial level is the low proton conductivity, high cost, and high fuel permeability of the graphene-Nafion–based membranes. However, it is well established that the characteristics of membranes, which are currently used in industries, can be significantly improved with the addition of graphene derivatives materials. The use of graphene as functional additives in the composition of the electrocatalytic layer of fuel cells will definitely improve their properties and will increase the activity and stability of the electrocatalyst. Lee *et al.* prepared a cross-linked membrane based on SPAES and vinyl FGO (VGO) as cross-linker and filler, where cross-linked SPAES90 having 1.0 wt% of VGO exhibited higher proton conductivities than Nafion 211 at 80°C under different range of RH conditions (50%–95%) [44]. Lee *et al.* prepared a composite based on sulfonated polytriazole grafted GO (SPTA-GO) as fillers in poly(arylene ether sulfone) (SPAES) and the maximum power density of the membrane reached up to 1.58 W cm^{-2} at 80°C at 100% RH condition, better than commercial Nafion 212 (1.33 W cm^{-2}) [45]. A. Ribes-Greus *et al.* developed a composite membrane containing GO as filler, a cross-linking agent such as SSA with PVA, and the sPVA/30SSA/GO composite membrane performed a better proton conductivity of 1.95 mS/cm at 25°C [46].

Mughal *et al.* prepared a nanocomposite membrane containing poly 2-N-acrylamido-2-methyl–1-propane sulfonic acid (PAMPS), carboxyl group containing poly(vinyl chloride) (CPVC), and 2-N-acrylamido-2-methyl–1-propane sulfonic acid having GO with proportions of CPVC (10%), PAMPS (20%), and PAMPS-mGO (20%), where the membrane offered superior performance with IEC of 1.3 mmol/g, oxidative stability of 97.2%, WU of 40.8%, proton conductivity of 151 S/cm, power density of 566.5 mW/cm^2 at 120°C, Young modulus of 797 MPa, water content of 17.43, current density of 1,537 mA/cm^2, tensile strength of 16.8 MPa, and elongation at break % of 2.7 MPa [47]. Pal *et al.* prepared composite consisting of matrix of SPEEK and nanoribbons of RGO wrapped titanium dioxide (TiO_2) microspheres as filler. The proton transport of the SPEEK/rGONR@TiO_2 (SPTC) nanohybrid PEM improved from 188% (0.62 S cm^{-1} for SPEEK) to 1.79 S cm^{-1} for SPTC at 100°C [48]. Loh and co-workers developed composite membrane based on polybenzimidazole (PBI) with the addition of a sulfonated GO (SGO)–based inorganic filler where the 2 wt% of PBI-SGO composite membrane showed the highest proton conductivity of 9.142 mS cm^{-1} at 25°C, which further increased to 29.30 mS cm^{-1} at 150°C [49]. Nicotera *et al.* developed and tested hybrid nanocomposite comprising TiO_2 nanoparticles grown on nanofillers of GO platelets in matrix of sulfonated polysulfone and the developed membrane (3 wt% of filler) showed excellent proton mobility and unusual high retention capacity of water at high temperatures [50]. In 2021, Liu *et al.* fabricated layered membrane based on GO sheets using a simple strategy by cross-linking two long-chain polymers (PDA and SPVA), and the resultant membrane was found to be enriched with high proton conductivity of 0.303 S cm^{-1} and high tensile strength value of 216.5 MPa at 80°C [51]. Table 3.1 shows the recent developments of graphene-based membranes for fuel cells.

Table 3.1 Recent developments of graphene based membranes for PEM fuel cells.

Composition of membrane	Proton conductivity (mS cm^{-1})	Temperature (°C)	References
GO-polySPM	29.8	80	[24]
AQSA-GO	30	25	[25]
SPEEK/ABPBI-GO	7.5	140	[26]
SPEEK/PVA-GO	83	130	[27]
SPAEK/PW-mGO	260.7	80	[28]
SGO/SPAEK	124	90	[29]
PVA/SSA/GO	3.06	30	[30]
NAFION/GO	57	80	[31]
SPAES-GO	183	120	[32]
SPEEK/RGO	79	50	[33]
GO-G-SPEEK/NAFION	219	90	[34]

(Continued)

Table 3.1 Recent developments of graphene based membranes for PEM fuel cells. (*Continued*)

Composition of membrane	**Proton conductivity (mS cm^{-1})**	**Temperature (°C)**	**References**
PBI-RGO/PPBI/PBI-RGO	113.8	170	[35]
s-PEEK/s-PBI/s-GO	217	80	[36]
NAFION/SGO	11.62	120	[37]
SG/SPEEK	63	54	[38]
GO	50	60	[39]
SPAES/SATS-GO	131.43	80	[40]
SPVdF-HFP/GO	0.243	100	[41]
PBI/GO	120	160	[42]
PU/GO/PDDA/GO	1.83×10^2	150	[43]
SPAES90/VGO	35.75	80	[44]
SPAES/SPTA-GO	412.5	80	[45]
sPVA/30SSA/GO	1.95	25	[46]
CPVC/PAMPS/PAMPS-mGO	1.51×10^3	120	[47]
SPEEK/rGONR@TiO_2	1790	100	[48]
PBI-SGO	29.30	150	[49]
SPSU/GO/TiO_2	98.91	80	[50]
sPVA/GO	303.0	80	[51]

3.3 Conclusion

Graphene has emerged as one of the best alternatives for future energy source specifically for PEMFC-based technology and, in the past few years, it has encouraged the research community due to exceptional characteristics and properties. The main obstacles for the successful commercialization of PEMFCs are hindered by the cost and durability, which can be overcome by the use of graphene-based membranes. This chapter has highlighted the influence of graphene in membrane modification for fuel cell technology. It has also represented the state-of-the-art and the recent development on graphene-based PEMFCs and the influence of the graphene-based nanocomposite membranes on the performance of PEMFCs.

Acknowledgements

We acknowledge the Department of Chemistry, National Institute of Technology Agartala.

References

1. Zarrin, H., Higgins, D., Jun, Y., Chen, Z., Fowler, M., Functionalized graphene oxide nanocomposite membrane for low humidity and high temperature proton exchange membrane fuel cells. *J. Phys. Chem. C.*, 115, 20774, 2011.
2. Adigoppula, V.K., Khan, W., Anwar, R., Argun, A.A., Asmatulu, R., Graphene based nafion® nanocomposite membranes for proton exchange membrane fuel cells. *Int. Mech. Eng Congress Expo.*, 54877, 325, 2011.
3. Enotiadis, A., Angjeli, K., Baldino, N., Nicotera, I., Gournis, D., Graphene-based nafion nanocomposite membranes: enhanced Proton transport and water retention by novel organo-functionalized graphene oxide nanosheets. *Small.*, 8, 3338, 2012.
4. Feng, K., Tang, B., Wu, P., Evaporating Graphene Oxide Sheets (GOSs) for rolled up GOSs and its applications in proton exchange membrane fuel cell. *ACS Appl. Mater. Interfaces.*, 5, 1481, 2013.
5. Xue, C., Zou, J., Sun, Z., Wang, F., Han, K., Zhu, H., Graphite oxide/functionalized graphene oxide and polybenzimidazole composite membranes for high temperature proton exchange membrane fuel cells. *Int. J. Hydrog. Energy.*, 39, 7931, 2014.
6. He, Y., Wang, J., Zhang, H., Zhang, T., Zhang, B., Cao, S., Liu, J., Polydopamine-modified graphene oxide nanocomposite membrane for proton exchange

membrane fuel cell under anhydrous conditions. *J. Mater. Chem. A.*, 2, 9548, 2014.

7. Shirdast, A., Sharif, A., Abdollahi, M., Prediction of proton conductivity of graphene oxide-containing polymeric membranes. *Int. J. Hydrog. Energy.*, 39, 1760, 2014.
8. Beydaghi, H., Javanbakht, M., Kowsari, E., Synthesis and characterization of poly(vinyl alcohol)/sulfonated graphene oxide nanocomposite membranes for use in PEMFCs. *Ind. Eng. Chem. Res.*, 53, 16621, 2014.
9. Ho, H., Matsuda, J., Yang, M., Wang, L., Rafailovich, M., Effects of Graphene Oxide on Proton Exchange Membrane Fuel Cells. *Ind. Eng. Chem. Res.*, 5, 16621, 2014.
10. Zhao, Y., Fu, Y., He, Y., Hu, B., Liu, L., Lü, J., Lü, C., Enhanced performance of poly (ether sulfone) based composite proton exchange membranes with sulfonated polymer brush functionalized graphene oxide. *RSC Adv.*, 5, 93480, 2015.
11. Ko, T., Kim, K., Lim, M.Y., Nam, S.Y., Kim, T.H., Kim, S.K., Lee, J.C., Sulfonated poly(arylene ether sulfone) composite membranes having poly(2,5-benzimidazole)-grafted graphene oxide for fuel cell applications. *J. Mater. Chem. A.*, 3, 20595, 2015.
12. He, G., Chang, C., Xu, M., Hu, S., Li, L., Zhao, J., Li, Z., Li, Z., Yin, Y., Gang, M., Wu, H., Yang, X., Guiver, M.D., Jiang, Z., Tunable nanochannels along graphene oxide/polymer core–shell nanosheets to enhance proton conductivity. *Adv. Funct. Mater.*, 25, 7502, 2015.
13. Yang, J., Liu, C., Gao, L., Wang, J., Xu, Y., He, R., Novel composite membranes of triazole modified graphene oxide and polybenzimidazole for high temperature polymer electrolyte membrane fuel cell applications. *RSC Adv.*, 122, 101049, 2015.
14. Wu, W., Li, Y., Chen, P., Liu, J.L., Wang, J., Zhang, H., Constructing Ionic Liquid-Filled Proton Transfer Channels within Nanocomposite Membrane by Using Functionalized Graphene Oxide. *ACS Appl. Mater. Interfaces.*, 8, 588, 2016.
15. Zhang, B., Cao, Y., Jiang, S., Li, Z., He, G., Wu, H., Enhanced proton conductivity of nafion nanohybrid membrane incorporated with phosphonic acid functionalized graphene oxide at elevated temperature and low humidity. *J. Membr. Sci.*, 518, 243, 2016.
16. Peng, K.J., Lai, J.Y., Liu, Y.L., Nanohybrids of graphene oxide chemically-bonded with nafion: Preparation and application for proton exchange membrane fuel cells. *J. Membr. Sci.*, 514, 86, 2016.
17. Thimmappa, R., Devendrachari, M.C., Shafi, S., Freunberger, S., Thotiyl, M.O., Proton conducting hollow graphene oxide cylinder as molecular fuel barrier for tubular H2-air fuel cell. *Int. J. Hydrog. Energy.*, 41, 22305, 2016.
18. Zhao, Y., Fu, Y., Hu, B., Lü., C., Quaternized graphene oxide modified ionic cross-linked sulfonated polymer electrolyte composite proton exchange membranes with enhanced properties. *Solid State Ion.*, 294, 43, 2016.

19. Thimmappa, R., Devendrachari, M.C., Kottaichamy, A.R., Tiwari, O., Gaikwad, P., Paswan, B., Thotiyl, M.O., Stereochemistry-dependent proton conduction in proton exchange membrane fuel cells. *Langmuir.*, 32, 359, 2016.
20. Vinothkannan, M., Kim, A.R., Nahm, K.S., Yoo, D.J., Ternary hybrid (SPEEK/SPVdF-HFP/GO) based membrane electrolyte for the applications of fuel cells: Profile of improved mechanical strength, thermal stability and proton conductivity. *RSC Adv.*, 6, 108851, 2016.
21. Yang, H.N., Lee, W.H., Choi, B.S., Kim, W.J., Preparation of nafion/Pt-containing TiO_2/graphene oxide composite membranes for self humidifying proton exchange membrane fuel cell. *J. Membr. Sci.*, 504, 20, 2016.
22. Cai, Y., Yue, Z., Xu, S., A novel polybenzimidazole composite modified by sulfonated graphene oxide for high temperature proton exchange membrane fuel cells in anhydrous atmosphere. *J. Appl. Polym. Sci.*, 133, 44986, 2017.
23. Wang, J., Bai, H., Zhang, J., Zhao, L., Chen, P., Li, Y., Liu, J., Acid-base block copolymer brushes grafted graphene oxide to enhance proton conduction of polymer electrolyte membrane. *J. Membr. Sci.*, 531, 47, 2017.
24. He, X., He, G., Zhao, A., Wang, F., Mao, X., Yin, Y., Cao, L., Zhang, B., Wu, H., Jiang, Z., Facilitating proton transport in nafion-based membranes at low humidity by incorporating multi-functional graphene oxide nanosheets. *ACS Appl. Mater. Interfaces.*, 9, 27676, 2017.
25. Gautama, M., Devendracharia, M.C., Thimmappaa, R., Kottaichamya, A.R., Shafia, S., Gaikwada, P., Kotreshb, H.M.N., Thotiyl, M.O., Polarity governed selective amplification of through plane proton shuttling in proton exchange membrane fuel cell. *Phys. Chem. Chem. Phys.*, 19, 7751, 2017.
26. Qiu, X., Ueda, M., Hu, H., Sui, Y., Zhang, X., Wang, L., Poly(2,5-benzimidazole)-grafted graphene oxide as an effective proton conductor for construction of nanocomposite proton exchange membrane. *ACS Appl. Mater. Interfaces.*, 9, 33049, 2017.
27. Rodriguez, J.L.R., Escorihuela, J., Gimenez, A.G.B.E., Feria, O.S., Compa, V., Proton conducting electrospun sulfonated polyether ether ketone graphene oxide composite membranes. *RSC Adv.*, 7, 53481, 2017.
28. Oha, K., Sonb, B., Sanetuntikulc, J., Shanmugam, S., Polyoxometalate decorated graphene oxide/sulfonated poly(arylene ether ketone) block copolymer composite membrane for proton exchange membrane fuel cell operating under low relative humidity. *J. Membr. Sci.*, 541, 386, 2017.
29. Jang, H.R., Vinothkannan, M., Kim, A.K., Yoo, D.J., Constructing proton-conducting channels within sulfonated(poly arylene ether ketone) using sulfonated graphene oxide: A nano-hybrid membrane for proton exchange membrane fuel cells. *Bull. Korean Chem. Soc*, 39, 715, 2018.
30. Guisasola, C.G., Greus, A.R., Dielectric relaxations and conductivity of cross-linked PVA/SSA/GO composite membranes for fuel cells. *Polym. Test.*, 67, 55, 2018.

31. Zhang, S., Li, D., Kang, J., Ma, G., Liu, Y., Electrospinning preparation of a graphene oxide nanohybrid proton-exchange membrane for fuel cells. *J. Appl. Polym. Sci.*, 135, 46443, 2018.
32. Chen, R.M., Xu, F.Z., Fu, K., Zhou, J.J., Shi, Q., Xue, C., Lyua, Y.C., Guo, B.K., Li, G., Enhanced proton conductivity and dimensional stability of proton exchange membrane based on sulfonated poly(arylene ether sulfone) and grapheme Oxide. *Mater. Res. Bull.*, 103, 142, 2018.
33. Wu, H., Wu, X., Wu, Q., Yan, W., High performance proton-conducting composite based on vanadium-substituted Dawson-type heteropoly acid for proton exchange membranes. *Compos. Sci. Technol.*, 162, 1, 2018.
34. Gao, S., Xu, H., Fang, Z., Ouadah, A., Chen, H., Chen, X., Shi, L., Ma, B., Jing, C., Zhu, C., Highly sulfonated poly(ether ether ketone) grafted on graphene oxide as nanohybrid proton exchange membrane applied in fuel cells. *Electrochim. Acta*, 283, 428, 2018.
35. Cai, Y., Yue, Z., Teng, X., Xu, S., Radiation grafting graphene oxide reinforced polybenzimidazole membrane with a sandwich structure for high temperature proton exchange membrane fuel cells in anhydrous atmosphere. *Eur. Polym. J.*, 103, 207, 2018.
36. Gao, S., Chen, X., Xu, H., Luo, T., Ouadah, A., Fang, Z., Li, Y., Wang, R., Jing, C., Zhu, C., Sulfonated graphene oxide-doped proton conductive membranes based on polymer blends of highly sulfonated poly(ether ether ketone) and sulfonated polybenzimidazole. *J. Appl. Polym. Sci.*, 135, 46547, 2018.
37. Vinothkannan, M., Kim, A.R., Kumar, G.G., Yoo, D.J., Sulfonated graphene oxide/Nafion composite membranes for high temperature and low humidity proton exchange membrane fuel cells. *RSC Adv.*, 8, 7494, 2018.
38. Cao, N., Zhou, C., Wang, Y., Ju, H., Tan, D., Li, J., Synthesis and characterization of sulfonated graphene oxide reinforced sulfonated poly (ether ether ketone) (SPEEK) composites for proton exchange membrane materials. *Mater.*, 11, 516, 2018.
39. Ahn, M., Liu, R., Lee, C., Lee, W., Designing carbon/oxygen ratios of graphene oxide membranes for proton exchange membrane Fuel Cells. *J. Nanomater.*, 2019, 6464713, 2019.
40. Lee, H., Han, J., Kim, K., Kim, J., Kim, E., Shin, H., Lee, J.C., Highly sulfonated polymer-grafted graphene oxide composite membranes for proton exchange membrane fuel cells. *J. Ind. Eng. Chem.*, 74, 223, 2019.
41. Lavanya, G., Paradesi, D., Hemalatha, P., Development of proton conducting polymer nanocomposite membranes based on SPVdF-HFP and graphene oxide for H2-O2 fuel cells. *J. Macromol. Sci. A.*, 57, 283, 2020.
42. Budak, Y. and Devrim, Y., Micro-cogeneration application of a high temperature PEM fuel cell stack operated with polybenzimidazole based membranes. *Int. J. Hydrog. Energy.*, 45, 35198, 2020.
43. Jia, T., Shen, S., Xiao, L., Jin, J., Zhao, J., Che., Q., Constructing multilayered membranes with layer-by-layer self-assembly technique based on graphene

oxide for anhydrous proton exchange membranes. *Eur. Polym. J.*, 122, 109362, 2020.

44. Kim, J., Kim, K., Han, J., Lee, H., Kim, H., Kim, S., Sung, Y.E., Lee, J.C., End group cross-linked membranes based on highly sulfonated poly(arylene ether sulfone) with vinyl functionalized graphene oxide as a cross-linker and a filler for proton exchange membrane fuel cell application. *J. Polym. Sci. Part A: Polym. Chem.*, 58, 3456, 2020.
45. Han, J., Lee, H., Kim, J., Kim, S., Kim, H., Kim, E., Sung, Y.E., Kim, K., Lee, J.C., Sulfonated poly(arylene ether sulfone) composite membrane having sulfonated polytriazole grafted graphene oxide for high-performance proton exchange membrane fuel cells. *J. Membr. Sci.*, 612, 118428, 2020.
46. Ballester, S.C.S., Soria, V., Rydzek, G., Ariga, K., Greus, A.R., Synthesis and characterization of bisulfonated poly(vinyl alcohol)/ graphene oxide composite membranes with improved proton exchange capabilities. *Polym. Test.*, 91, 106752, 2020.
47. Mughal, Z.N., Shaikh, H., Memon, S., Raza, R., Shah, R., Bhanger, M.I., Synthesis and characterization of poly 2-N-acrylamido-2-methyl1-propane sulfonic acid functionalized graphene oxide embedded electrolyte membrane using DOE for PEMFC. *Int. J. Energy Res.*, 44, 10354, 2020.
48. Roy, T., Wanchoo, S.K., Pal, K., Novel sulfonated poly (ether ether ketone)/ rGONR@TiO_2 nanohybrid membrane for proton exchange membrane fuel cells. *Solid State Ion.*, 349, 115296, 2020.
49. Yusoff, Y.N., Loh, K.S., Wong, W.Y., Daud, W.R.W., Lee, T.K., Sulfonated graphene oxide as an inorganic filler in promoting the properties of a polybenzimidazole membrane as a high temperature proton exchange membrane. *Int. J. Hydrog. Energy.*, 45, 27510, 2020.
50. Simari, C., Lufrano, E., Godbert, N., Gournis, D., Coppola, L., Nicotera, I., Titanium dioxide grafted on graphene oxide: Hybrid nanofiller for effective and low-cost proton exchange membranes. *J. Nanomater.*, 10, 1572, 2020.
51. Cai, Y.Y., Yang, Q., Sun, L.X., Zhu, Z.Y., Zhu, A.M., Liu, Q.L., Bioinspired layered proton-exchange membranes with high strength and proton conductivity. *Int. J. Hydrog. Energy.*, 46, 4087, 2021.

4

Carbon Nanotube–Based Membranes for Proton Exchange Membrane Fuel Cells

Umesh Fegade[1]* and K. E. Suryawanshi[2]

[1]Department of Chemistry, Bhusawal Arts Science and P. O. Nahata Commerce College, Bhusawal, Maharashtra, India
[2]Department of Applied Science and Humanities, R. C. Patel Institute of Technology, Shirpur, Dist. Dhule (M.S.), India

Abstract
In recent years, many fuel cell devices are produced due to their property of conversing chemical to electrical energy through reactions. From all the kinds of fuel cells, carbon nanotube (CNT)–based proton exchange membrane (PEM) fuel cells are the best in their stability, process at little temperature, maximum efficiency, and medium power generation.

The Cu-CNFs/PBI/IrO_2 membrane is developed to convert CO_2 to important products. In this membrane, low current density allowed the transformation of carbon dioxide into organic products with 85% selectivity and high current density increased catalytic activity and produce lighter saturated products.

Nowadays, replacement of high-cost catalysts with low-cost, high-efficient catalysts are a major challenge for the scientist. For this purpose, CNT-based PEM fuel cells are the most interesting material. In this chapter overview of the application-based approach, CNT-based PEM fuel cells are explained due to their auspicious properties, for example, huge surface area, maximum conductivity, and resist to corrosion.

Keywords: Carbon nanotubes, PEM fuel cells

**Corresponding author*: umeshfegade@gmail.com

Inamuddin, Omid Moradi and Mohd Imran Ahamed (eds.) Proton Exchange Membrane Fuel Cells: Electrochemical Methods and Computational Fluid Dynamics, (51–72) © 2023 Scrivener Publishing LLC

4.1 Introduction

The "fossil fuel uses" in the past few decades increased due to high demand, and it creates the "global warming" problem. Currently, the utilization and energy transformation are increased over the world [1]. To overcome this problem, carbon-based PEM fuels cells are a more promising material.

In 1991, Iijima discover carbon nanotubes (CNTs) with their 1D structure [2]. He classifies CNTs into three categories (Figure 4.1), i.e., single-, double-, and multi-layer CNTs [3–6]. CNTs increase the rate of electron transfer from electro-active species to electrodes that is why it acts as new material for the production of devices [7, 8]. CNTs are used in biotechnology, photonics, nanoelectronics, etc., due to electrical, chemical, physical, and thermal properties. As compared to regular materials, it is observed that CNTs are more interesting material due to their high clarity [9], electrical conductivity [10, 11], stability [12, 13], agricultural [14], drug transport, genetic work [15], and biomedical applications [16].

The CNTs are synthesized by different procedures such as solubilization [17], chemical vapor deposition, ball mill, or flame procedure [18]. Because of the insolubility in some solvents, uses of CNTs are limited, and, for that purpose, its surface modification is required [19–21].

The proton-conducting property of electrolytes gives the name PEM cell. The PEM fuel cell is the best capable material due to its rapid start-up and shut-down, great power density, and its solid nature. Because of the low-temperature and rigid nature, there is a lot of interest to produce PEM fuel cells [22]. The PEM fuel cell [23] mainly consists of the following:

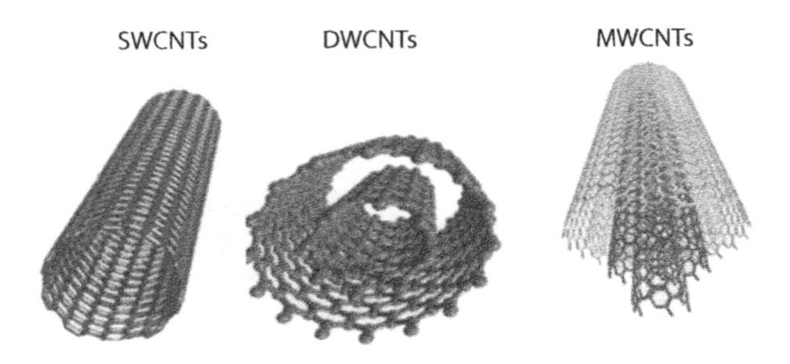

Figure 4.1 Single-, double-, and multi-layer CNTs [3].

a) Proton exchange membrane (PEM): It acts as the electrolyte in a fuel cell. Currently, commercial perfluorosulfonic acid membranes are used due to their high H^+ ion-conducting property, chemically inertness, and stability [24]. This membrane acts as an insulator for electricity, prevents gases entry, and is always hydrated. However, this type of membrane has certain disadvantages such as high cost, low tolerance to certain gases, and decreases conductivity at a certain temperature.
b) Electrocatalyst: A good electrocatalyst contains a huge surface area, high stability, and electrical properties. A carbon material with the support of platinum is used as a catalyst. The platinum must be uniformly dispersed over the surface, and electron transformation is easily occurs.
c) Carbon electrodes: It is present in the form of porous paper or cloth on the anode and cathode sides. It is also called as "gas diffusion layer" (GDL) due to its property of uniform distribution over electrocatalyst. Carbon electrodes show hydrophobic nature. The hot press combination of membrane, electrocatalyst, and carbon electrodes is known as membrane electrode setup.
d) Flow field plates: Graphite plates are used as flow channels for hydrogen and oxygen. It is used for the constant spreading of reactants over the body of the fuel cell. Graphite is used to help heat away from fuel cells' operation.
e) Current collecting plates: Gold-coated copper plates are used for the collection of current. To provide power to device, current is taken from fuel cells.
f) End plates: Aluminum end plates are used to clamp all fuel cells setup via nuts and bolts.

In the past few decades, there is great attention to produce PEM cells by using organic hybrid membranes at high temperatures [25]. To attain high temperature, three aspects are considered to prepare PEMs [26]: a) replace water through the volatile fluid; b) mixing of hard proton conductors; and c) mixing of organic fillers in the membrane. In this chapter overview of CNT-based membranes, PEM fuel cells are discussed.

4.2 Overview of Carbon Nanotube–Based Membranes PEM Cells

The CNTs containing Nafion material membranes are synthesis by melt-blending at 250°C. It is observed that, due to the incorporation of CNTs, proton conductivity decreases and, therefore, it is necessary to reduce filler amount to increasing conductivity [27]. The incorporation of CNTs with other elements modifies the mechanical and electronic properties of fuel cells [28]. The Pt/CNT layer is synthesized through *in situ* development of CNT on carbon paper and direct sputter deposition of Pt compound. The SEM and TEM of the deposited layer show uniform distributed CNT porous structure and Pt nano-dots on the surface. Because of low Pt doping, it gives a high power density of 595 mW cm^{-2} [29]. The FePc/CNTs show high ORR performance in alkaline solution. It also observed that CNTs play an important role in ORR, and MWCNT catalyst gives high current densities and low ORR potential [30].

In TEM analysis, it is observed that uniform distribution of Pt particles on CNx rather than CNTs. Pt/CNx shows high catalytic activity, high electrochemical surface area than Pt/CNTs, and it is observed by single-cell fuel tests [31]. The reduction of sodium borohydride and polyol methods is used to synthesize Pt/f-MWNTs and Pt_3Co/f-MWNTs. The maximum power density of 798 mW cm^2 is obtained for Pt_3Co/f-MWNTs [32]. The $TiSi_2Ox$-coated nitrogen-fixed CNTs are synthesized by the CVD method, and it is observed that, at the low electrochemical surface zone, Pt/An-$TiSi_2Ox$-NCNTs show good catalytic movement to ORR than Pt/NCNT catalysts [33]. The Pt/VACNT shows the highest power density of 697 mW/cm^2 through Pt doping of 102 µg/cm^2 at the cathode and 108 µg/cm^2 at the anode, which is 21% more [34].

In current days, researchers have a lot of interest to produce CNT-based PEM fuel cell. Kaewsai *et al.* synthesized a Pt deposit CNT covered with polybenzimidazole (PBI) and Py-m-PBIs (PBI/CNT and Py-m-PBI/CNTs) catalysts. It is observed that synthesized catalyst shows improved performance in PEM fuel cell at 160°C because of polymer-N content wrapped with CNT surface and bringing together of imidazole and pyridine-N content [35]. The plasma-enriched CVD technique is used to synthesize PEM fuel cells with CNTs developed GDL. The synthesized fuel cell shows high performance as compared to GDLs, which employ Vulcan XC-72 as the mixed microporous layer (MPL). It happens because of the huge macro-pore dimensions size for gas transport, and it shows 822 mW cm^{-2} using H_2/Air at 80% RH power density of fuel cell [36].

Figure 4.2 shows that CNTs are developed on carbon cloth and PBI are incorporate on CNTs to produce PBI-CNT hybrid electrode earlier deposition of Pt on PBI-CNT. The H_2/O_2 HT-PEM fuel cell shows that the low Pt-deposited Pt/PBI-CNT/CC electrode shows high power density (620 mW cm^{-2}) as related to the high Pt-deposited Pt/PBI-CNT/CC electrode at 160°C [37].

The sol-gel technique is used to obtain silica-based CNTs, which are used in the preparation of chitosan (CS)/SCNTs membrane. The prepared CS/SCNTs membrane shows high thermal and mechanical stability, and it is observed that proton conductivity of membrane grows from 0.015 to 0.025 S cm^{-1} due to electrostatic interfaces and H-bonding among chitson and SCNTs [38].

The PBI-based CNTs doped with H_3PO_4 are used to synthesize PEM. The doping of PBI into CNT membrane shows high proton conductivity 7.4×10^{-2} S cm^{-1} at 180°C due to high mechanical stability and tensile strength [39].

The solution casting technique is used to develop functionalized carbon nanotube (f-CNT)/SPEEK nanohybrid PEMs fuel cell. The prepared fuel cell shows high proton conductivity at elevated temperature, and activation energy is decreased due to the addition of f-CNT [40].

In phosphorus-fixed CNT, it is observed that doping of P has a large positive effect for increasing adsorption of Pt related to doping of N. It happens because Pt atom directly attaches with P-doped CNT and in N-doped

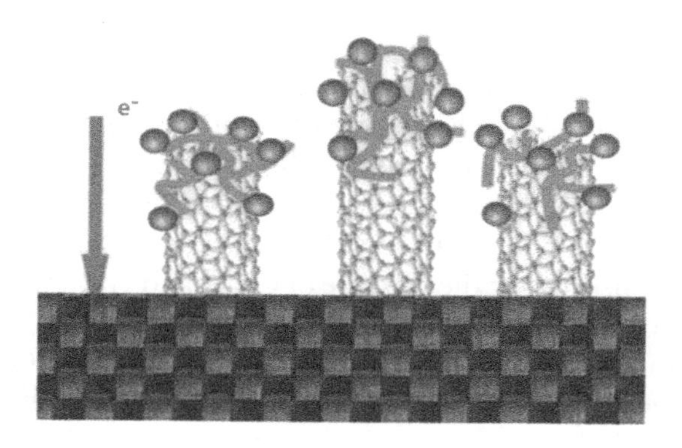

Figure 4.2 Schematic presentation of Pt/PBI-CNT electrode [37].

CNT interaction that occurs between Pt and carbon atoms [41]. Many researchers produce commercial HT-PEM fuel cells at operating temperature 80°C–130°C. Above this temperature, acid-doped membrane controls the field, completely scrap the system, and more acceptance to contaminated fuel with increasing in start-up time [42]. The addition of manganese oxide on carbon nano-tubes increased the performance of fuel cells with double power output [43].

The CNT-based PEM fuel cell increases thermal conductivity, electrical conductivity, and mechanical strength. The presence of CNTs raises the performance of fuel cells and reduces platinum catalysts in fuel cells [44].

PBI membrane is obtained to synthesize high-temperature PEM fuel cells because its high thermal and proton conductivity property are used in automotive industries. HT-PEMFC is used to migrate carbon monoxide poising, moisture, and heat in the system [45]. The PEM fuel cell is the best significant applicant in the automobile industry due to its quick start-up, process at low temperature, max energy productivity, and bulk power density [46]. CNTs are mainly used to prepare fuel cells because of its promising properties, for example, increasing cell performance, water permeability, and improving thermal and mechanical properties [47]. The PPy/carbon material with different carbon ratios is prepared by oxidative reaction of pyrrole with carbon black, i.e., chemical oxidative polymerization process. It is noted that a larger percentage of carbon expands the fuel cells performance [48].

The Cu-CNFs/PBI/IrO_2 membrane is developed to convert CO_2 to important products. In this membrane, low current density allowed the transformation of carbon dioxide into organic products with 85% selectivity and high current density increased catalytic activity and produced lighter saturated products [49]. The carbon-based supporting materials are used to produce 3D cells because of its durability, large surface area–to-volume ratio, increased electron transfer, and electrolyte diffusion and also reduce toxic products [50]. The synthesized GO/CNTs/SPEN fuel cell gives high proton conductivity 0.1197 S/cm at 20°C. The proton conductivity increases because of the mixture of CNTs and GO and low methanol penetrability of 2.015×10^{-7} cm^2 S^{-1} [51].

The polyaniline-covered CNT-held PtCo (PtCo/xPAN-CNT) catalysts are synthesized for O_2 reduction in PEM fuel cells. The synthesized catalyst shows high catalytic movement by a current density of 36.9 mA cm^{-2} in a 0.5 M H_2SO_4 PEM fuel cell [52]. The hybrid clay/organo-modified CNTs with sulfonated groups are used for the prepared nanomaterial membrane.

This membrane was obtained by the mixture of 1D nanotubes and 2D silicates structure of smectic clay [53].

The CNT wrapped phosphonic acid (CNT-POH) with sulfonated poly(ether ether ketone) (SPEEK) is used to synthesize PEMs. Mixing CNT-POH into SPEEK increases proton conductivity, which is recognized as the identical spreading of extremely hydrophilic phosphonic acid clusters and forms proton passage or canals in the membrane [54].

CNT fluids add to the CS matrix by the solution casting technique to produce PEM fuel cells. The CNT fluids improve the mechanical, thermal, oxidation strengths, and proton conductivity of PEM fuel cells due to the uniform distribution of fluids in CS matrixes and deliver a novel channel for proton transfer. CS/CNT fluid PEM fuel gives the maximum proton conductivity of 0.044 S cm^{-1} at 80°C [55]. Microwave-assisted technique is used to synthesize Pt-Pd/CNT catalysts by polyol path in the existence of SDS (Pt-Pd/CNT-S) or CTAB (Pt-Pd/CNT-C). The prepared Pt-Pd/CNT-S shows high fuel cell performance due to less Pt composition (12 wt%) as compared to clean Pt or Pt/C (20 wt%), and this fuel cell has scope for the commercial market due to great activity, high stability, and low cost [56].

For better performance of PEM fuel cells, it needs good water balance through the cell and good working conditions. In PEM fuel cells, short side-chainionomers, a more aqua holding capacity, high cell performance, and large and dynamic surface area are observed than traditional long side-chain ionomers [57]. Because of high Pt doping (0.5mg/cm^2), catalyst-glazed membrane lacking hot press that includes gas diffusion sheets gives high power density of 0.95 W/cm^2 [58]. Figure 4.3 shows that, in the PEM fuel cell device, hydrogen oxidized at the anode to generate electrons and proton drives to the cathode by the circuit and membrane. The oxygen is reduced on the cathode by reacting with electrons and protons to generate water [59].

Marquardt *et al.* produced thermocell created on PEM fuel cells with H_2 gas electrodes (Figure 4.4). This thermocell is similar to PEM fuel cells, and the only difference is that both electrodes supply hydrogen in which, if the single electrode is warmed up, then other is cooled. Both electrodes produce the change in chemical and electrode potential gradient for generating electricity [60].

The performance of CNTs catalyst-based PEM fuel cells is improved by its graphitization effect. The GCNT-held Pt demonstrates high efficiency, constancy, and high performance. The PEM fuel cell prepared by microwave synthesis technique shows current density 0.36, 0.30, and 0.20 A cm^2

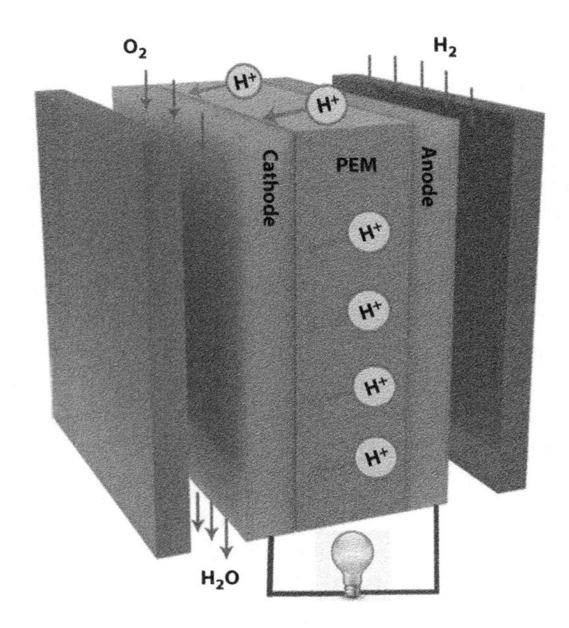

Figure 4.3 The device structure of PEM fuel cells [59].

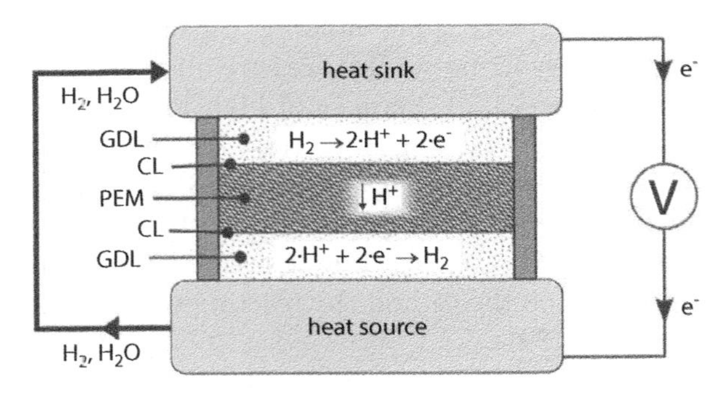

Figure 4.4 Thermocell based on a PEM [60].

for membrane made up from Pt/GCNT, Pt/CNT, and Pt/C catalysts, respectively [61]. It is observed that the fuel cell performance is improved if Pt catalysts are used in a low percentage [62]. The screen printing and doctor-blade technique are used to synthesize freeze-drying of cathode electrodes and it raises porosity and influences interaction with Pt/C catalyst as compared to oven or vacuum drying [63]. It is observed that the interaction study on the result of input factors gives greater power to the

CNT membrane, and it is found by preparing a larger CNT membrane [64].

The CS-built PEM fuel cell with polydopamine (PDA) shows high mechanical, thermal, and oxidation stability; it happens because H-bonding is present among the CS matrix and PDA covering. The CS/PDA@CNT PEM fuel cell shows 0.028 S cm^{-1} proton conductivity, which is 33.3% greater than pure CS membrane [65]. It is observed that when polytetrafluoroethylen merged into CNT-based electrocatalyst layers used in PEM fuel cells; it increases the power density due to the development of hydrophobicity in low humidity [66]. The Vulcan/PBI/Pt PEM fuel cells give 1.16 kWg^{-1} power density, which is 20% higher than the Vulcan/Pt (0.97 kWg^{-1}). This improvement is observed due to the polymer layer, which prevents Pt nanoparticles to go to Vulcan and improves coating Naflon ionomers [67]. The platinum-coated multi-walled CNT is used to prepare Pt/MWCNT PEM fuel cells, and it shows a larger power density (0.360 W/cm^2) than Pt/C (0.310 W/cm^2) at 160°C [68]. Ge *et al.* observed that, if current densities increase, then it shows membrane dehydration in fuel cells [69].

The power density is high for CNT2-based anode, due to its high surface concentration of nitrophenyl groups and surface N/C ratio. The MFC performance increased due to the incorporation of nitrogen on anode surface and holding on to nitrogen composition [70].

Eren *et al.* prepared PBI-covered MWCNTs as supporting solid for Pt-containing fuel cell and found that, at high-temperature (180°C), Pt-PBI/MWCNT shows power density improved from 38 to 47 mW cm^{-2} [71]. Figure 4.5 shows that PEM fuel cell is mainly prepared from bipolar plates and membrane electrode assembly. The MEA is mainly prepared through PEM, GDL, and catalyst film. PEM and CL are the main parts of PEM fuel cells. The GDL is a porous device structure, and it scatters uniform gas and acts as drainage for water [72].

In recent days, polymer PBI is mainly used with fillers to increase proton movement because of the large thermal and chemical strength in PEM fuel cells [73]. The bio-polymer CS, cellulose, and carrageenan are mostly used to increase proton conductivity, eco-friendly, and chief [74]. The power densities of NMWCNT@GONR anode and the MWCNT@GONR anode were 3,444 and 3,291 mW m^{-2}, respectively, which are three times larger than MWCNT modified anode. The power density increased because of a functional group [75].

The volume of Pt/CA-200 improved the pore availability to nanoform catalyst, gas molecules, and Nafion ionomer molecules, and thus, it increases the performance of PEM fuel five to seven times [76]. The 100 mM PtNP

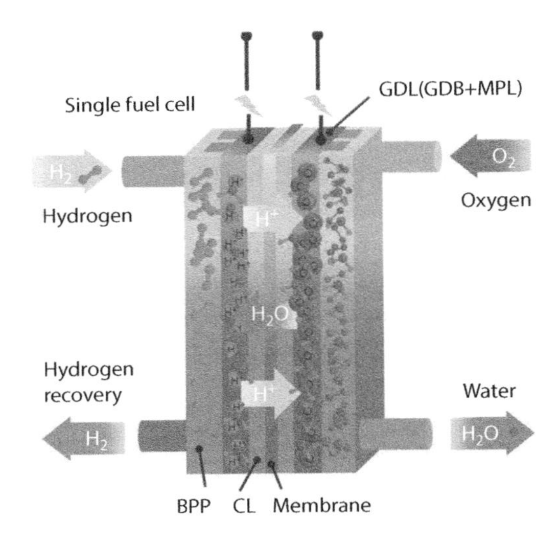

Figure 4.5 Single-cell PEM fuel cell [72].

material ink with 10 wt% CNT shows high ECSA catalytic activity for H_2, and it shows the highest value 37.9 cm^2/mg Pt. [77]. The thermocell containing a PEM fuel cell with H_2 electrodes shows 45.3 mW/cm^2 power density at a temperature of 323 K [78].

The polymer-based MWCNT is prepared by the solution growth technique, and it shows a large surface area and high current density. Figure 4.6 shows the synthesized MWCNT-based catalyst shows a mesh-like structure and light grey surface in SEM analysis. In XRD analysis, two highest peaks at 26° and 43° were observed and it shows planes 002 and 100 [79].

The CS-containing polymer membrane is mainly used to synthesize many types of fuel cells such as PEM, DMFC, and MFC. The CS is made up of a combination of fillers and a mixture of CS and polymer, chemical changes to increases its property. In addition, the CS membrane was upgraded in PEM cells by adding proton conductors [80]. The graphene MPL is shown more stability after 30,000 cycles of AST test and depletion observed in the first 10,000 cycles. It concluded that, if the number of cycles increases, then the performance of the cell decreases. The synthesized fuel cell shows high performance at the broad range of moisture conditions [81].

The greater surface area of the Pt/CNT catalyst shows the long duration stability and performance. The XRD analysis gives the size of platinum from a range of 4.9–5.0 nm, and EDS results show the presence of platinum

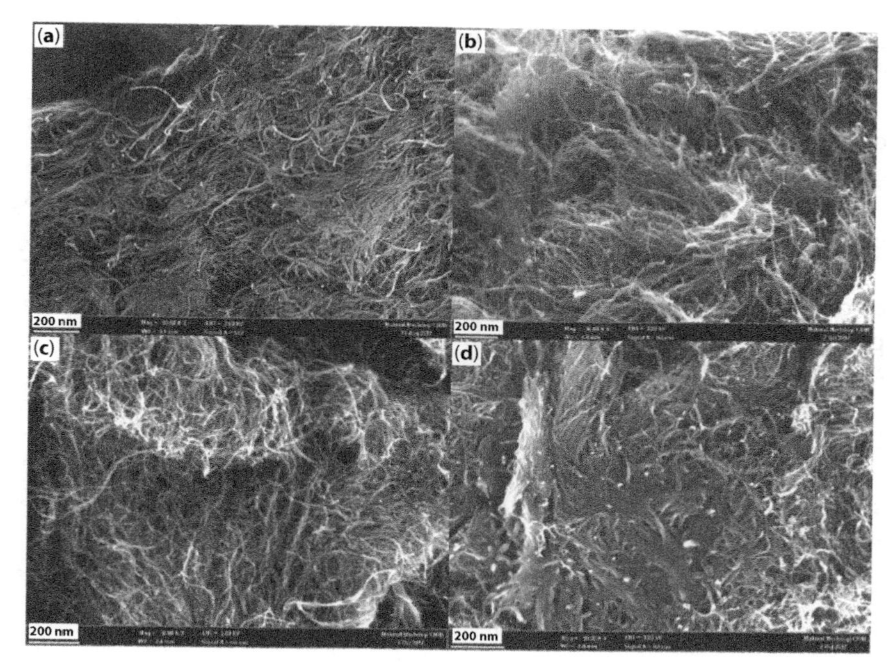

Figure 4.6 SEM images of MWCNT [79].

and carbon [82]. The SWCNTs are synthesized by mixing with surfactants such as SDBS, SDS, and SC. In thermoelectric analysis, n-type Seeback-coefficients are observed for SDBS/SWCNT at 350°C [83].

A general CNT membrane consists of four stages (Figure 4.7): a) transfer of mass to the electrode, b) adsorption-desorption on CNT membrane,

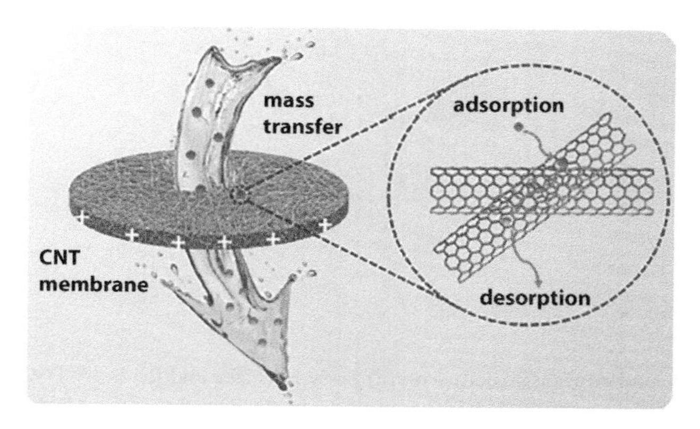

Figure 4.7 Filtration mechanism of CNT membrane [84].

c) electron transfer through membrane, and d) large chemical reaction process followed by electron transfer [84].

Ghosh *et al.* [85] observed that conducting polymers show high conductivity and stability when they act as a catalyst in PEM and MFCs fuel cells. Cheng *et al.* [86] synthesized polymer electrolyte membrane fuel cells with carbon support and they observed that these supporting materials enhance oxygen-carrying resistance power.

HPTS techniques are used to synthesize Nafion membranes with ionic electroactive polymer actuators, and it is achieved that mixing of MWCNTs produces high stability, power, chemical properties, and uniform thickness of Nafion membrane [87].

The Fe-N-SWCNT and Fe-N-DWCNT are synthesized for CO_2 adsorption, and it forms that energy for Fe-N-DWCNT is greater than Fe-N-SWCNT with more stability of adsorption of CO_2. Figure 4.8 shows the structure of CO_2 adsorbed on an iron atom, bidentate adsorption occurs with carbon and oxygen [88].

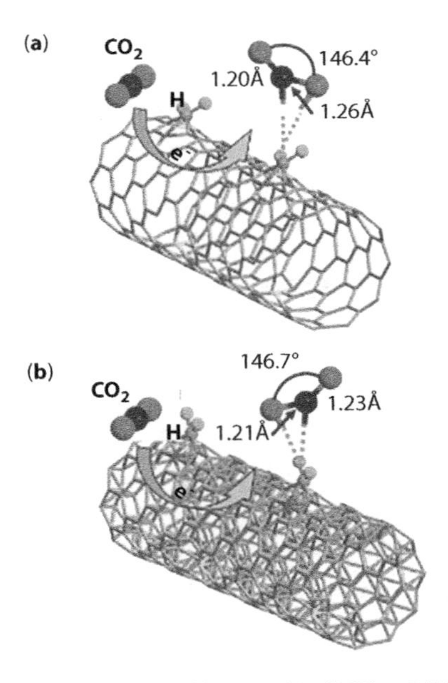

Figure 4.8 CO_2 adsorption structure on (a) Fe-N-SWCNT and (b) Fe-N-DWCNT surface [CC BY license (http://creativecommons.org/licenses/by/4.0/) of Ref. 88].

The PVDF/PSSA/CNF catalyst is synthesized by mixing of solution of poly(4-styrene sulfonic acid) and C nanofiber in a polyvinylidene fluoride polymer matrix. The synthesized catalyst shows improved properties such as gauge factor, piezoresistive sensing by adding C nanofiber. The XRD analysis of shows the shifting of phases from α to β, and crystalline nature is analyzed by the DSC technique [89].

The CNT bucky papers are used to prepare CNTs/epoxy composites, and it shows the conversion of mechanical performance from low to high, electrical insulator to semiconductor, and thermal insulator to the conductor as compared to glass fiber/epoxy composites [90].

PFSA Nafion®-type membranes with CNTs are synthesized by using ultrasonication techniques, and it shows localization of CNTs in hydrophilic holes and flow of membrane. The ionic transport occurs due to the carboxylic group and the proportion of CNTs increases the water intakes and volume of holes reduced [91].

Arora *et al.* [92] said that CNT nanomaterial is used in many processes such as ultra and nanofiltration, making of membranes, separation, and catalysis due to its unique properties. Liu *et al.* [93] synthesized a new method to produce carbon fabrics made up of electrospun carbon fibers containing CNT bristles that contain metal particles.

Figure 4.9 shows SEM images of microorganisms on an anode surface. Figures 4.9a and b clearly show that microorganisms stick to in and outside of carbon fibers. Figures 4.9c and d show that many microorganisms present on the surface, and some of them exist inside the carbon fiber felt. Figures 4.9e and f show that more microorganisms stick to the surface of the carbon fiber brush [94].

For the synthesis of PEM fuel cells, the mixture of platinum and palladium support with graphene is used due to their more active property. It also observed that, when the concentration of CNTs on surface adsorption increases, power also increases [95]. Basheer *et al.* observed that polymer on the surface of CNTs shows low electrical conductivity, and it shows shielding efficiency [96].

The carbon nanofibers are synthesized by loading acetylene gas on Fe-Co-kaolin catalyst and CNFs/Pt was synthesized by wet impregnation technique by mixing of K_2PtCl_4 onto carbon nanofibers. The XRD analysis gives 5.54- to 6.69-nm particle size for platinum and, if carbon nanofibers increase, then particle size is decreased [97].

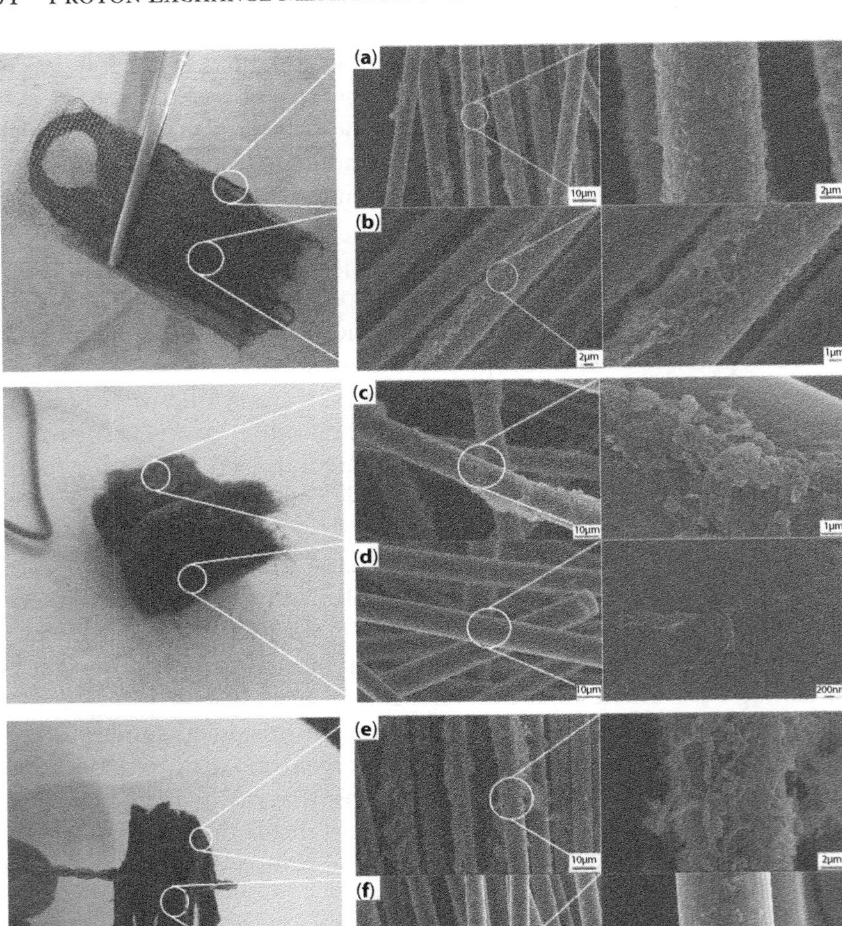

Figure 4.9 SEM images of the carbon fibers (a, b), carbon fiber felt (c, d), and carbon fiber brush (e, f) [94].

References

1. Bhogilla, S.S., Ito, H., Segawa, T., Kato, A., Nakano, A., Experimental study laboratory scale totalized hydrogen energy utilization system using wind power data. *Int. J. Hydrogen. Energy*, 42, 13827, 2017.
2. Iijima, S., Helical microtubules of graphitic carbon. *Nature*, 354, 56, 1991.

3. Azizi-Lalabadi, M., Hashemi, H., Feng, J., Jafari, S.M., Carbon nanomaterials against pathogens; the antimicrobial activity of carbon nanotubes, graphene/graphene oxide, fullerenes, and their nanocomposites. *Adv. Colloid Interface Sci.*, 284, 102250, 2020.
4. Harris, P., Carbon nanotubes and other graphitic structures as contaminants on evaporated carbon films. *J. Microsc.*, 186, 88, 1997.
5. Bethune, D.S., Kiang, C.H., De Vries, M.S., Gorman, G., Savoy, R., Vazquez, J., Beyers, R., Cobalt-catalysed growth of carbon nanotubes with single-atomic-layer walls. *Nature*, 363, 605, 1993.
6. Neves, W.Q., Alencar, R.S., Ferreira, R.S., Torres-Dias, A.C., Andrade, N.F., San-Miguel, A., Kim, Y.A., Endo, M., Kim, D.W., Muramatsu, H., Aguiar, A.L., Effects of pressure on the structural and electronic properties of linear carbon chains encapsulated in double wall carbon nanotubes. *Carbon*, 133, 446, 2018.
7. Zhou, J., Chen, Y., Lan, L., Zhang, C., Pan, M., Wang, Y., Han, B., Wang, Z., Jiao, J., Chen, Q., A novel catalase mimicking nanocomposite of Mn (II)-poly-L-histidine-carboxylated multi walled carbon nanotubes and the application to hydrogen peroxide sensing. *Anal. Biochem.*, 567, 51, 2019.
8. Alizadeh, T., Atashi, F., Ganjali, M.R., Molecularly imprinted polymer nanosphere/multi-walled carbon nanotube coated glassy carbon electrode as an ultrasensitive voltammetric sensor for picomolar level determination of rdx. *Talanta*, 194, 415, 2019.
9. Li, W., Liang, C., Qiu, J., Zhou, W., Han, H., Wei, Z., Sun, G., Xin, Q., Carbon nanotubes as support for cathode catalyst of a direct methanol fuel cell. *Carbon*, 40, 791, 2002.
10. Li, W., Liang, C., Zhou, W., Qiu, J., Zhou, Z., Sun, G., Xin, Q., Preparation and characterization of multiwalled carbon nanotube-supported platinum for cathode catalysts of direct methanol fuel cells. *J. Phys. Chem. B.*, 107, 6292, 2003.
11. Tian, Z.Q., Jiang, S.P., Liang, Y.M., Shen, P.K., Synthesis and characterization of platinum catalysts on multiwalled carbon nanotubes by intermittent microwave irradiation for fuel cell applications. *J. Phys. Chem. B.*, 110, 5343, 2006.
12. Shao, Y., Yin, G., Gao, Y., Shi, P., Durability study of Pt/C and Pt/CNTs catalysts under simulated PEM fuel cell conditions. *J. Electrochem. Soc*, 153, A1093, 2006.
13. Wang, X., Li, W., Chen, Z., Waje, M., Yan, Y., Durability investigation of carbon nanotube as catalyst support for proton exchange membrane fuel cell. *J. Power Sources.*, 158, 154, 2006.
14. Pereira, A.E., Grillo, R., Mello, N.F., Rosa, A.H., Fraceto, L.F., Application of poly (epsiloncaprolactone) nanoparticles containing atrazine herbicide as an alternative technique to control weeds and reduce damage to the environment. *J. Hazard Mater.*, 268, 207, 2014.

15. Chen, J., Chen, S., Zhao, X., Kuznetsova, L.V., Wong, S.S., Ojima, I., Functionalized single walled carbon nanotubes as rationally designed vehicles for tumor-targeted drug delivery. *J. Am. Chem. Soc*, 130, 16778, 2008.
16. Foroughi, J., Spinks, G.M., Aziz, S., Mirabedini, A., Jeiranikhameneh, A., Wallace, G.G., Kozlov, M.E., Baughman, R.H., Knitted carbon-nanotube-sheath/spandex-core elastomeric yarns for artificial muscles and strain sensing. *ACS Nano.*, 10, 9129, 2016.
17. Fujigaya, T. and Nakashima, N., Soluble carbon nanotubes and nanotube-polymer composites. *J. Nanosci. Nanotechnol.*, 12, 1717, 2012.
18. Yellampalli, S., Carbon nanotubes: Synthesis, characterization, applications, BoD–books on Demand, 2011.
19. Katouzian, I. and Jafari, S.M., Protein nanotubes as state-of-the-art nanocarriers: synthesis methods, simulation and applications. *J. Control Release.*, 303, 302, 2019.
20. Murphy, M., Theyagarajan, K., Ganesan, P., Senthilkumar, S., Thenmozhi, K., Electrochemical biosensor for the detection of hydrogen peroxide using cytochrome c covalently immobilized on carboxyl functionalized ionic liquid/multiwalled carbon nanotube hybrid. *Appl. Surf. Sci.*, 492, 718, 2019.
21. Wang, L., Wang, Y., Zhuang, Q., Simple self-referenced ratiometric electrochemical sensor for dopamine detection using electrochemically pretreated glassy carbon electrode modified by acid-treated multiwalled carbon nanotube. *J. Electroanal. Chem.*, 851, 113446, 2019.
22. Peighambardoust, S.J., Rowshanzamir, S., Amjadi, M., Review of the proton exchange membranes for fuel cell applications. *Int. J. Hydrog. Energy.*, 35, 9349, 2010.
23. Karthikeyan, K., Karuppanan, M.K., Panthalinga, P., Nanoscale, Catalyst Support Materials for Proton-Exchange Membrane Fuel Cells, in: *Handbook of Nanomaterials for Industrial Applications*, pp. 468–495, 2018.
24. Nasef, M.M., Radiation-grafted membranes for polymer electrolyte fuel cells: current trends and future directions. *Chem. Rev.*, 114, 12278, 2014.
25. Kim, D.J., Jo, M.J., Nam, S.Y., A review of polymer nanocomposite electrolyte membranes for fuel cell application. *J. Ind. Eng. Chem.*, 21, 36, 2015.
26. Bose, S., Kuila, T., Nguyen, T.X.H., Kim, N.H., Lau, K.T., Lee, J.H., Polymer membranes for high temperature proton exchange membrane fuel cell: Recent advances and challenges. *Prog. Polym. Sci.*, 36, 813, 2011.
27. Cele, N.P., Ray, S.S., Pillai, S.K., Ndwandwe, M., Nonjola, S., Sikhwivhilu, L., Mathe, M.K., Carbon Nanotubes Based Nafion Composite Membranes for Fuel Cell Applications. *Fuel Cells*, 10, 64, 2010.
28. Saha, M.S. and Kundu, A., Functionalizing carbon nanotubes for proton exchange membrane fuel cells electrode. *J. Power Sources*, 195, 6255, 2010.
29. Tang, Z., Poh, C.K., Lee, K.K., Tian, Z., Chua, D.H.C., Lin, J., Enhanced catalytic properties from platinum nanodots covered carbon nanotubes for proton-exchange membrane fuel cells. *J. Power Sources*, 195, 155, 2010.

30. Morozan, A., Campidelli, S., Filoramo, A., Jousselme, B., Palacin, S., Catalytic activity of cobalt and iron phthalocyaninesor porphyrins supported on different carbon nanotubes towards oxygen reduction reaction. *Carbon*, 49, 4839, 2011.
31. Chen, Y., Wang, J., Liu, H., NorouziBanis, M., Li, R., Sun, X., Sham, T.K., Ye, S., Knights, S., Nitrogen doping effects on carbon nanotubes and the origin of the enhanced electrocatalytic activity of supported Pt for proton-exchange membrane fuel cells. *J. Phys. Chem. C*, 115, 3769, 2011.
32. Vinayan, B.P., Jafri, R.I., Nagar, R., Rajalakshmi, N., Sethupathi, K., Ramaprabhu, S., Catalytic activity of platinum cobalt alloy nanoparticles decorated functionalized multiwalled carbon nanotubes for oxygen reduction reaction in PEMFC. *Int. J. Hydrog. Energy*, 37, 412, 2012.
33. NorouziBanis, M., Sun, S., Meng, X., Zhang, Y., Wang, Z., Li, R., Cai, M., Sham, T.K., Sun, X., $TiSi_2Ox$ coated N-doped carbon nanotubes as Pt catalyst support for the oxygen reduction reaction in PEMFCs. *J. Phys. Chem. C*, 117, 15457, 2013.
34. Shen, Y., Xia, Z., Wang, Y., KokPoh, C., Lin, J., Pt coated vertically aligned carbon nanotubes as electrodes for proton exchange membrane fuel cells. *Proc. Eng.*, 93, 34, 2014.
35. Kaewsai, D., Lin, H.L., Liu, Y.C., Yu, T.L., Platinum on pyridine-polybenzimidazole wrapped carbon nanotube supports for high temperature proton exchange membrane fuel cells. *Int. J. Hydrog. Energy*, 41, 10430, 2016.
36. Xie, Z., Chen, G., Yu, X., Hou, M., Shao, Z., Hong, S., Mu, C., Carbon nanotubes grown *in situ* on carbon paper as a microporous layer for proton exchange membrane fuel cells. *Int. J. Hydrog. Energy*, 40, 8958, 2015.
37. Du, H.Y., Yang, C.S., Hsu, H.C., Huang, H.C., Chang, S.T., Wang, C., Chen, J.C., Chen, K.H., Chen, L.C., Pulsed electrochemical deposition of Pt NPs on polybenzimidazole-CNT hybrid electrode for high-temperature proton exchange membrane fuel cells. *Int. J. Hydrog. Energy*, 40, 14398, 2015.
38. Liu, H., Gonga, C., Wang, J., Liu, X., Liu, H., Cheng, F., Wang, G., Zheng, G., Qina, C., Wen, S., Chitosan/silica coated carbon nanotubes composite proton exchange membranes for fuel cell applications. *Carbohydr. Polym.*, 136, 1379, 2016.
39. Moreno, N.G., Gervasio, D., García, A.G., Robles, J.F.P., Polybenzimidazole-multiwall carbon nanotubes composite membranes for polymer electrolyte membrane fuel cells. *J. Power Sources*, 300, 229, 2015.
40. Gahlot, S. and Kulshrestha, V., Dramatic improvement in water retention and proton conductivity in electrically aligned functionalized CNT/SPEEK nanohybrid PEM. *ACS Appl. Mater. Interfaces*, 7, 1, 264–272, 2015.
41. Tong, Y., Zhang, X.Y., Wang, Q., Xu, X., The adsorption mechanism of platinum on phosphorus-doped single walled carbon nanotube. *Comput. Theor. Chem.*, 1059, 1, 2015.

42. Chandan, A., Hattenberger, M., Kharouf, A., Du, S., Dhir, A., Self, V., Pollet, B.G., Ingram, A., Bujalski, W., High temperature (HT) polymer electrolyte membrane fuel cells (PEMFC) - A review. *J. Power Sources*, 231, 264, 2013.
43. Amade, R., Vila-Costa, M., Hussain, S., Casamayor, E.O., Bertran, E., Vertically aligned carbon nanotubes coated with manganese dioxide as cathode material for microbial fuel cells. *J. Mater. Sci.*, 50, 1214, 2015.
44. Mukherjee, S., Bates, A., Lee, S.C., Lee, D.H., Park, S., A review of application of CNTs in PEM Fuel Cells. *Int. J. Green Energy*, 12, 787, 2014.
45. Haque, M.A., Sulong, A.B., Loh, K.S., HeriantoMajlan, E., Husaini, T., Rosli, R.E., Acid doped polybenzimidazoles based membrane electrode assembly for high temperature proton exchange membrane fuel cell: A review. *Int. J. Hydrog. Energy*, 42, 9156, 2016.
46. Wang, C., Wang, S., Peng, L., Zhang, J., Shao, Z., Huang, J., Sun, C., Ouyang, M., He, X., Recent progress on the key materials and components for proton exchange membrane fuel cells in vehicle applications. *Energies*, 9, 603, 2016.
47. Goh, K., Karahan, H.E., Wei, L., Bae, T.H., Fane, A.G., Wang, R., Chen, Y., Carbon nanomaterials for advancing separation membranes: A strategic perspective. *Carbon*, 109, 694, 2016.
48. Das, E. and Yurtcan, A.B., Effect of carbon ratio in the polypyrrole/carbon composite catalyst support on PEM fuel cell performance. *Int. J. Hydrogen Energy*, 41, 13171, 2016.
49. Gutiérrez-Guerra, N., Valverde, J.L., Romero, A., Serrano Ruiz, J.C., Lucas-Consuegra, A., Electrocatalytic conversion of CO_2 to added-value chemicals in a high-temperature proton-exchange membrane reactor. *Electrochem. Commun.*, 81, 128, 2017.
50. Hoa, L.Q., Vestergaard, M.C., Tamiya, E., Carbon-based nanomaterials in biomass-based fuel-fed fuel cells. *Sensors*, 17, 2587, 2017.
51. Feng, M., Huang, Y., Cheng, T., Liu, X., Synergistic effect of graphene oxide and carbon nanotubes on sulfonated poly(arylene ether nitrile)-based proton conducting membranes. *Int. J. Hydrogen Energy*, 42, 8224, 2017.
52. Kaewsai, D., Piumsomboon, P., Pruksathorn, K., Hunsom, M., Synthesis of polyaniline-wrapped carbon nanotube-supported PtCo catalysts for proton exchange membrane fuel cells: activity and stability tests. *RSC Adv.*, 7, 20801, 2017.
53. Simari, C., Baglio, V., Vecchio, C., Aricò, A.S., Agostino, R.G., Coppola, L., Rossi, C.O., Nicotera, I., Reduced methanol crossover and enhanced proton transport in nanocomposite membranes based on clay-CNTs hybrid materials for direct methanol fuel cells. *Ionics*, 23, 2113, 2017.
54. Zhang, W., Zheng, H., Zhang, C., Li, B., Fang, F., Wang, Y., Strengthen the performance of sulfonated poly (ether ether ketone) as proton exchange membranes with phosphonic acid functionalized carbon nanotubes. *Ionics*, 23, 2103–2112, 2017.

55. Wang, J., Gong, C., Wen, S., Liu, H., Qin, C., Xiong, C., Dong, L., Proton exchange membrane based on chitosan and solvent-free carbon nanotube fluids for fuel cells applications. *Carbohydr. Polym.*, 186, 200, 2018.
56. Bharti, A. and Cheruvally, G., Surfactant assisted synthesis of Pt-Pd/MWCNT and evaluation as cathode catalyst for proton exchange membrane fuel cell. *Int. J. Hydrog. Energy*, 43, 31, 14729–14741, 2018.
57. Shahgaldi, S., Alaefour, I., Zhao, J., Li, X., Impact of ionomer in the catalyst layers on proton exchange membrane fuel cell performance under different reactant flows and pressures. *Fuel*, 227, 35, 2018.
58. Shahgaldi, S., Alaefour, I., Li, X., Impact of manufacturing processes on proton exchange membrane fuel cell performance. *Appl. Energy*, 225, 1022, 2018.
59. Zeng, Z., Xu, R., Zhao, H., Zhang, H., Liu, L., Xu, S., Lei, Y., Exploration of nanowire- and nanotube-based electrocatalysts for oxygen reduction and oxygen evolution reaction. *Mater. Today Nano*, 3, 54, 2018.
60. Marquardt, T., Valadez Huerta, G., Kabelac, S., Modeling a thermo cell with proton exchange membrane and hydrogen electrodes. *Int. J. Hydrog. Energy*, 43, 43, 19841–19850, 2018.
61. Devrim, Y. and Arıca, E.D., Investigation of the effect of graphitized carbon nanotube catalyst support for high temperature PEM fuel cells. *Int. J. Hydrog. Energy*, 45, 5, 3609–3617, 2019.
62. Ganesan, A. and Narayanasamy, M., Ultralow loading of platinum in proton exchange membranebased fuel cells: a brief review. *Mater. Renew. Sustain. Energy.*, 8, 18, 2019.
63. Talukdar, K., Delgado, S., Lagarteira, T., Gazdzicki, P., Friedrich, K.A., Minimizing mass-transport loss in proton exchange membrane fuel cell by freeze-drying of cathode catalyst layers. *J. Power Sources*, 427, 309, 2019.
64. Vijayaraghavan, V., Garg, A., Gao, L., Multiphysics-based statistical model for investigating the mechanics of carbon nanotubes membranes for proton-exchange membrane fuel cell applications. *J. Electrochem. Energy Convers. Storage*, 16, 031005, 2019.
65. Wang, J., Gong, C., Wen, S., Liu, H., Qin, C., Xiong, C., Dong, L., A facile approach of fabricating proton exchange membranes by incorporating polydopamine functionalized carbon nanotubes into chitosan. *Int. J. Hydrog. Energy*, 44, 6909, 2019.
66. Weerathunga, D.T.D., Jayawickrama, S.M., Phua, Y.K., Nobori, K., Fujigaya, T., effect of polytetrafluoroethylene particles in cathode catalyst layer based on carbon nanotube for polymer electrolyte membrane fuel cells. *Bull. Chem. Soc Jpn.*, 92, 2038, 2019.
67. Jayawickrama, S.M., Han, Z., Kido, S., Nakashima, N., Fujigaya, T., Enhanced platinum utilization efficiency of polymer-coated carbon black as an electrocatalyst in polymer electrolyte membrane fuel cells. *Electrochim. Acta*, 312, 349, 2019.

68. Devrim, E. and DamlaArıca, Y., Multi-walled carbon nanotubes decorated by platinum catalyst for high temperature PEM fuel cell. *Int. J. Hydrog. Energy*, 44, 18951, 2019.
69. Ge, N., Banerjee, R., Muirhead, D., Lee, J., Liu, H., Shrestha, P., Wong, A.K.C., Jankovic, J., Tam, M., Susac, D., Stumper, J., Bazylak, A., Membrane dehydration with increasing current density at high inlet gas relative humidity in polymer electrolyte membrane fuel cells. *J. Power Sources*, 422, 163, 2019.
70. Iftimie, S. and Dumitru, A., Enhancing the performance of microbial fuel cells (MFCs) with nitrophenyl modified carbon nanotubes-based anodes. *Appl. Surf. Sci.*, 492, 661, 2019.
71. Eren, E., Ozkan, N., Devrim, Y., Polybenzimidazole-modified carbon nanotubes as a support material for platinum-based high-temperature proton exchange membrane fuel cell electrocatalysts. *Int. J. Hydrog. Energy*, 46, 57, 29556–29567, 2021.
72. Yang, Y., Zhou, X., Li, B., Zhang, C., Recent progress of the gas diffusion layer in proton exchange membrane fuel cells: Material and structure designs of microporous layer. *Int. J. Hydrog. Energy*, 46, 5, 4259–4282, 2021.
73. Escorihuela, J., Olvera-Mancilla, J., Alexandrova, L., Castillo, L.F., Compañ, V., Recent progress in the development of composite membranes based on polybenzimidazole for high-temperature proton exchange membrane (PEM) fuel cell applications. *Polymers*, 12, 1861, 2020.
74. Rosli, N.A.H., Loh, K.S., Wong, W.Y., Yunus, R.M., Lee, T.K., Ahmad, A., Chong, S.T., Review of chitosan-based polymers as proton exchange membranes and roles of chitosan-supported ionic liquids. *Int. J. Mol. Sci.*, 21, 632, 2020.
75. Liu, Y.C., Hung, Y.H., Liu, S.F., Guo, C.H., Liu, T.Y., Sun, C.L., Chen, H.Y., Core–shell structured multiwall carbon nanotube-graphene oxide nanoribbon and its N-doped variant as anodes for high-power microbial fuel cells. *Sustain. Energy Fuels*, 4, 5339, 2020.
76. Gu, K., Kim, E.J., Sharma, S.K., Sharma, P.R., Bliznakov, S., Hsiao, B.S., Rafailovich, M.H., Mesoporous carbon aerogel with tunable porosity as catalyst support for enhanced proton exchange membrane fuel cell performance. *Mater. Today Energy*, 9, 100560, 2020.
77. Nagelli, E.A., Burpo, F.J., Marbach, D.A., Romero, A.N., Rabbia, D.J., Mahr, H.W., Jaskot, M.H., Murray, A.N., Chu, D.D., Scalable carbon nanotube/Platinum nanoparticle composite inks from salt templates for oxygen reduction reaction electrocatalysis for PEM fuel cells. *J. Compos. Sci.*, 4, 160, 2020.
78. Marquardt, T., Kube, J., Radici, P., Kabelac, S., Experimental investigation of a thermocell with proton exchange membrane and hydrogen electrodes. *Int. J. Hydrog. Energy*, 45, 1268, 2020.
79. Haque, M.A., Sulong, A.B., Shyuan, L.K., Majlan, E.H., Husaini, T., Rosli, R.E., Synthesis of polymer/MWCNT nanocomposite catalyst supporting

materials for high temperature PEM fuel cells. *Int. J. Hydrog. Energy*, 46, 4339, 2020.
80. Rosli, N.A.H., Loh, K.S., Wong, W.Y., Yunus, R.M., Lee, T.K., Ahmad, A., Chong, S.T., Review of chitosan-based polymers as proton exchange membranes and roles of chitosan-supported ionic liquids. *Int. J. Mol. Sci.*, 21, 632, 2020.
81. Shahgaldi, S., Ozden, A., Li, X., Hamdullahpur, F., A scaled-up proton exchange membrane fuel cell with enhanced performance and durability. *Appl. Energy*, 268, 114956, 2020.
82. Kim, T., Kwon, Y., Kwon, S., Seo, J.G., Substrate effect of platinum-decorated carbon on enhanced hydrogen oxidation in PEMFC. *ACS Omega*, 5, 26902, 2020.
83. Seki, Y., Nagata, K., Takashiri, M., Facile preparation of air-stable n-type thermoelectric single-wall carbon nanotube films with anionic surfactants. *Sci. Rep.*, 10, 1, 2020.
84. Liu, Y., Gao, G., Vecitis, C.D., Prospects of an electroactive carbon nanotube membrane toward environmental applications. *Acc. Chem. Res.*, 15, 53, 12, 2892–2902, 2020 Dec.
85. Ghosh, S., Das, S., Mosquera, M.E.G., Conducting polymer-based nanohybrids for fuel cell application. *Polymers*, 12, 2993, 2020.
86. Chenga, X., Wei, G., Wang, C., Shen, S., Zhang, J., Experimental probing of effects of carbon support on bulk and local oxygen transport resistance in ultra-low Pt PEMFCs. *Int. J. Heat Mass Transf.*, 164, 120549, 2021.
87. Park, K. and Jeon, Effect of ionic polymer membrane with multiwalled carbon nanotubes on the mechanical performance of ionic electroactive polymer actuators. *Polymers*, 12, 396, 2020.
88. Yoon, S.H., Park, H., Elbashir, N.O., Han, D.S., Effect of Fe/N-doped carbon nanotube (CNT) wall thickness on CO_2 conversion: A DFT study. *Sustain. Mater. Technol.*, 26, e00224, 2020.
89. Chaturvedi, M., Panwar, V., Prasad, B., Piezoresistive sensitivity tuning using polyelectrolyte as interface linker in carbon based polymer composites. *Sens. Actuator A Phys.*, 312, 112151, 2020.
90. Trakakis, G., Tomara, G., Datsyuk, V., Sygellou, L., Bakolas, A., Tasis, D., Papagelis, K., Mechanical, electrical, and thermal properties of carbon nanotube buckypapers/epoxy nanocomposites produced by oxidized and epoxidized nanotubes. *Materials*, 13, 4308, 2020.
91. Safronova, E., Parshina, A., Kolganova, T., Yelnikova, A., Bobreshova, O., Pourcelly, G., Yaroslavtsev, A., Potentiometric multisensory system based on perfluorosulfonic acid membranes and carbon nanotubes for sulfacetamide determination in pharmaceuticals. *J. Electroanal. Chem.*, 873, 114435, 2020.
92. Arora, B. and Attri, P., Carbon Nanotubes (CNTs): A potential nanomaterial for water purification. *J. Compos. Sci.*, 4, 135, 2020.
93. Liu, X., Ouyang, M., Orzech, M.W., Niu, Y., Tang, W., Chen, J., Marlow, M.N., Puhan, D., Zhao, Y., Tan, R., Colin, B., *In-situ* fabrication of carbon-metal

fabrics as freestanding electrodes for high-performance flexible energy storage devices. *Energy Storage Mater.*, *30*, 329, 2020.

94. Xu, H., Zhang, M., Ma, Z., Zhao, N., Zhang, K., Song, H., Li, X., Improving electron transport efficiency and power density by continuous carbon fibers as anode in the microbial fuel cell. *J. Electroanal. Chem.*, 857, 113743, 2019.
95. Jeong, G.U. and Lee, B.J., Interatomic potentials for Pt-C and Pd-C systems and a study of structure-adsorption relationship in large Pt/Graphene system. *Comput. Mater. Sci.*, 185, 109946, 2020.
96. Basheer, B.V., George, J.J., Siengchin, S., Parameswaranpillai, J., Polymer grafted carbon nanotubes-Synthesis, properties, and applications: A review. *Nano-Struct. Nano-Objects*, 22, 100429, 2020.
97. Mudi, K.Y., Abdulkareem, A.S., Kovo, A.S., Azeez, O.S., Tijani, J.O., Eterigho, E.J., Development of carbon nanofibers/Pt nanocomposites for fuel cell application. *Arab. J. Sci. Eng.*, *45*, 7329, 2020.

5

Nanocomposite Membranes for Proton Exchange Membrane Fuel Cells

P. Satishkumar[1], Arun M. Isloor[1,2]* and Ramin Farnood[3]

[1]Membrane and Separation Technology Laboratory, Department of Chemistry, National Institute of Technology Karnataka, Surathkal, Mangalore, India
[2]Apahatech Solutions, Science and Technology Entrepreneur's Park, National Institute of Technology, Karnataka, Surathkal, Mangalore, India
[3]Department of Chemical Engineering and Applied Chemistry, University of Toronto, Ontario, Canada

Abstract
The development of green technologies like fuel cell is need of the day because of their zero emission and as an efficient technology to produce electrical energy. Among the different varieties of fuel cells, enhancing the performance of proton exchange membrane (PEM) fuel cell is emphasized because of their numerous advantages such as easy portability, less corrosive nature, and leakage-free convenient setup. Generally used Nafion membranes in PEM fuel cells show few limitations such as the inability to work at high temperature and low relative humidity. Nanocomposite membranes play an indispensable role in overcoming these flaws. Incorporating numerous nanoadditives like silica, titanium dioxide, zirconium dioxide, graphene oxide, zirconium phosphate, heteropolyacids, and metal-organic frameworks into the variety of polymer matrix such as Nafion, sulfonated polybenzimidazole, polysulfone, sulfonated poly(ether ether ketone), and biopolymers involving polyvinyl alcohol, chitosan is assessed with its characteristic properties of proton conductivity, mechanical stability, oxidative stability, and power density. Nanocomposite membranes aid to increase the mechanical stability of the PEMs by the combination of two or more polymer layers and especially with a solid support layer. Development of natural, biodegradable polymer-based PEMs with enhanced proton conducting ability and chemical stability was possible only because of the nanocomposite model; otherwise, it was not possible. Certain hygroscopic inorganic additives improved the water uptake capacity of the

**Corresponding author*: isloor@yahoo.com

Inamuddin, Omid Moradi and Mohd Imran Ahamed (eds.) Proton Exchange Membrane Fuel Cells: Electrochemical Methods and Computational Fluid Dynamics, (73–110) © 2023 Scrivener Publishing LLC

nanocomposite membranes even at elevated temperatures. A large pool of nanocomposite membranes that can meet the desired characteristics of PEMs for fuel cell applications is reviewed in detail.

Keywords: Proton exchange membrane (PEM), fuel cell, nanocomposite membranes, low humidity fuel cells, high-temperature fuel cells

5.1 Introduction

Increase in population and urbanization results in an escalation of global energy consumption [1]. A prodigious amount of draining fossil fuels and natural gases are burnt daily to meet this demand. This is taking a huge toll on the environment and human health by the emission of a variety of potential greenhouse gases and toxic compounds into the atmosphere [2]. In response to this alarming scenario, scientific communities have come up with many alternative energy technologies like wind generators, turbine generators, solar cells, and fuel cells. Out of which fuel cell technology stands out as the promising one because of its high energy conversion, zero emission, and portable power applications [3]. Unlike fossil fuel fed engines, fuel cell reduces power drop by eliminating the intermediate steps,

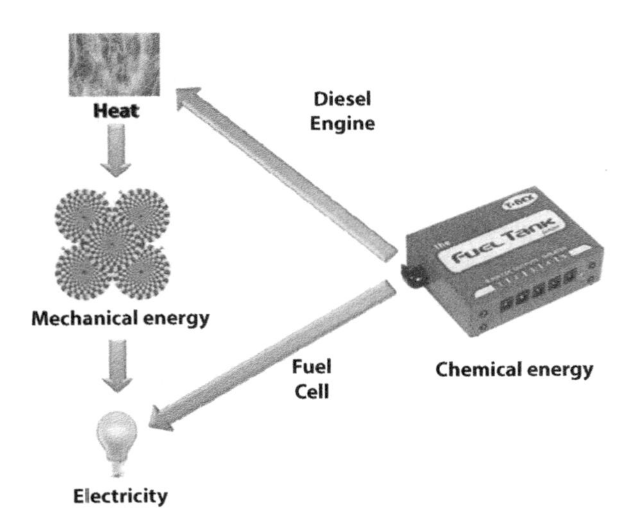

Figure 5.1 Comparison of fuel cell and diesel engine in converting chemical energy in to electrical energy. Reprinted with permission from [4], Copyright 2011, Progress in Polymer Science, Elsevier.

and electrical energy is directly obtained from chemical energy which is shown in Figure 5.1 [4].

Fuel cell was first developed by Robert Grove in 1839 [5], and this encouraged further advancement of the invention through years and led to different types of fuel cells like proton exchange membrane (PEM) fuel cell, direct

Table 5.1 Categories of fuel cells with operational specifications. Modified from [7], Copyright 2014, Elsevier.

Type of fuel cell	Typical electrolyte	Charge carrier	Electrical efficiency (%)	Operational temperature (°C)
Proton exchange membrane fuel cell (PEMFC)	Perfluorosulfonic acid	H^+	60	50–80
Direct methanol fuel cell (DMFC)	Perfluorosulfonic acid	H^+	60	90–120
Alkaline fuel cell (AFC)	Aqueous potassium hydroxide (KOH)	OH^-	50	50–200
Molten carbonate fuel cell (MCFC)	Solution of Na_2CO_3, K_2CO_3, and Li_2CO_3 in Lithium aluminate ($LiAlO_2$)	CO_3^{2-}	45–50	600–700
Solid oxide fuel cell (SOFC)	Zirconia stabilized by Yttria	O^{2-}	60	800–1,000
Phosphoric acid fuel cell (PAFC)	Phosphoric acid (H_3PO_4) in silicon carbide	H^+	40	160–220
Microbial fuel cell (MFC)	Ion exchange membrane	H^+	65	20–60

methanol fuel cell, phosphoric acid (PA) fuel cell, alkaline fuel cell, solid oxide fuel cell, microbial fuel cell, and molten carbonate fuel cell [6]. Cardinal points of different categories of fuel cells are mentioned in Table 5.1 [7].

Among the different categories of fuel cells, PEMFCs have got emphasizing attention because of their distinctive features like quick startup, lightweight, facile portability, and broad temperature range, and it is less prone to corrosion when compared to other fuel cells with liquid electrolyte [8]. In PEMFC, anode and cathode are separated by PEM which plays a pivotal role in transporting H^+ ions from an oxidizing electrode to reducing electrode while concomitantly clogging the transmission of electrons and fuels [9, 10]. Figure 5.2 is a schematic representation of PEMFC showing different components and process involved in electricity generation along with water as a by-product.

$$\textit{At Anode}: 2\mathrm{H}_2 \rightarrow 4\mathrm{H}^+ + 4\mathrm{e}^-$$

$$\textit{At Cathode}: O_2 + 4H^+ + 4e^- \rightarrow 2H_2O$$

$$\textit{Overall}: 2\mathrm{H}_2 + O_2 \rightarrow 2H_2O + \textit{electricity}$$

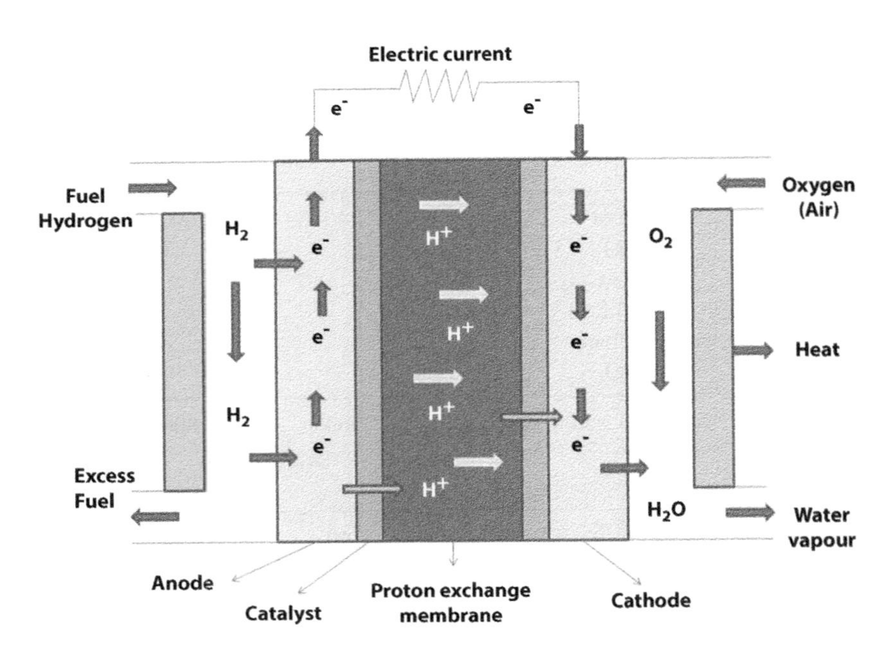

Figure 5.2 Schematic representation of PEMFC showing different components and process involved in electricity generation along with water as a by-product.

The vital characteristics of a good PEM include the following [11].

1. Excellent proton conductivity
2. High water uptake capacity
3. Good mechanical stability
4. Wide temperature range
5. Oxidative stability
6. Resistant to fuel crossover
7. Low cost and economical
8. High power density

Nafion is ubiquitous and is a commercially available PEM because of its good proton conductivity and chemical stability [12]. However, it has its limitations like fuel permeability, high cost, under par performance at low humidity, and inability to work at high temperatures [13]. In order to overcome these snags, different polymeric membranes are used such as poly(benzimidazole) (PBI) [14], sulfonated poly(ether ether ketone) (SPEEK) [15], poly(vinyl alcohol) (PVA) [16], sulfonated poly(ether sulfone) (SPSF) [17] and chitosan (CS) [11] (Table 5.2). However, these membranes individually failed to meet all the criteria of a good PEM, and it led to the development of nanocomposite membranes where a polymeric membrane property is customized by the addition of a variety of inorganic nanoparticles as per the need, operating condition, and application [18].

5.2 Nanocomposite Membranes for PEMFC

To achieve unique functional properties, an organic polymer matrix is added with nanosized inorganic moieties. The added inorganic particles have no less than one dimensions that are of the order of nanometer [19]. It may be isodimensional nanoparticles or two-dimensional (2D) nanotubes or one-dimensional silicate layers [20]. Filing of inorganic particles into the organic matrix alters membrane proton conducting ability, mechanical strength, resistance to oxidation, capacity to withstand high temperature, water uptake, and much more [21]. Figure 5.3 highlights the key factors which are influenced by both phases [22]. Out from the large pool of inorganic nanoparticles, major ones include heteropolyacids [23], titanium dioxide [24], zirconium dioxide [25], boehmite [26], cerium dioxide [27], and silicates [28]. Many times there is a possibility of additives agglomeration instead of uniform distribution, which, in turn, fails to achieve the

Table 5.2 Structures of different polymeric membranes.

Polymer	Structure
Nafion	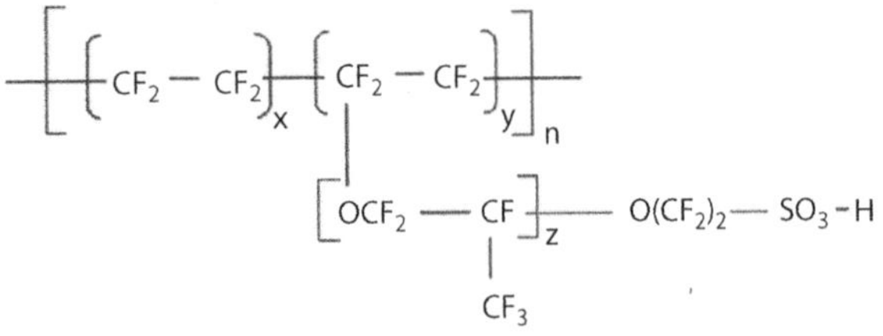
PBI	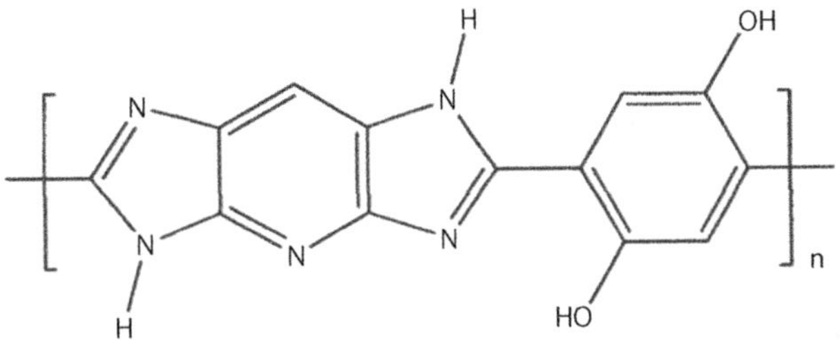

(Continued)

Table 5.2 Structures of different polymeric membranes. (*Continued*)

Polymer	Structure
SPEEK	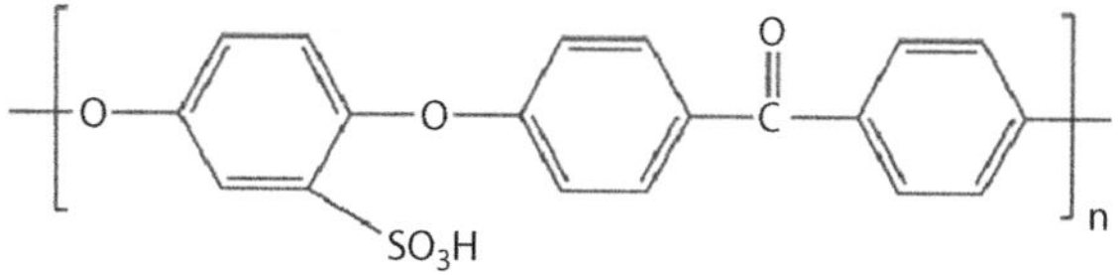 O, O, O, C, SO_3H, n
PVA	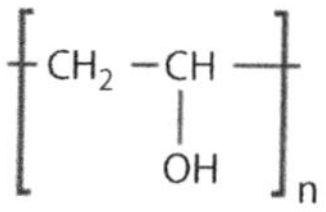 CH_2 –CH, OH, n
SPSF	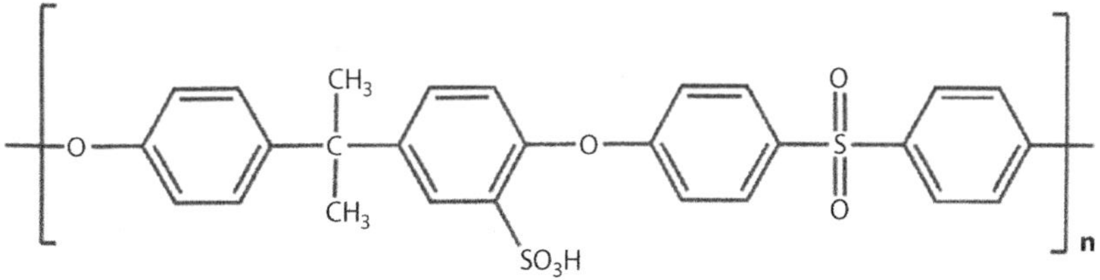 O, CH_3, C, CH_3, O, O, S, O, SO_3H, n
Chitosan	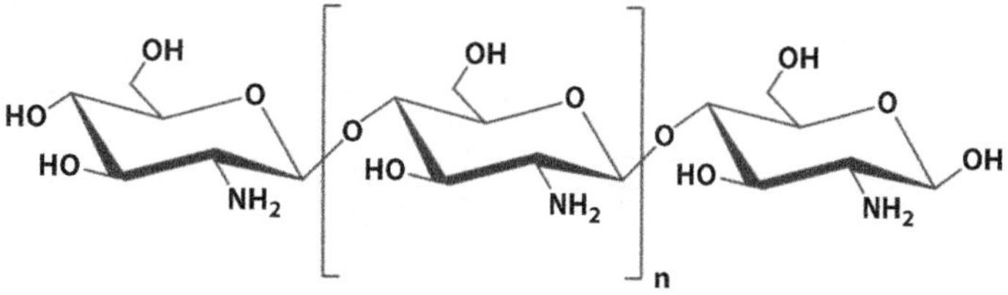 OH, O, HO, HO, NH_2, O, OH, O, HO, NH_2, O, OH, O, HO, OH, NH_2, n

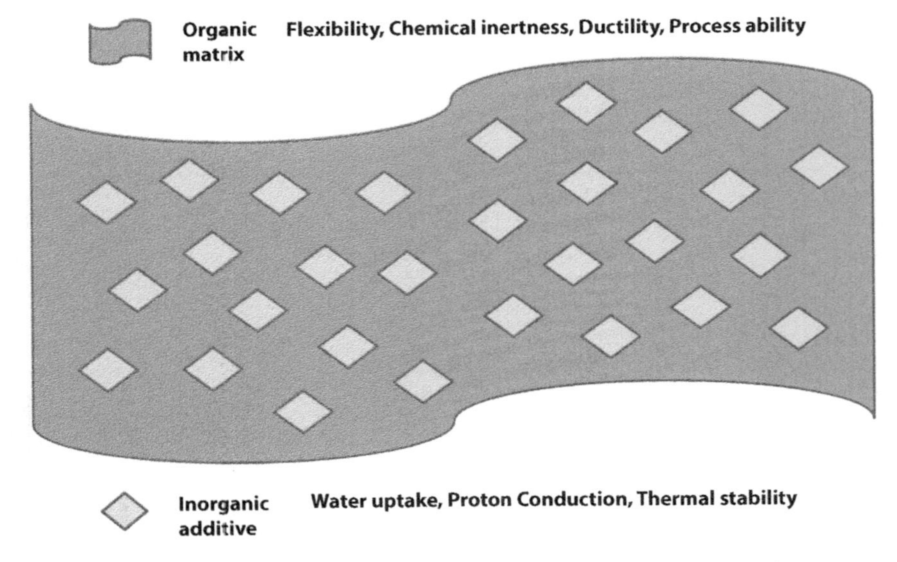

Figure 5.3 Depicts different characteristic properties influenced by organic polymer framework and inorganic materials in the nanocomposite membrane.

intended property of the membrane. To get better of this limitation, surface modification of inorganic additives is done [29].

Different routes are accessible to make nanocomposite membranes. Among them, the sol gel process, doping of inorganic particles, and direct blending of inorganic components into the polymer arrangement are widely used because of their facile preparation route [30].

5.3 Evaluation Methods of Proton Exchange Membrane Properties

5.3.1 Proton Conductivity Measurement

Proton conductivity is measured either by four-point probe or two-point probe method impedance spectroscopy technique, in which membrane is placed in between two electrodes of an electrochemical cell and direction of measurement may be in-plane or through plane [31]. Proton conductivity is determined (in siemens per centimeter) using Equation 5.1.

$$\sigma = \frac{d}{i\,S_a} \tag{5.1}$$

where d is the distance in centimeters between the two electrodes, i is impedance, and S_a is surface area of the membrane.

5.3.2 Water Uptake Measurement

The water uptake capacity of the membrane is nothing but the mass ratio of absorbed water to that of the dry membrane [32]. The prepared membrane is dried in an oven and immersed in a deionized water bath at room temperature for 24 h. The water uptake % is calculated using Equation 5.2.

$$WaUp\,\% = \frac{\left(WNCM_W - DNCM_W\right)}{DNCM_W} \times 100 \tag{5.2}$$

where $WNCM_w$ and $DNCM_w$ are weight of wet and dry nanocomposite membranes, respectively.

5.3.3 Oxidative Stability Measurement

Membrane was soaked in Fenton's solution containing 3% H_2O_2 and 2ppm $FeSO_4$ at 80°C. Depending on the time taken by the membrane to degrade in to pieces and by weight loss of the membrane for 1 h, its oxidative stability was determined [33].

5.3.4 Thermal and Mechanical Properties Measurement

The thermogravimetric analysis method is generally employed to evaluate the thermal stability and weight loss at different temperatures involving evaporation of water, rupture of functional groups, and degradation of polymer matrix [34]. Mechanical strength of the membrane is analyzed by dynamical mechanical analysis and by plotting stress sustained against elongation at break.

5.4 Nafion-Based Membrane

Titanium dioxide added Nafion nanocomposite membranes were fabricated by Amjadi *et al.* via sol-gel method [35]. As per their results, 3 wt% loading of TiO_2 into the pure Nafion matrix enhances its water uptake by 51% because of the hygroscopic nature of the TiO_2 nanoparticles but, interestingly, instead of an increase in water uptake, proton conductivity decreased with the addition of TiO_2. Proton conductivity of 2 and 5 wt% TiO_2-doped Nafion membrane at room temperature with low relative humidity of 20% is found to be 0.00267 S cm^{-1} and 0.00167 S cm^{-1}, respectively. The reason for this is discontinued proton conduction path due to the incorporation of TiO_2 in between the sulfonic acid clusters. T_g of the polymer is increased by the Insertion of TiO_2 and which will aid the membrane to operate at an elevated temperature of 110°C. Heteropolyacids are inorganic oxyacids that exhibit powerful acidity because of their keggin structure [36]. Amirinejada *et al.* incorporated Cesium hydrogen salt of heteropolyacids (CsHPs) involving two combinations, i.e., $Cs_{2.5}H_{0.5}PMo_{12}O_{40}$ (CsPMo) and $Cs_{2.5}H_{0.5}PW_{12}O_{40}$ (CsPW), into the Nafion polymer, which resulted in higher water retention but build a slight hindrance to the activity of the sulfonic group [23]. The maximum power density values of CsPMo and CsPW were found to be 420 and 351 mW cm^{-2}, respectively at 80°C, while undoped Nafion membranes only able to produce 266 mW cm^{-2}. The conductivity of 15 wt% CsPW is higher than that of pristine Nafion membrane and its value promotes from 5.80×10^{-3} to 1.42×10^{-2} S cm^{-1} with an increase in the relative humidity from 30% to 100% [33]. The structure of CsPW is depicted in Figure 5.4.

Up to 300°C stable sulfated titania-silica (SO_4^{2-}/TiO_2-SiO_2)(STS) Nafion nanocomposite membranes were fabricated by Sayeed *et al.* with 2 wt% loading of STS into the Nafion matrix via sol gel method and its water uptake ability increased with increasing the addition of STS up to 5 wt% [37]. Devrim *et al.* prepared Nafion–titanium silicon oxide ($TiSiO_4$) nanocomposite membranes in which connectivity of ionic clusters of the polymer are improved because of the presence of nanosized $TiSiO_4$, and this inorganic nanoparticle brings forth a favorable pathway for proton movement, which is responsible for increased membrane proton conductivity [38]. Water uptake and membrane degradation temperature enhanced along with $TiSiO_4$ doping. The highest power density of 420 mA cm^{-2} was observed for Nafion-$TiSiO_4$ membrane at 80°C, which is

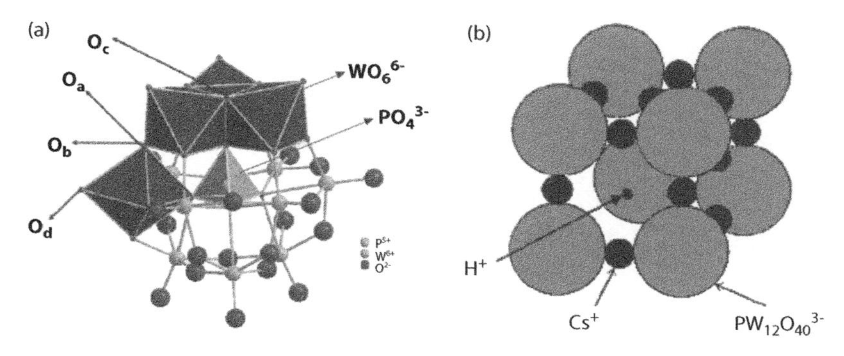

Figure 5.4 Primary (a) and secondary (b) structure of CsPW. Reprinted from [22], Copyright 2010, Journal of Power Sources, Elsevier.

operating at 0.6 V. For high temperature (120°C) and low humidity (25%) fuel cell application, Zarrin *et al.* fabricated sulfonic acid functionalized graphene oxide (GO) Nafion nanocomposite (F-GO/Nafion) membrane with 10 wt% F-GO, which showed proton conductivity of 0.047 S cm^{-1} and is four times greater than recast Nafion membrane (0.012 S cm^{-1}) [39]. The same membrane is successful in providing a power density of 0.15 W cm^{-2} and is around 3.6 times higher than that of the recast Nafion membrane (0.042 W cm^{-2}).

Peng *et al.* reacted C=C of GO with C-F linkage of Nafion through an atom transfer radical addition (ATRA) reaction to obtain nanohybrids of GO-Nafion, which showed a 1.6 times surge in proton conductivity when compared to recast Nafion membrane [40]. The incorporation of 0.10 wt% of GO by this method showed 0.04008 S cm^{-1} proton conductance at 20°C. Schematic representation of ATRA reaction and formation of nanohybrid is shown in Figure 5.5.

Chemical stability of the Nafion membrane is enhanced by Taghizadeh and Vatanparast in 2016 by incorporating ZrO_2 nanoparticles into the polymer matrix, which was confirmed by lower fluoride release and weight loss in Fenton's test [41]. Figure 5.6 shows the graphical representation and surface images of the pristine and nanocomposite membrane. Since ZrO_2 is hydrophilic in nature, its addition significantly increased water uptake percentage but is not supposed to better the thermal strength of the membrane. The incorporation of ZrO_2 lessens the proton conducting ability of the membrane. Pure Nafion showed the conductance of 0.018 S cm^{-1},

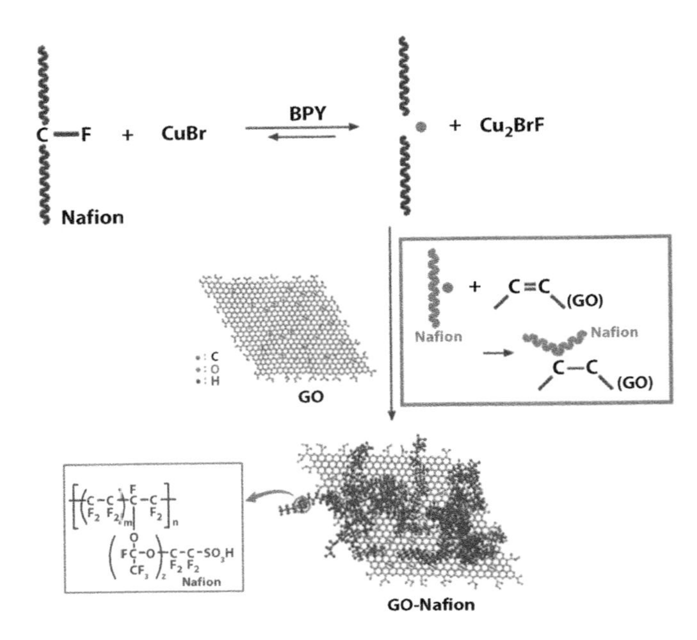

Figure 5.5 ATRA reaction of GO surface with Nafion to give hybrid GO-Nafion membrane. Reprinted from [40], Copyright 2016, Journal of Membrane Science, Elsevier.

while 2, 4, and 8 wt% ZrO_2-incorporated membranes showed the value of 0.01623, 0.01457, and 0.00868 S cm^{-1}, respectively.

In 2018, Parnian *et al.* fabricated ZrO_2-added recast Nafion nanocomposite membrane, which clearly showed the increase in tensile strength upon ZrO_2 doping [42]. Improved mechanical robustness is an indication of enhanced intermolecular forces between the filler and the polymer matrix and also uniform dispersion of ZrO_2 nanoparticles in the membrane. Chai *et al.* reported the preparation of Nafion carbon nanocomposite membrane by hydrothermal treatment with the infusion of glucose solution [43]. Nafion carbon nanocomposite membrane exhibited smaller weight loss in TGA analysis at 320°C to 550°C because of the remarkable thermal stability of carbon nanoparticles and carbon loading of 3.6 wt% showed highest proton conductivity of 0.126 S cm^{-1}. Water uptake capacity of pure Nafion 117 membrane is 38%, and it appreciably increased to 70.6% with an increase in carbon doping of 9.8 wt%.

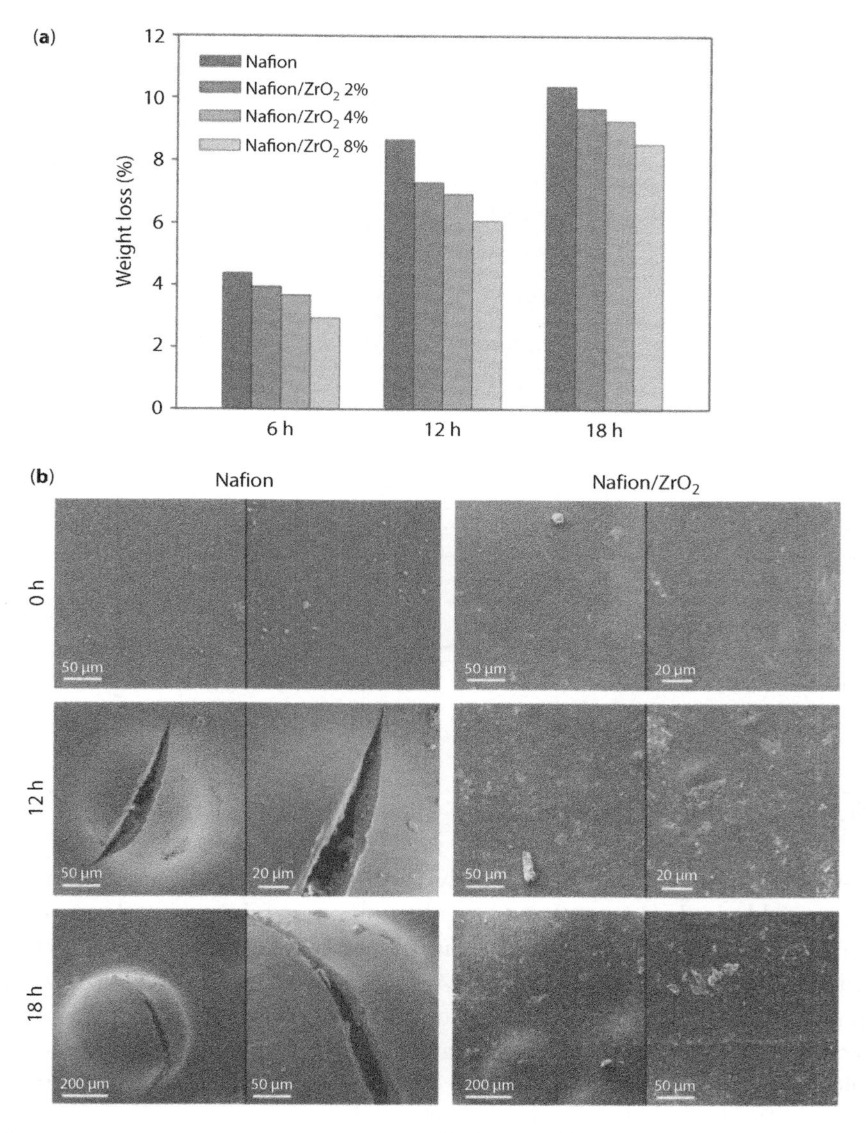

Figure 5.6 (a) Fenton's test result showing weight loss% with respect to time for plain, 2%, 4%, and 8% ZrO_2-loaded Nafion nanocomposite membrane. (b) Image of broken pristine Nafion membrane and unbroken ZrO_2 nanoparticles added Nafion nanocomposite membrane after Fenton's test for a specified time. Reprinted from [41], Copyright 2016, Journal of Colloid and Interface Science, Elsevier.

Table 5.3 Different Nafion based nanocomposite membranes showing their proton conductivity along with their additive percentage and temperature.

S. no.	Membrane	Additives composition	Proton conductivity in S cm^{-1}	Temperature in °C	Reference
1	Nafion-CsPMo	$Cs_{2.5}H_{0.5}PMo_{12}O_{40}$	0.0580	25	[23]
2	Nafion-CsPWA	15wt% CsPWA	0.0142	120	[33]
3	Nafion-TiO_2	2 wt% TiO_2	0.0021	25	[35]
4	Nafion- $TiSiO_4$	10 wt% $TiSiO_4$	0.300	100	[38]
5	Nafion-Functionalized GO	10 wt% F-GO	0.0470	120	[39]
6	Nafion-GO	0.10 wt% GO	0.0823	95	[40]
7	Nafion-ZrO_2	4 wt% ZrO_2	0.0145	25	[41]
8	Nafion/ZrO_2	7.5 wt% ZrO_2	0.050	100	[42]
9	Nafion-Carbon	3.6 wt% C	0.126	25	[43]

5.5 Poly(Benzimidazole)-Based Membrane

The water uptake capacity of a poly(benzimidazole) (PBI) membrane is appreciably high (14 wt%) compared to other membranes such as polycarbonate (0.3 wt%), polyether ketone (0.5 wt%), and polyimide (1.2 wt%) when submerged in a water bath [44]. Hooshyari *et al.* used $BaZrO_3$ as nanoadditives to enhance proton conductivity of the pristine PBI membrane because it possesses low activation enthalpy of proton mobility, simple cubic perovskite structure, high coordination number, and its aptness to hold PA doping [45]. PA doping helps in better proton conductivity and is reflected in PA-doped $BaZrO_3$-PBI nanocomposite membrane, which displayed maximum proton conductivity of 0.125 S cm^{-1} at 180°C with 5% relative humidity. $BaZrO_3$-PBI nanocomposite membrane exhibited a maximum power density of 0.56 W cm^{-2} at 180°C and 0.5 V, which makes it a good membrane for high-temperature PEMFC. Schematic representation of proton transfer in $BaZrO_3$-PBI nanocomposite membrane is shown in Figure 5.7.

Suryani *et al.* fabricated a PBI nanocomposite membrane by incorporating sulfonated silica nanoparticles (S-SNP) in it [46]. In this case, silica modification (Figure 5.8) plays a crucial role in enhancing the interaction between S-SNP and PBI, which, in turn, increased the tensile strength of the membrane. Nanocomposite PBI membranes with 10 wt% of S-SNP exhibited three-fold proton conductivity compared to plain PBI membrane.

In 2012, (p-carboxyphenyl)maleimide (pCPM)–functionalized SNPs (SNP-pCPM) are utilized to fabricate PBI nanocomposite membrane [28]. SNP-pCPM PBI membrane of 10 wt% showed 25% enhancement in its

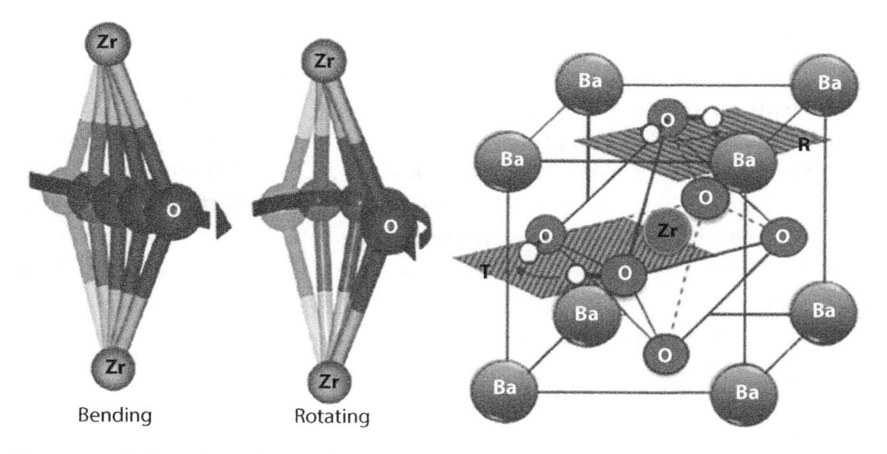

Figure 5.7 Enlarged Bending and rotating motion Zr-O-Zr bond in $BaZrO_3$ perovskite structure along with proton transfer (T) and reorientation (R) pathways where energy minimum of the proton is depicted by white balls. Reprinted from [45], Copyright 2014, Journal of Power Sources, Elsevier.

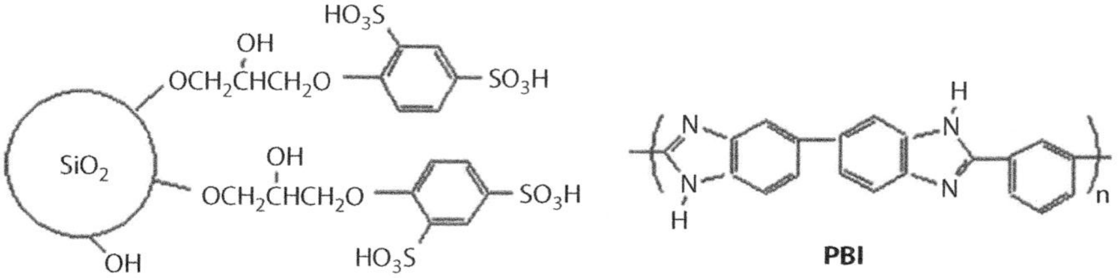

Figure 5.8 Structures of sulfonated silica nanoparticle (S-SNP) and PBI membrane. Reprinted from [46], Copyright 2009, Journal of Membrane Science, Elsevier.

proton conductivity value (0.0500 S cm^{-1}) at 160°C when compared to pristine PBI membrane. SNP-pCPM PBI membrane of 20 wt% exhibited a tensile strength of 98 MPa, which is 20 MPa greater compared to plain PBI membrane. Plain PBI showed a maximum power density of 530 mW cm^{-2} and is increased to 650 mW cm^{-2} for SNP-pCPM PBI nanocomposite membrane. An efficient PBI-based PEM capable of working at high temperature with minimum (1%) aid of humidity is fabricated by Linlina *et al.* [47]. The acid doped PBI nanocomposite membrane with 10 wt% modified silica (m-SNP) showed H^+ ion conductivity of 0.038 S cm^{-1} at 140°C, while plain PBI membrane under identical condition showed proton conductivity of 0.015 S cm^{-1}. Pristine PBI membrane showed water uptake of 14.3%, while the addition of hygroscopic silica nanoparticles increases its value up to 23.6% at 80°C. Virgin PBI membrane PA doping level is found to be 60.8%, and it is increased up to 70.92% for PBI nanocomposite membrane incorporated with 2 wt% unmodified silica (US) nanoparticles. The reason for this is the increased number of hydrogen bonding between OH groups of US and PA. Hence, PA doping level for 2 wt% modified silica–added nanocomposite is found to be only 28.65%. Nawn *et al.* utilized ZrO_2 as nanoadditives for poly[2,2'-(m-phenylene)-5,5'-bibenzimidazole] (PBI4N) membranes and it increased doping of PA to get a composition of $[PBI4N(ZrO_2)_{0.231}](H_3PO_4)_{13}$, which showed notable proton conductivity of 0.104 S cm^{-1} at 185°C [48]. The stability of the PA-doped membranes lasts only up to 180°C, while the undoped membranes were found to be stable up to 400°C. It indicates a decrease in thermal stability with an increase in PA doping. Fe_2TiO_5 nanoparticles of 4 wt% are uniformly dispersed in the PBI matrix by Shabanikia *et al.*, which significantly enhanced the proton conductance up to 0.078 S cm^{-1} at 180°C [49]. Fe_2TiO_5 nanoparticle inclusion resulted in an exceptional increase in PA acid uptake and tensile strength compared to the pristine membrane. In dry condition, Fe_2TiO_5-PBI nanocomposite membrane produced a maximum power density of 380 mW cm^{-2} at 180°C at 0.5 V. Shabanikia *et al.* in another work used $SrCeO_3$ as an additive in the preparation PBI-based nanocomposite membrane, which possesses a simple perovskite structure and good hygroscopic nature [50]. Membrane with 8 wt% $SrCeO_3$ showed proton conductivity of 0.105 S cm^{-1} and power density of 0.440 W cm^{-2} in 0.5 V at 180°C. Devrim *et al.* were successful in getting a proton conductivity value of 0.1027 S cm^{-1} at 180°C from the PBI-based nanocomposite membrane using SiO_2 as nanofiller [51]. However, its maximum power density was found to be less compared to other membranes and is only 0.250 W cm^{-2} at 165°C. A comparative study of three different PBI-based nanocomposite membranes incorporated with TiO_2, SiO_2, and ZrP was carried out by Ozdemir *et al.* [24]. In their study, 5 wt% ZrP embedded PBI nanocomposite membrane is emphasized because it showed significant PA doping of 85% and highest proton conductivity of 0.200 S cm^{-1} at 180°C, while 5 wt% TiO_2-added PBI membrane showed proton conductivity of 0.044 S cm^{-1}, which is less than the pristine PBI membrane. Table 5.4 represents a comparative study of three different PBI-based nanocomposite membranes involving proton conductivity, acid uptake, and mechanical strength.

TiO_2-incorporated PBI nanocomposite membrane proton conductivity was enhanced above pristine PBI membrane by Esmaeilzadea *et al.* with certain alteration [29]. In this case, TiO_2 nanoparticles were modified by reacting it with cellulose (CL) (Figure 5.9), which is further doped with acid and incorporated in sulfonated PBI (S-PBI) matrix. TiO_2-CL-SO_3H-/S-PBI nanocomposite membranes of 16 wt% displayed proton conductivity of 0.006 S cm^{-1}, which was an outstanding improvement over virgin S-PBI membrane (0.0015 S cm^{-1}) at 120°C. However, the incorporation of this modified titania failed to increase the thermal stability of the membrane.

Recently, in 2021, Jheng and co-workers utilized modified carbon nanofibers in hexafluoroisopropylidene containing polybenzimidazole (6FPBI) nanocomposite membranes [52]. The modification involves acetylation in presence of polyphosphoric acid, which is depicted in Figure 5.10. Surface modification of carbon nanofibers with amino benzyl (AB) groups resulted in detaining PA molecules, which, in turn, facilitate proton movement. Notable improvement was observed in tensile strength (78.9 MPa) and acid doping level (12%) of nanocomposite membrane with 0.3 wt% CNF-AB–added 6FPBI. At 160°C, appreciable enhancement in proton

Table 5.4 Mechanical strength comparison of plain and TiO_2-, SiO_2-, ZrP-added PBI-based nanocomposite membranes. Reprinted from [24], Copyright 2021, International Journal of Hydrogen Energy, Elsevier.

Membrane type	Inorganic content (wt%)	Acid uptake (%)	Tensile strength (MPa)	Elongation at break (%)
PBI	O	13.4	127	59
PBI/SiO_2	5	13.6	102	72
PBI/TiO_2	5	13.4	85.6	34
PBI/ZrP	5	15.4	119	47

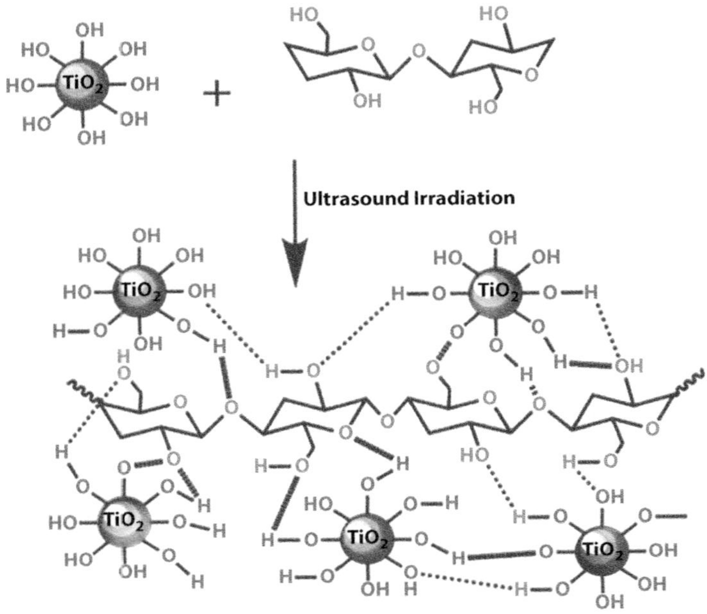

Figure 5.9 Surface modification of TiO_2 nanoparticles with cellulose. Reprinted from [29], Copyright 2018, Ultrasonics Sonochemistry, Elsevier.

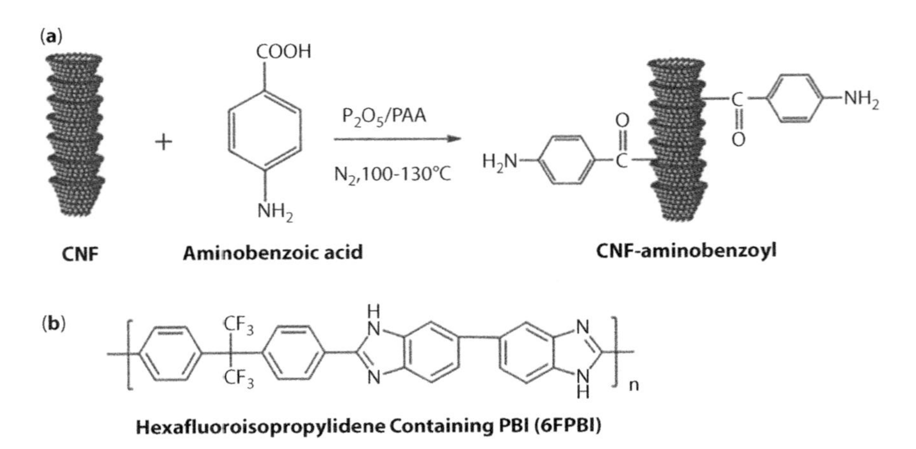

Figure 5.10 (a) Modification of CNF using aminobenzoic acid and (b) structure of hexafluoroisopropylidene containing polybenzimidazole. Reprinted from [52], Copyright 2021, RSC advances.

Table 5.5 Different PBI based nanocomposite membranes showing their proton conductivity along with their additive percentage and temperature.

S. no.	Membrane	Additives composition	Proton conductivity in S cm^{-1}	Temperature in °C	Reference
1	PBI-ZrP-PA	5 wt% ZrP	0.200	140	[24]
2	PBI-TiO_2	5 wt% TiO_2	0.044	180	[24]
3	PBI-SNP-pCPM	10 wt% SNP-pCPM	0.0500	160	[28]
4	S-PBI-TiO_2-cellulose-SO_3H	16% TiO_2-cellulose-SO_3H	0.028	85	[29]
5	PBI-$BaZrO_3$	4% wt% $BaZrO_3$	0.125	180	[45]
6	PBI-(m-SNP)	10 wt% m-SNP	0.0038	140	[47]
7	PBI- ZrO_2-PA	PBI4N$(ZrO_2)_{0.231}$]$(H_3PO_4)_{13}$	0.104	185	[48]
8	PBI-Fe_2TiO_5	4 wt% of Fe_2TiO_5	0.078	180	[49]
9	PBI/ $SrCeO_3$	8 wt % $SrCeO_3$	0.105	180	[50]
10	PBI/SiO_2	5 wt % SiO_2	0.1027	180	[51]
11	6FPBI- CNF-AB	0.3 wt% CNF-AB	0.200	160	[52]

conductivity was observed with a value of 0.200 S cm^{-1} and also power density of 461 mW cm^{-2} for 0.3 wt% CNF-aminobenzoyl/6FPBI nanocomposite membrane.

5.6 Sulfonated Poly(Ether Ether Ketone)-Based Membranes

Peighambardoust and group members fabricated SPEEK-based nanocomposite membranes by incorporating 15 wt% $Cs_{2.5}H_{0.5}PW_{12}O_{40}$ (heteropolyacid) supported with 1.25 wt% Pt catalyst, which exhibited good proton conductivity of 0.0682 S cm^{-1} and is higher than the value of pure Nafion 117 membrane under similar experimental conditions [53]. Water uptake capacity improved up to 45.86% with the addition of heteropolyacid because of its hydrophilic nature and its ability to provide additional hydrogen bonding sites for water molecules. The durability of this membrane is studied by Rowshanzamir *et al.*, in which it is proposed that heteropolyacid converts hydrogen peroxide and crossover hydrogen to water (Equations 5.1 and 5.2) and is the reason for less chemical degradation of the membrane [54]. The chemical reaction of permeable H_2 and O_2 results in the formation of water molecules on the platinum surface and absorption of those water molecules by $Cs_{2.5}H_{0.5}PW_{12}O_{40}$ explains self-humidifying nature of the membrane (Figure 5.11).

$$Cs_{2.5}H_{0.5}PW_{12}O_{40} + H_2O_2 \rightarrow Cs_{2.5}H_{0.5}PW_{12}O_{40}(O) + H_2O \quad (5.3)$$

$$Cs_{2.5}H_{0.5}PW_{12}O_{40}(O) + H_2 \rightarrow Cs_{2.5}H_{0.5}PW_{12}O_{40} + H_2O \quad (5.4)$$

Mossayebi *et al.* prepared SPEEK nanocomposite membrane by adding sulfated zirconia (ZS) nanoparticles [55]. The addition of 5.94% SZ significantly enhanced chemical stability (up to 102 min) and produced maximum proton conductivity of 0.0038 S cm^{-1} at 100°C. In this work,

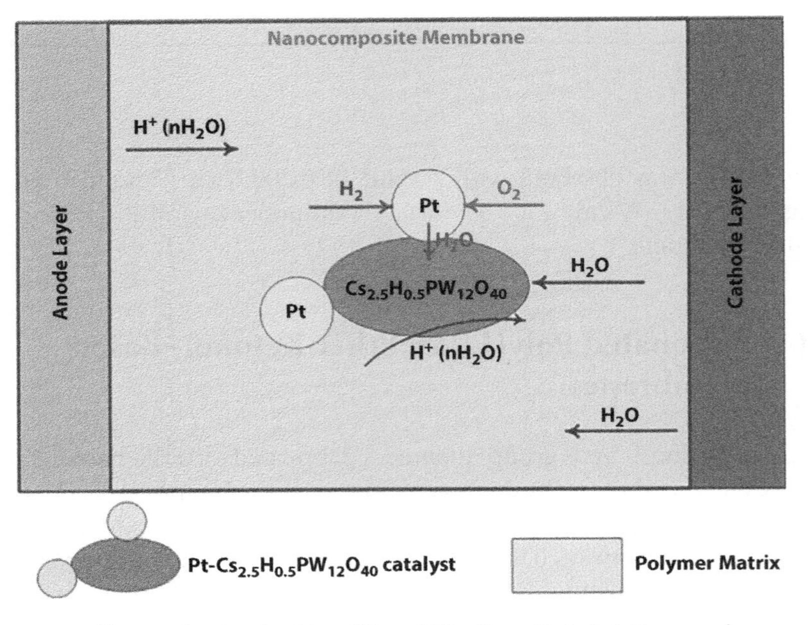

Figure 5.11 The reaction involved in self-humidification effect of platinum and heteropolyacid - $Cs_{2.5}H_{0.5}PW_{12}O_{40}$. Reprinted from [54], Copyright 2014, International Journal of Hydrogen Energy, Elsevier.

sulfonation of zirconia plays a key role in stabilizing the low temperature tetragonal structure of Zr (depicted in Figure 5.12) and reduction in particle size by preventing particle agglomeration. Mechanical strength of nanocomposite membrane increased by the incorporation of SZ nanoparticles in the vacant spaces of SPEEK matrix and water retention ability multiplied 1.5 times over virgin SPEEK membrane.

Song *et al.* treated 12-Tungstophosphoric acid (HPW) with tetraethoxysilane (TEOS) to synthesize hetero poly tungstic acid (HPW)/meso-SiO_2 nanoparticles, which are then embedded with cross-linked SPEEK membrane [56]. HPW/meso-SiO_2 cross-linked nanocomposite membrane of 20 wt% exhibited proton conductivity of 1.9 10^{-3} S cm^{-1} at 120°C under 30% RH, which is 10 times higher than the pristine SPEEK membrane under similar conditions. The appreciable increase in proton conductivity is because of nanoadditives, which assisted in developing ionic clusters pathways and also in preventing water evaporation at elevated temperatures. Sulfonation of PEEK polymer decreases its thermal stability. To enhance the thermal stability, Gashoul *et al.* used ZrO_2 nanoparticles as additives in preparation of SPEEK-ZrO_2 nanocomposite membrane, which is stable up to 200°C and suitable PEM fuel cells [25]. Along with thermal stability, other crucial factors such as water uptake capacity, mechanical stability, and oxidative stability were improved to an appreciable extent with 10 wt% loadings of ZrO_2 nanoparticles in the SPEEK matrix. Proton conductivity of ZrO_2-SPEEK nanocomposite membrane showed improved value over plain membrane only above 80°C. Table 5.6 gives information about the chemical stability of different nanocomposite membranes with different wt% of ZrO_2 nanoparticles.

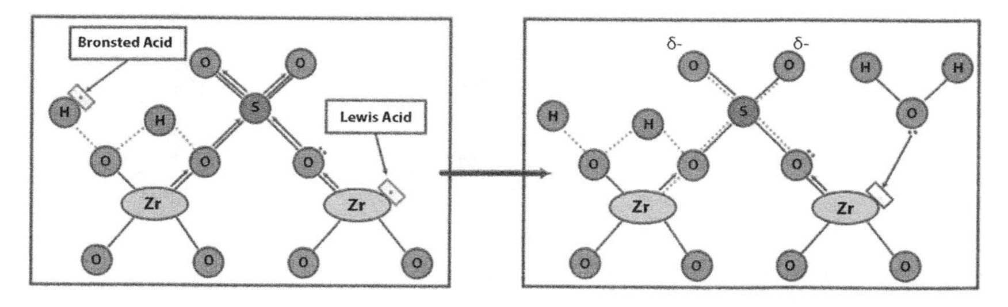

Figure 5.12 Model showing water adhesion on the surface of sulfated zirconia. Reprinted from [55], Copyright 2016, International Journal of Hydrogen Energy, Elsevier.

Table 5.6 Outcome of chemical stability tests of plain and ZrO_2-added nanocomposite membranes. Reprinted from [25], Copyright 2016, Elsevier.

Samples	Hydrolytic stability		Oxidative stability	
	Residual weight (%)	Enhancement (%)	Degradation time (min)	Enhancement (%)
SPEEK	65	-	39	-
SPEEK-2.5% ZrO_2	69	6.1%	49	25.6
SPEEK-5% ZrO_2	71	9.2%	54	38.5
SPEEK-7.5% ZrO_2	73	12.3%	60	53.9
SPEEK-10% ZrO_2	76	16.9%	69	76.9

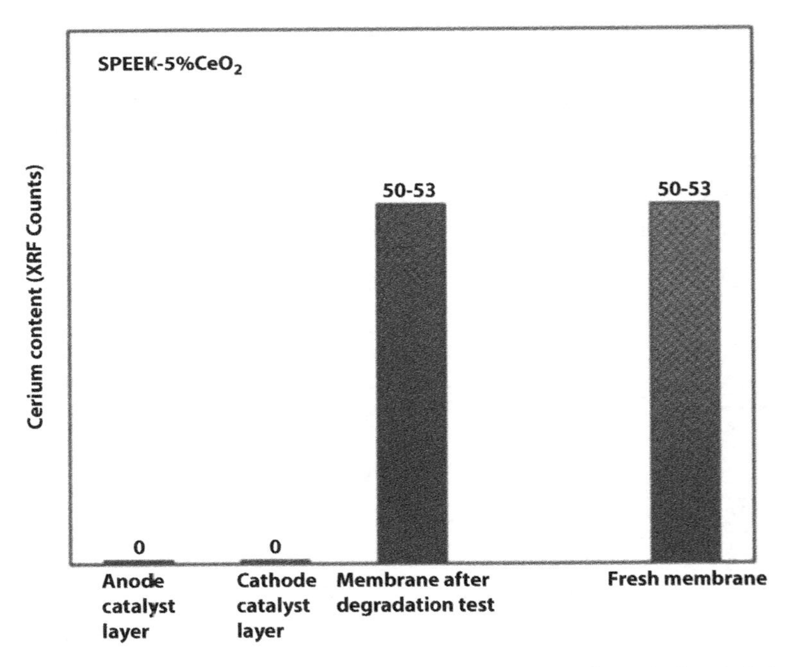

Figure 5.13 XRF result of 5% CeO_2-SPEEK nanocomposite membrane before and after the degradation test. Reprinted from [27], Copyright 2018, Journal of membrane Science, Elsevier.

Table 5.7 Different SPEEK based nanocomposite membranes showing their proton conductivity along with their additive percentage and temperature.

S. no.	Membrane	Additives composition	Proton conductivity in S cm^{-1}	Temperature in °C	Reference
1	SPEEK-$Cs_{2.5}H_{0.5}PW_{12}O_{40}$-Pt	15 wt % $Cs_{2.5}H_{0.5}PW_{12}O_{40}$ + 2wt% Pt	0.0682	25	[53]
2	SPEEK-SZ	5.94 wt % SZ	0.0038	100	[55]
3	SPEEK-(HPW-meso-SiO_2)	20 wt% HPW-meso-SiO_2	0.0019	120	[56]

Parnian *et al.* fabricated 2.5 wt% CeO_2-containing SPEEK nanocomposite membrane, which exhibited good mechanical stability, thermal stability, and chemical stability [57]. Nanocomposite membrane showed lower hydrogen crossover and better proton conductivity than plain membrane. Virgin SPEEK membrane electrode assemblies showed an open-circuit voltage (OCV) decay rate of 0.52 mV h^{-1} after 230 h of accelerated stress test to cause chemical degradation. While 10 wt% CeO_2 embedded SPEEK membrane showed 0.09 mV h^{-1} [27]. The reason for this is scavenged free radicals by ceria nanoparticles, which enhanced membrane durability. In the Fenton test, 10 wt% CeO_2-SPEEK membrane took 87 min for degradation and is far better over pristine membrane, which has taken only 39 min. Stewart *et al.* reported that CeO_2 ions are mobile in the Nafion matrix, and they move toward both the electrodes, which result in the reduction of its efficiency [58]. However, in the case of SPEEK membranes, CeO_2 is very much stable and is confirmed by XRF results (Figure 5.13).

5.7 Poly(Vinyl Alcohol)–Based Membranes

Attaran and co-workers fabricated two kinds of PVA-based nanocomposite membranes involving PVA-$BaZrO_3$ and PVA-poly(vinyl pyrrolidone)–BaZrO3 (PVA-P-$BaZrO_3$) nanocomposite membranes [59]. Proton conductivity of PVA-P-$BaZrO_3$ nanocomposite membrane with 1 wt% $BaZrO_3$ at 70°C was found to be 0.0601 S cm^{-1} and showed highest peak power density of 28.98 mW cm^{-2}. Table 5.8 gives the comparison of two PVA-based nanocomposite membrane and plain membrane along with

Table 5.8 Comparison of water uptake, proton conductivity, and mechanical strength of plain PVA, PVA-$BaZrO_3$, and PVA-P-$BaZrO_3$ nanocomposite membranes at 25°C. Reprinted from [59], Copyright 2014, Elsevier.

Membranes	Water uptake %	Proton conductivity in S cm^{-1}	Elongation at break (%)	Tensile strength (MPa)
PVA	180	0.0005	5.23	25.29
PVA-$BaZrO_3$	195	0.018	6.27	30.53
PVA-P-$BaZrO_3$	220	0.03	21.55	100.5

their mechanical strength, water uptake capacity, and proton conductivity measurements.

Intending to increase proton conductivity of PVA membrane, Hooshyari *et al.* incorporated $La_2Ce_2O_7$ nanoparticles into the PVA matrix using glutaraldehyde as a cross-linking agent to prepare PVA-$La_2Ce_2O_7$ nanocomposite membrane [60]. It significantly increased proton conductivity up to 0.019 S cm^{-1} with 1 wt% $La_2Ce_2O_7$ loading. $La_2Ce_2O_7$ lattice structure and water molecule incorporation in its vacancies are represented in Figure 5.14.

3-Mercaptopropyltrimethoxysilane was grafted on nanoporous silica (SBA-15) by Beydaghi *et al.* and then oxidized to get SBA-15-SO_3H nanoparticles (Figure 5.15), which were incorporated in PVA matrix with

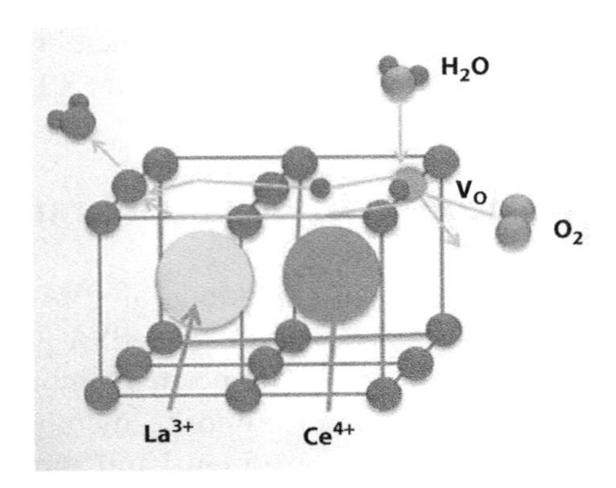

Figure 5.14 Schematic representation of water incorporation in $La_2Ce_2O_7$ lattice structure. Reprinted from [60], Copyright 2014, Iranian Journal of Hydrogen and Fuel Cell.

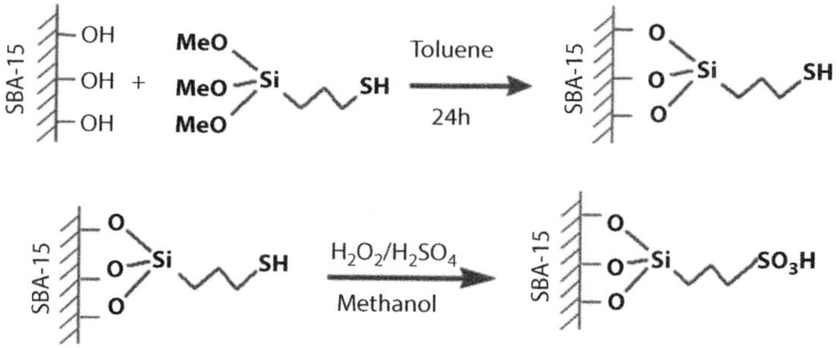

Figure 5.15 Surface modification of SBA-15 using 3-Mercaptopropyltrimethoxysilane and subsequent oxidation to get sulfonated SBA-15 nanoparticles. Reprinted from [61], Copyright 2014, Springer Nature.

Table 5.9 Different PVA based nanocomposite membranes showing their proton conductivity along with their additive percentage and temperature.

S. no.	Membrane	Additives composition	Proton conductivity in S cm^{-1}	Temperature in °C	Reference
1	PVA-P-$BaZrO_3$	1 wt% $BaZrO_3$	0.0601	70	[59]
2	PVA-$La_2Ce_2O_7$	1wt% $La_2Ce_2O_7$	0.019	25	[60]
3	PVA-SBA-15-SO_3H	5wt% SBA-15-SO_3H	0.006	25	[61]
4	PVA-SGO	5 wt% SGO	0.050	25	[62]

cross-linking agent glutaraldehyde (GLA) [61]. Hence, formed 5 wt% SBA-15-SO_3H induced PVA-SBA-15-SO_3H nanocomposite membrane showed significant enhancement in water uptake up to 80% and thermal stability up to 180°C. Proton conductivity enhancement is not appreciable and found to be 0.006 S cm^{-1}.

Beydaghi *et al.* incorporated 5 wt% aryl sulfonated GO (SGO) nanoparticles in PVA matrix to prepare PVA-SGO nanocomposite membrane, which exhibited excellent thermal stability up to 223°C and possess a high tensile strength of 67.8 MPa [62]. Prepared nanocomposite membrane showed good proton conductivity of 0.050 S cm^{-1}.

5.8 Sulfonated Polysulfone–Based Membranes

Polysulfone membranes are extensively used in fabricating nanocomposite membranes for a variety of applications including ultrafiltration membranes [63] because of their good film-forming strength [64]. In addition to that PSF membranes show excellent thermal stability, mechanical strength, oxidative stability, and ability to withstand a wide pH range [65, 66]. Sulfonated polysulfone (SPSF) membranes emerged as an alternative for extensively used Nafion membranes because of their low cost and facile fabrication process [67]. Leila Ahmadian-Alam and Hossein Mahdavimixed imidazole encapsulated metal-organic framework (MOF) NH_2MIL-53(Al) and sulfonic acid modified silica to produce efficient proton conducting nanoparticles and embedded them in SPSF matrix to produce nanocomposite membrane [68]. Modified silica–added MOF-SPSF nanocomposite membrane of 5 wt% displayed H^+ ion conductivity value of 0.017 S cm^{-1} at 70°C. The addition of silica and MOF significantly enhanced its tensile strength from 19 to 30 Mpa and exhibited a power density of 40.80 mW cm^{-2}. Simari *et al.* utilized Mg/Al-NO^{-3} layered double hydroxides (LDHs) as nanoadditives in the ratio of 2:1 in SPSF matrix, in which sulfonation of PSF polymer matrix was done by trimethylsilyl chlorosulfonate [67]. Hence, formed SPSF-LDH nanocomposite membrane significantly enhanced proton conductivity value up to 0.0137 S cm^{-1} at 120°C with a relative humidity of 30% and is almost 30% greater than recast Nafion membrane. This nanocomposite membrane showed outstanding water retention capacity, and it leads to produce a power density of 204.5 mW cm^{-2}, which was double the power density produced by recast Nafion membrane at elevated temperature (110°C) with RH of only 25%. In another work, Simari *et al.* fabricated SGO incorporated SPSF nanocomposite membrane (SPSF-SGO), in which aminopropansulfonic acid is utilized for the sulfonation of GO additive [69]. SGO is firmly attached with the polymer matrix and homogenous membrane morphology accounts for the enhanced mechanical and thermal stability of the membrane. GO comprise reactive carbonyl groups and hydroxyl groups and exhibit excellent hydrophilicity along with good proton conduction capacity [70]. SGO incorporated PSF nanocomposite membrane showed higher proton conductivity than Nafion 212 membrane at less than 50% relative humidity, while GO embedded PSF and plain PSF membrane showed much less proton conductivity at all different RH condition (Figure 5.16). PSF-SGO nanocomposite membrane displayed a power density of 182.6 mW cm^{-2} at 110°C at a low RH of 25%.

Amino acid modified CL whiskers (Figure 5.17) were embedded in the PSF matrix by Xu *et al.*, and it resulted in PEMs with outstanding proton

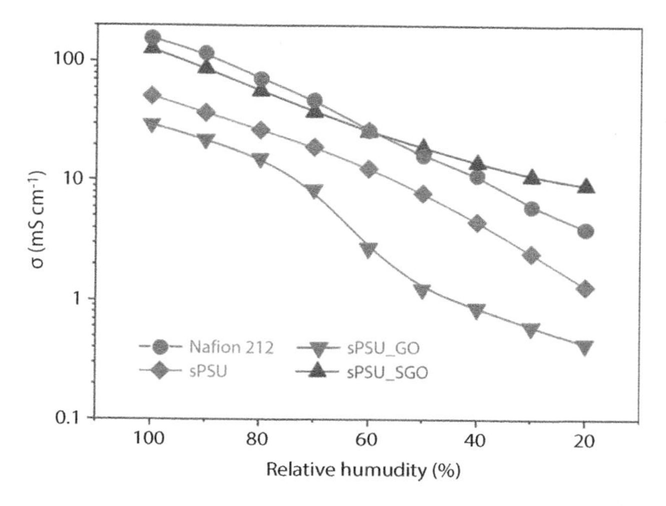

Figure 5.16 Proton conductivity plot of sulfonated polysulfone (sPSU), GO-Sulfonated polysulfone (sPSU_GO), SGO-sulfonated polysulfone (sPSU_SGO), and plain Nafion 212 membranes at different relative humidity %. Reprinted from [69], Copyright 2021, Elsevier.

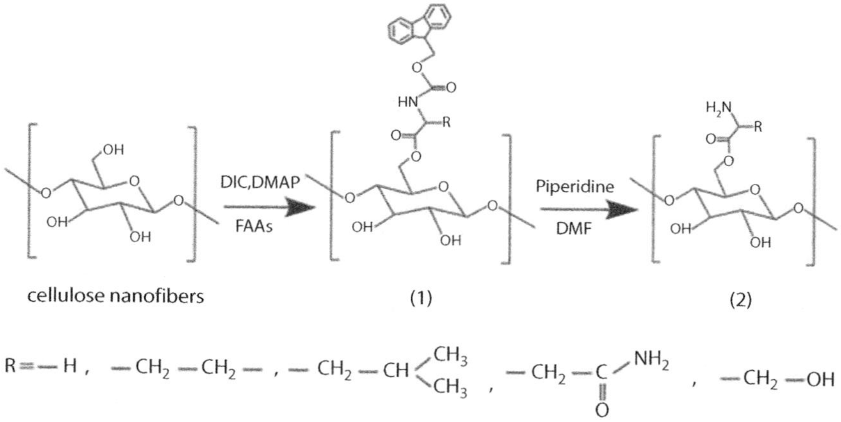

DIC,DMAP
FAAs
Piperidine
DMF
cellulose nanofibers
(1)
(2)

R = —H , —CH_2—CH_2— , —CH_2—$CH(CH_3)_2$, —CH_2—C(=O)—NH_2 , —CH_2—OH

Figure 5.17 Reaction showing an amino acid modification of cellulose whiskers (R-amino acid residue). Reprinted from [71], Copyright 2019, Elsevier.

conductivity [71]. At 80°C 4 wt% L-serine–modified nanocomposite membrane SPSF-SER-4 exhibited proton conductivity of 0.209 S cm^{-1}, while 10 wt% incorporated SPSF-SER-10 showed further enhancement in proton conductivity of 0.234 S cm^{-1}.

Table 5.10 Different Sulfonated polysulfone based nanocomposite membranes showing their proton conductivity along with their additive percentage and temperature.

S. no.	Membrane	Additives composition	Proton conductivity in S cm^{-1}	Temperature in °C	Reference
1	SPSF-LDH	3 wt% LDH	0.0137	120	[67]
2	SPSF-(NH_2MIL-53(Al)-mSiO_2)	5 wt% mSiO_2	0.017	70	[68]
3	SPSF-SER-10	10wt% Cellulose-SER	0.234	80	[71]

5.9 Chitosan-Based Membranes

Biopolymer CS is nothing but a long chain of D-glucosamine obtained from N-deacetylation of chitin using an alkali [72, 73]. CS-based membranes are of a great deal of interest because of their biodegradability, non-toxicity, hydrophilic nature, and good film-forming ability [65, 74, 75]. CS is highly hydrophilic and water insoluble [73], and its mechanical strength can be enhanced by the addition of silica, TiO_2, silver, etc. [76]. Gomez *et al.* prepared a polymer matrix by mixing CS and polyvinyl alcohol in the ratio of 80:20 and so the formed matrix is incorporated with 1,000-ppm TiO_2 nanoparticles [77]. CS-PVA-TiO_2 nanocomposite membrane doped with 2M KOH showed increased thermal stability up to 140°C with excellent water uptake capacity. Mechanical strength also improved with a 50% enhancement in tensile strength. Divya *et al.* incorporated 0.75 wt% exfoliated MoS_2 in CS matrix to fabricate CS-MoS_2 nanocomposite membrane, and it increased the membrane proton conductivity to 0.0032 S cm^{-1} at 80°C [78]. CS-MoS_2 nanocomposite membrane significantly enhanced tensile strength up to 55 MPa, while plain CS membrane was stable only up to 30MPa and nanocomposite membrane exhibited outstanding thermal stability of 400°C. Vijayalekshmi *et al.* protonated polyaniline using sulfuric acid and treated it with hydrophilic silica to get PANI-SiO_2 nanoparticles [79]. PANI-SiO_2-embedded CS-PANI-SiO_2 nanocomposite membrane of 3 wt% showed the highest proton conductivity of 0.00839S cm^{-1} at 80°C, which is due to the doped sulfate ions, hydroxyl groups on silica, and protonated amine groups of the polymer matrix. Nanocomposite membrane improved water uptake capacity by nearly up to 90%. Tensile strength increased with an increase in nanoparticle addition, and it was found that 7 wt% PANI-SiO_2–added CS composite membrane displayed tensile strength of 102 MPa, while plain CS membrane showed a tensile strength of only 20 MPa. Boehmite nanoadditives (OS) are coated with p-toluene sulfonic acid and incorporated into CS matrix by Ahmed *et al.* to prepare CS-OS nanocomposite membrane [26]. OS-embedded nanocomposite membrane of 5 wt% displayed good proton conductivity of 0.032 S cm^{-1} and is approximately equal to the conductivity of plain Nafion membrane. In a nanocomposite membrane preparation, Ahmed *et al.* utilized sulfonated TiO_2 nanoparticles (STiO_2) as additives in CS matrix [80]. STiO_2-added CS-STiO_2 nanocomposite membrane of 5 wt% exhibited proton conductivity of 0.035 S cm^{-1}. The reason for the enhanced thermal stability and tensile strength of the composite membrane is due to the interaction between $-SO_3H$ groups of TiO_2 and $-NH_2$ groups of CS which resist the movement of the CS chain. Cui *et al.* used three different components involving one-dimensional sodium lignin sulfonate functionalized carbon nanotubes (SCNT), 2D GO, and three-dimensional zirconium-based MOFs (UiO-66) to fabricate 7 wt% SCNT-GO-U–added CS-SCNT-GO-U nanocomposite membrane, which showed good proton conductivity of 0.064 S cm^{-1} at 70°C [81]. Selectivity and mechanical strength of composite membrane also increased appreciably. Wang and group members fabricated nanocomposite membranes of amino acid functionalized CS nanofiber mats (A-CSF) incorporated with SPSF in between the inter fiber voids of the A-CSF matrix [82]. SEM images of the plain and composite membrane are displayed in Figure 5.18. Excellent proton conductivity of 0.192 S cm^{-1} was displayed by serine functionalized SPSF-filled CSF

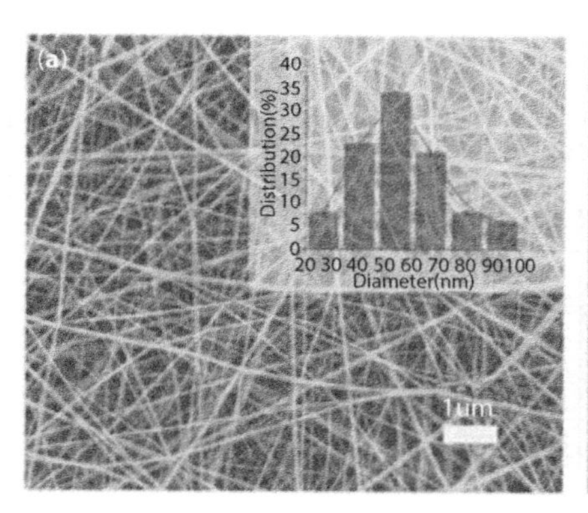

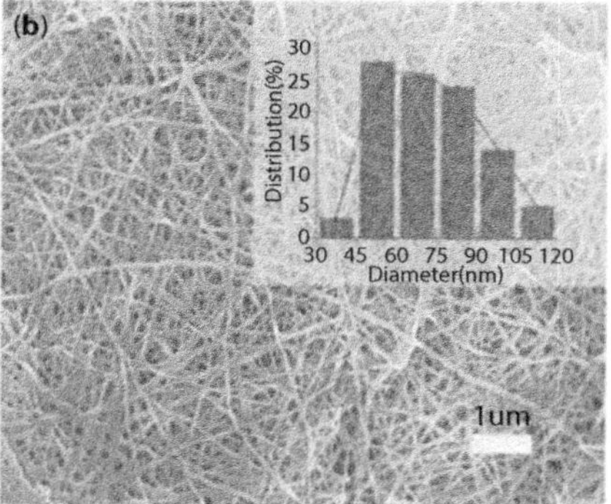

Figure 5.18 SEM images of (a) A-CSF and (b) SPSF-Ser-CSF. Reprinted from [71], Copyright 2019, European Polymer Journal, Elsevier.

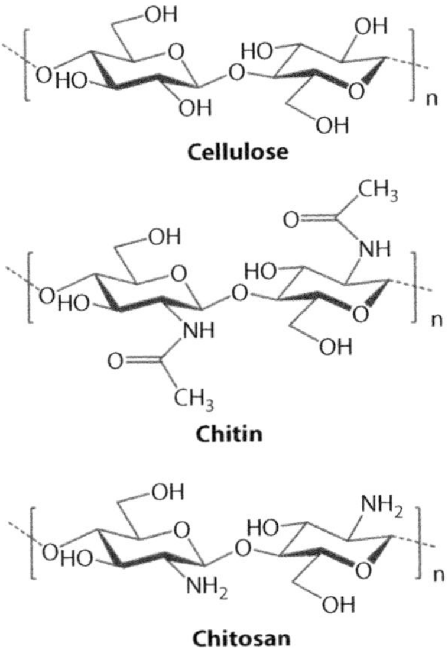

Figure 5.19 Structure of cellulose, chitin and chitosan. Reprinted from [83], Copyright 2021, Elsevier.

Table 5.11 Different chitosan based nanocomposite membranes showing their proton conductivity along with their additive percentage and temperature.

S. no.	Membrane	Additives composition	Proton conductivity in S cm^{-1}	Temperature in °C	Reference
1	Chitosan- OS_1	5 wt% OS_1	0.032	25	[26]
2	Chitosan-PVA-TiO_2	1000 ppm TiO_2	10^{-6}	25	[77]
3	Chitosan-E-MoS_2	0.75 wt% E-MoS_2	0.00361	80	[78]
4	Chitosan-PAni-SiO_2	3 wt % PAni-SiO_2	0.00839	80	[79]
5	CS-$STiO_2$	5wt% $STiO_2$	0.035	25	[80]
6	CS-SCNT-GO-U	7wt% SCNT-GO-U	0.064	70	[81]
5	Chitosan-SPSF-Ser-CSF	5 wt % SPSF	0.192	80	[82]
6	Chitosan-CL-SSA-0.05	0.05 M SSA	4.76x10^{-5}	25	[84]

(SPSF-Ser-CSF) nanocomposite membrane 80°C. Virgin SPSF membrane showed of tensile strength of 21.7 MPa. While SPSF-Ser-CSF nanocomposite membrane increased it up to 26.8 MPa.

CL is a natural polymer and its structure is close to that of the CS, and only the difference is that, instead of $-NH_2$, $-OH$ group is attached at C_2 carbon (Figure 5.19) [83]. CS and CL are mixed in a weight ratio of 1:0.18 by Rahman *et al.* and fabricated two nanocomposite membranes with 0.05 and 0.10 M sulfosuccinic acid (SSA) to give CS-CL-SSA-0.05 and CS-CL-SSA-0.10 nanocomposite membrane, respectively [84]. Among these two membranes, CS-CL-SSA-0.05 exhibited the highest proton conductivity of 4.73×10^{-5} S cm^{-1}, which is greater than that of the plain Nafion 212 membrane. CS-CL-SSA-0.10 showed excellent water uptake of 104% at 90°C while, at the same, temperature pristine Nafion 212 membrane showed water uptake of only 16.3%.

5.10 Conclusions

Nanocomposite PEMs boosted the development of efficient fuel cells and were successful in improving proton conductivity, high-temperature stability, and working ability at a wide relative humidity range. Uniform dispersion of fillers in the polymer matrix and strong interaction between nanoadditives and polymer matrix leads to membranes with excellent mechanical stability with high tensile strength. Spotlight on surface modification of nanoadditives is mainly to achieve this goal and to prevent agglomeration as well. The incorporation of hydrophilic additives amplifies the water uptake capacity of the membrane and, in turn, facilitates proton transport, but, in few cases, it may decrease the proton conductivity by disconnecting proton conducting pathways of the base polymer matrix. Advancements in the biodegradable, nontoxic, and bountiful natural polymer (PVA and CS)–based PEMs are mainly because of nanocomposite membranes, in which additives play a crucial role in increasing its proton conductivity and chemical stability. The finest way to get custom-tailored properties, low cost, and quality PEM fuel cells is by nanocomposite membranes.

References

1. Ahmad, T. and Zhang, D., A critical review of comparative global historical energy consumption and future demand: The story told so far. *Energy Rep.*, 6, 1973, 2020.

2. Jain, V., Fossil fuels, GHG emissions and clean energy development: Asian giants in a comparative perspective. *Millenn. Asia*, 10, 1, 2019.
3. Ogungbemi, E., Wilberforce, T., Ijaodola, O., Thompson, J., Olabi, A.G., Review of operating condition, design parameters and material properties for proton exchange membrane fuel cells. *Int. J. Energ. Res.*, 45, 1227, 2021.
4. Mishra, A.K., Bose, S., Kuila, T., Kim, N.H., Lee, J.H., Silicate-based polymer-nanocomposite membranes for polymer electrolyte membrane fuel cells. *Prog. Polym. Sci.*, 37, 842, 2012.
5. Grove, W.R., On a small voltaic battery of great energy; some observations on voltaic combinations and forms of arrangement; and on the inactivity of a copper positive electrode in nitro-sulphuric acid. *Philos. Mag.*, 15, 287, 1839.
6. Pourzare, K., Mansourpanah, Y., Farhadi, S., Advanced nanocomposite membranes for fuel cell applications: A comprehensive review. *Biofuel Res. J.*, 3, 496, 2016.
7. Sharaf, O.Z. and Orhan, M.F., An overview of fuel cell technology: Fundamentals and applications. *Renew. Sust. Energ. Rev.*, 32, 810, 2014.
8. Narayanamoorthy, B., Datta, K.K.R., Eswaramoorthy, M., Balaji, S., Improved oxygen reduction reaction catalyzed by Pt/Clay/Nafion nanocomposite for PEM fuel cells. *ACS Appl. Mater. Interfaces*, 4, 3620, 2012.
9. Dushyant, S., Spivey, J.J., Berry, D.A., Introduction to fuel processing, in: *Fuel cells: technologies for fuel processing*, S. Dushyant, J.J. Spivey, D.A. Berry, (Eds.), pp. 1–9, Elsevier, Amsterdam, Netherlands, 2012.
10. Authayanun, S., Im-orb, K., Arpornwichanop, A., A review of the development of high temperature proton exchange membrane fuel cells. *Chin. J. Catal.*, 36, 473, 2015.
11. Muhmed, S.A., Nor, N.A.M., Jaafar, J., Ismail, A.F., Othman, M.H.D., Rahman, M.A., Aziz, F., Yusof, N., Emerging chitosan and cellulose green materials for ion exchange membrane fuel cell: A review. *Energy Ecol. Environ.*, 5, 85, 2020.
12. Hickner, M.A., Ghassemi, H., Kim, Y.S., Einsla, B.R., McGrath, J.E., Alternative polymer systems for Proton Exchange Membranes (PEMs). *Chem. Rev.*, 104, 4587, 2004.
13. Elakkiya, S., Arthanareeswaran, G., Venkatesh, K., Kweon, J., Enhancement of fuel cell properties in polyethersulfone and sulfonated poly (ether ether ketone) membranes using metal oxide nanoparticles for proton exchange membrane fuel cell. *Int. J. Hydrog. Energy*, 43, 21750, 2018.
14. Zhou, Z., Zholobko, O., Wu, X.-F., Aulich, T., Thakare, J., Hurley, J., Polybenzimidazole-based polymer electrolyte membranes for High-temperature fuel Cells: Current status and prospects. *Energies*, 14, 135, 2021.
15. Jones, D.J. and Rozière, J., Recent advances in the functionalisation of polybenzimidazole and polyetherketone for fuel cell applications. *J. Membr. Sci.*, 185, 41, 2001.
16. Wong, C.Y., Wong, W.Y., Loh, K.S., Daud, W.R.W., Lim, K.L., Khalid, M., Walvekar, R., Development of poly(vinyl alcohol)-based polymers as proton

exchange membranes and challenges in fuel cell application: A review. *Polym. Rev.*, 60, 171, 2020.
17. Kerres, J., Cui, W., Reichle, S., New sulfonated engineering polymers via the metalation route. I. Sulfonated poly(ethersulfone) PSU Udel® via metalation-sulfination-oxidation. *J. Polym. Sci. A Polym. Chem.*, 34, 2421, 1996.
18. Laberty-Robert, C., Vallé, K., Pereira, F., Sanchez, C., Design and properties of functional hybrid organic–inorganic membranes for fuel cells. *Chem. Soc Rev.*, 40, 961, 2011.
19. Karthikeyan, C.S., Nunes, S.P., Prado, L.A.S.A., Ponce, M.L., Silva, H., Ruffmann, B., Schulte, K., Polymer nanocomposite membranes for DMFC application. *J. Membr. Sci.*, 254, 139, 2005.
20. Herring, A.M., Inorganic–polymer composite membranes for proton exchange membrane fuel cells. *Polym. Rev.*, 46, 245, 2006.
21. Carraro, M. and Gross, S., Hybrid materials based on the embedding of organically modified transition metal oxoclusters or polyoxometalates into polymers for functional applications: A review. *Materials*, 7, 3956, 2014.
22. Siddiqui, S., II and Chaudhry, S.A., Organic/inorganic and sulfated zirconia nanocomposite membranes for proton-exchange membrane fuel cells, in: *Organic-Inorganic Composite Polymer Electrolyte Membranes: Preparation, Properties, and Fuel Cell Applications*, D. Inamuddin, A. Mohammad, A. Asiri, (Eds.), pp. 219–240, Springer Cham, Switzerland, 2017.
23. Amirinejad, M., Madaeni, S.S., Rafiee, E., Amirinejad, S., Cesium hydrogen salt of heteropolyacids/Nafion nanocomposite membranes for proton exchange membrane fuel cells. *J. Membr. Sci.*, 377, 89, 2011.
24. Özdemir, Y., Üregen, N., Devrim, Y., Polybenzimidazole based nanocomposite membranes with enhanced proton conductivity for high temperature PEM fuel cells. *Int. J. Hydrog. Energy*, 42, 2648, 2017.
25. Gashoul, F., Parnian, M.J., Rowshanzamir, S., A new study on improving the physicochemical and electrochemical properties of SPEEK nanocomposite membranes for medium temperature proton exchange membrane fuel cells using different loading of zirconium oxide nanoparticles. *Int. J. Hydrog. Energy*, 42, 590, 2017.
26. Ahmed, S., Cai, Y., Ali, M., Khanal, S., Xu, S., Preparation and performance of nanoparticle-reinforced chitosan proton-exchange membranes for fuel-cell applications. *J. Appl. Polym. Sci.*, 136, 46904, 2019.
27. Parnian, M.J., Rowshanzamir, S., Prasad, A., Advani, S., Effect of ceria loading on performance and durability of sulfonated poly (ether ether ketone) nanocomposite membranes for proton exchange membrane fuel cell applications. *J. Membr. Sci.*, 565, 342, 2018.
28. Suryani, Chang, Y.-N., Lai, J.-Y., Liu, Y.-L., Polybenzimidazole (PBI)-functionalized silica nanoparticles modified PBI nanocomposite membranes for proton exchange membranes fuel cells. *J. Membr. Sci.*, 403, 1, 2012.

29. Esmaeilzade, B., Esmaielzadeh, S., Ahmadizadegan, H., Ultrasonic irradiation to modify the functionalized bionanocomposite in sulfonated polybenzimidazole membrane for fuel cells applications and antibacterial activity. *Ultrason. Sonochem.*, 42, 260, 2018.
30. Tripathi, B.P. and Shahi, V.K., Organic–inorganic nanocomposite polymer electrolyte membranes for fuel cell applications. *Prog. Polym. Sci.*, 36, 945, 2011.
31. Tang, J., Yuan, W.Z., Wang, J., Tang, J., Li, H., Zhang, Y., Perfluorosulfonate ionomer membranes with improved through-plane proton conductivity fabricated under magnetic field. *J. Membr. Sci.*, 423, 267, 2012.
32. Xi, J., Wu, Z., Qiu, X., Chen, L., Nafion/SiO_2 hybrid membrane for vanadium redox flow battery. *J. Power Sources*, 166, 531, 2007.
33. Amirinejad, M., Madaeni, S.S., Navarra, M.A., Rafiee, E., Scrosati, B., Preparation and characterization of phosphotungstic acid-derived salt/ Nafion nanocomposite membranes for proton exchange membrane fuel cells. *J. Power Sources*, 196, 988, 2011.
34. Zakaria, Z., Kamarudin, S.K., Timmiati, S.N., Membranes for direct ethanol fuel cells: An overview. *Appl. Energy*, 163, 334, 2016.
35. Amjadi, M., Rowshanzamir, S., Peighambardoust, S.J., Hosseini, M.G., Eikani, M.H., Investigation of physical properties and cell performance of Nafion/TiO_2 nanocomposite membranes for high temperature PEM fuel cells. *Int. J. Hydrog. Energy*, 35, 9252, 2010.
36. Vuillaume, P.Y., Mokrini, A., Siu, A., Théberge, K., Robitaille, L., Heteropolyacid/saponite-like clay complexes and their blends in amphiphilic SEBS. *Eur. Polym. J.*, 45, 1641, 2009.
37. Sayeed, M., Park, Y., Gopalan, A.I., Kim, Y.H., Lee, K.-P., Choi, S.-J., Sulfated titania–silica reinforced Nafion® nanocomposite membranes for proton exchange membrane fuel cells. *J. Nanosci. Nanotechnol.*, 15, 7054, 2015.
38. Devrim, Y., Erkan, S., Baç, N., Eroglu, I., Nafion/titanium silicon oxide nanocomposite membranes for PEM fuel cells. *Int. J. Energy Res.*, 37, 435, 2013.
39. Zarrin, H., Higgins, D., Jun, Y., Chen, Z., Fowler, M., Functionalized graphene oxide nanocomposite membrane for low humidity and high temperature proton exchange membrane fuel cells. *J. Phys. Chem. C*, 115, 20774, 2011.
40. Peng, K.-J., Lai, J.-Y., Liu, Y.-L., Nanohybrids of graphene oxide chemically-bonded with Nafion: Preparation and application for proton exchange membrane fuel cells. *J. Membr. Sci.*, 514, 86, 2016.
41. Taghizadeh, M.T. and Vatanparast, M., Ultrasonic-assisted synthesis of ZrO_2 nanoparticles and their application to improve the chemical stability of Nafion membrane in proton exchange membrane (PEM) fuel cells. *J. Colloid Interface Sci.*, 483, 1, 2016.
42. Parnian, M.J., Rowshanzamir, S., Alipour Moghaddam, J., Investigation of physicochemical and electrochemical properties of recast Nafion nanocomposite membranes using different loading of zirconia nanoparticles for

proton exchange membrane fuel cell applications. *Mater. Sci. Technol.*, 1, 146, 2018.

43. Chai, Z., Wang, C., Zhang, H., Doherty, C.M., Ladewig, B.P., Hill, A.J., Wang, H., Nafion–carbon nanocomposite membranes prepared using hydrothermal carbonization for proton-exchange-membrane fuel cells. *Adv. Funct. Mater.*, 20, 4394, 2010.
44. Li, Q., He, R., Berg, R.W., Hjuler, H.A., Bjerrum, N.J., Water uptake and acid doping of polybenzimidazoles as electrolyte membranes for fuel cells. *Solid State Ion.*, 168, 177, 2004.
45. Hooshyari, K., Javanbakht, M., Shabanikia, A., Enhessari, M., Fabrication $BaZrO_3$/PBI-based nanocomposite as a new proton conducting membrane for high temperature proton exchange membrane fuel cells. *J. Power Sources*, 276, 62, 2015.
46. Suryani, and Liu, Y.-L., Preparation and properties of nanocomposite membranes of polybenzimidazole/sulfonated silica nanoparticles for proton exchange membranes. *J. Membr. Sci.*, 332, 121, 2009.
47. Linlin, M., Mishra, A.K., Kim, N.H., Lee, J.H., Poly(2,5-benzimidazole)–silica nanocomposite membranes for high temperature proton exchange membrane fuel cell. *J. Membr. Sci.*, 411, 91, 2012.
48. Nawn, G., Pace, G., Lavina, S., Vezzù, K., Negro, E., Bertasi, F., Polizzi, S., Di Noto, V., Nanocomposite membranes based on polybenzimidazole and zro_2 for high-temperature proton exchange membrane fuel cells. *ChemSusChem*, 8, 1381, 2015.
49. Shabanikia, A., Javanbakht, M., Amoli, H.S., Hooshyari, K., Enhessari, M., Novel nanocomposite membranes based on polybenzimidazole and Fe_2TiO_5 nanoparticles for proton exchange membrane fuel cells. *Ionics*, 21, 2227, 2015.
50. Shabanikia, A., Javanbakht, M., Amoli, H.S., Hooshyari, K., Enhessari, M., Polybenzimidazole/strontium cerate nanocomposites with enhanced proton conductivity for proton exchange membrane fuel cells operating at high temperature. *Electrochim. Acta*, 154, 370, 2015.
51. Devrim, Y., Devrim, H., Eroglu, I., Polybenzimidazole/SiO_2 hybrid membranes for high temperature proton exchange membrane fuel cells. *Int. J. Hydrog. Energy*, 41, 10044, 2016.
52. Jheng, L.C., Rosidah, A.A., Hsu, S.L.C., Ho, K.-S., Pan, C.-J., Cheng, C.-W., Nanocomposite membranes of polybenzimidazole and amine-functionalized carbon nanofibers for high temperature proton exchange membrane fuel cells. *RSC Adv.*, 11, 9964, 2021.
53. Peighambardoust, S.J., Rowshanzamir, S., Hosseini, M.G., Yazdanpour, M., Self-humidifying nanocomposite membranes based on sulfonated poly(ether ether ketone) and heteropolyacid supported Pt catalyst for fuel cells. *Int. J. Hydrog. Energy*, 36, 10940, 2011.
54. Rowshanzamir, S., Peighambardoust, S.J., Parnian, M.J., Amirkhanlou, G.R., Rahnavard, A., Effect of Pt-$Cs_{2.5}H_{0.5}PW_{12}O_{40}$ catalyst addition on durability

of self-humidifying nanocomposite membranes based on sulfonated poly (ether ether ketone) for proton exchange membrane fuel cell applications. *Int. J. Hydrog. Energy*, 40, 549, 2015.

55. Mossayebi, Z., Saririchi, T., Rowshanzamir, S., Parnian, M.J., Investigation and optimization of physicochemical properties of sulfated zirconia/ sulfonated poly (ether ether ketone) nanocomposite membranes for medium temperature proton exchange membrane fuel cells. *Int. J. Hydrog. Energy*, 41, 12293, 2016.
56. Song, J.-M., Woo, H.-S., Sohn, J.-Y., Shin, J., 12HPW/meso-SiO_2 nanocomposite CSPEEK membranes for proton exchange membrane fuel cells. *J. Ind. Eng. Chem.*, 36, 132, 2016.
57. Parnian, M.J., Rowshanzamir, S., Prasad, A.K., Advani, S.G., High durability sulfonated poly (ether ether ketone)-ceria nanocomposite membranes for proton exchange membrane fuel cell applications. *J. Membr. Sci.*, 556, 12, 2018.
58. Stewart, S., Spernjak, D., Borup, R., Datye, A., Garzon, F., Cerium migration through hydrogen fuel cells during accelerated stress testing. *ECS Electrochem. Lett.*, 3, 19, 2014.
59. Attaran, A.M., Javanbakht, M., Hooshyari, K., Enhessari, M., New proton conducting nanocomposite membranes based on poly vinyl alcohol/poly vinyl pyrrolidone/$BaZrO_3$ for proton exchange membrane fuel cells. *Solid State Ion.*, 269, 98, 2015.
60. Javanbakht, M., Hooshyari, K., Enhessari, M., Beydaghi, H., Novel PVA/ $La_2Ce_2O_7$ hybrid nanocomposite membranes for application in proton exchange membrane fuel cells. *IJHFC*, 1, 105, 2014.
61. Beydaghi, H., Javanbakht, M., Badiei, A., Cross-linked poly(vinyl alcohol)/ sulfonated nanoporous silica hybrid membranes for proton exchange membrane fuel cell. *J. Nanostructure Chem.*, 4, 97, 2014.
62. Beydaghi, H., Javanbakht, M., Kowsari, E., Synthesis and characterization of poly(vinyl alcohol)/sulfonated graphene oxide nanocomposite membranes for use in proton exchange membrane fuel cells (PEMFCs). *Ind. Eng. Chem. Res.*, 53, 16621, 2014.
63. Padaki, M., Isloor, A.M., Wanichapichart, P., Polysulfone/N-phthaloylchitosan novel composite membranes for salt rejection application. *Desalination*, 279, 409, 2011.
64. Pereira, V.R., Isloor, A.M., Bhat, U.K., Ismail, A.F., Obaid, A., Fun, H.-K., Preparation and performance studies of polysulfone-sulfated nano-titania (S-TiO_2) nanofiltration membranes for dye removal. *RSC Adv.*, 5, 53874, 2015.
65. Ibrahim, G.P.S., Isloor, A., Ahmed, A., Lakshmi, B., Fabrication and characterization of polysulfone-zeolite ZSM-5 mixed matrix membrane for heavy metal ion removal application. *JAMST*, 18, 1, 2017.

66. Kumar, R., Isloor, A.M., Ismail, A.F., Rashid, S.A., Matsuura, T., Polysulfone–chitosan blend ultrafiltration membranes: Preparation, characterization, permeation and antifouling properties. *RSC Adv.*, 3, 7855, 2013.
67. Simari, C., Lufrano, E., Brunetti, A., Barbieri, G., Nicotera, I., Highly-performing and low-cost nanostructured membranes based on Polysulfone and layered doubled hydroxide for high-temperature proton exchange membrane fuel cells. *J. Power Sources*, 471, 228440, 2020.
68. Ahmadian-Alam, L. and Mahdavi, H., A novel polysulfone-based ternary nanocomposite membrane consisting of metal-organic framework and silica nanoparticles: As proton exchange membrane for polymer electrolyte fuel cells. *Renew. Energy*, 126, 630, 2018.
69. Simari, C., Lufrano, E., Brunetti, A., Barbieri, G., Nicotera, I., Polysulfone and organo-modified graphene oxide for new hybrid proton exchange membranes: A green alternative for high-efficiency PEMFCs. *Electrochim. Acta*, 380, 138214, 2021.
70. Syed Ibrahim, G.P., Isloor, A.M., Ismail, A.F., Farnood, R., One-step synthesis of zwitterionic graphene oxide nanohybrid: Application to polysulfone tight ultrafiltration hollow fiber membrane. *Sci. Rep.*, 10, 6880, 2020.
71. Xu, X., Zhao, G., Wang, H., Li, X., Feng, X., Cheng, B., Shi, L., Kang, W., Zhuang, X., Yin, Y., Bio-inspired amino-acid-functionalized cellulose whiskers incorporated into sulfonated polysulfone for proton exchange membrane. *J. Power Sources*, 409, 123, 2019.
72. Kumar, R., Isloor, A.M., Ismail, A.F., Matsuura, T., Performance improvement of polysulfone ultrafiltration membrane using N-succinyl chitosan as additive. *Desalination*, 318, 1, 2013.
73. Kumar, R., Isloor, A.M., Ismail, A.F., Matsuura, T., Synthesis and characterization of novel water soluble derivative of chitosan as an additive for polysulfone ultrafiltration membrane. *J. Membr. Sci.*, 440, 140, 2013.
74. Kolangare, I.M., Isloor, A.M., Karim, Z.A., Kulal, A., Ismail, A.F., , Inamuddin Antibiofouling hollow-fiber membranes for dye rejection by embedding chitosan and silver-loaded chitosan nanoparticles. *Environ. Chem. Lett.*, 17, 581, 2019.
75. Kumar, R., Isloor, A.M., Ismail, A.F., Rashid, S.A., Ahmed, A.A., Permeation, antifouling and desalination performance of TiO_2 nanotube incorporated PSf/CS blend membranes. *Desalination*, 316, 76, 2013.
76. Kumar, B.Y.S., Isloor, A.M., Kumar, G.C.M., Inamuddin, Nanohydroxyapatite reinforced chitosan composite hydrogel with tunable mechanical and biological properties for cartilage regeneration. *Sci. Rep.*, 9, 15957, 2019.
77. Ruiz Gómez, E.E., Mina Hernández, J.H., Diosa Astaiza, J.E., Development of a chitosan/PVA/TiO_2 Nanocomposite for application as a solid polymeric Electrolyte in Fuel Cells. *Polymers*, 12, 1691, 2020.
78. Divya, K., Rana, D., Alwarappan, S., Abirami Saraswathi, M.S.S., Nagendran, A., Investigating the usefulness of chitosan based proton exchange

membranes tailored with exfoliated molybdenum disulfide nanosheets for clean energy applications. *Carbohydr. Polym.*, 208, 504, 2019.

79. Vijayakumar, V. and Khastgir, D., Hybrid composite membranes of chitosan/sulfonated polyaniline/silica as polymer electrolyte membrane for fuel cells. *Carbohydr. Polym.*, 179, 152, 2018.
80. Ahmed, S., Arshad, T., Zada, A., Afzal, A., Khan, M., Hussain, A., Hassan, M., Ali, M., Xu, S., Preparation and Characterization of a Novel Sulfonated Titanium Oxide Incorporated Chitosan Nanocomposite Membranes for Fuel Cell Application. *Membranes*, 11, 450, 2021.
81. Cui, F., Wang, W., Liu, C., Chen, X., Li, N., Carbon nanocomposites self-assembly UiO-66-doped chitosan proton exchange membrane with enhanced proton conductivity. *Int. J. Energy Res.*, 44, 4426, 2020.
82. Wang, S., Shi, L., Zhang, S., Wang, H., Cheng, B., Zhuang, X., Li, Z., Proton-conducting amino acid-modified chitosan nanofibers for nanocomposite proton exchange membranes. *Eur. Polym. J.*, 119, 327, 2019.
83. Sun, X., Wu, Q., Picha, D.H., Ferguson, M.H., Ndukwe, I.E., Azadi, P., Comparative performance of bio-based coatings formulated with cellulose, chitin, and chitosan nanomaterials suitable for fruit preservation. *Carbohydr. Polym.*, 259, 117764, 2021.
84. Rahman, N.F., Loh, K.S., Mohamad, A.B., Kadhum, A.A.H., Lim, K.L., Synthesis and characterisation of chitosan-cellulose biocomposite membrane for fuel cell applications. *MJAS*, 20, 885, 2016.

6

Organic-Inorganic Composite Membranes for Proton Exchange Membrane Fuel Cells

Guocai Tian

State Key Laboratory of Complex Non-Ferrous Metal Resource Clean Utilization, Faculty of Metallurgical and Energy Engineering, Kunming University of Science and Technology, Kunming, China

Abstract

Fuel cell is known as one of the most promising and potential clean energy technologies. Proton exchange membrane (PEM), as the core part of fuel cell, plays an important role in the performance of fuel cell. In view of the shortcomings of perfluorosulfonic acid PEM in high-temperature and low-humidity working environment, preparing inorganic composite PEM with low-cost and high performance is an effective solution. In recent years, many methods have been tried to synthesize and improve inorganic composite PEMs. Based on the classification of the main inorganic fillers for the preparation of inorganic organic composite PEMs, in present chapter, we focused on the research progress and achievements of inorganic-organic composite PEMs. The relationship between various inorganic fillers and the properties of composite PEMs was reviewed, and the researchers' development direction in the future was prospected.

Keywords: Fuel cell, proton exchange membrane, organic-inorganic composite proton exchange membrane, research status, proton conductivity

6.1 Introduction

Proton exchange membranes fuel cell (PEMFC) has excellent performance and outstanding advantages [1–11], such as large working current, high specific power and specific energy density, high-energy efficiency,

Email: tiangc@kust.edu.cn; tiangc@iccas.ac.cn; tiangc01@gmail.com

Inamuddin, Omid Moradi and Mohd Imran Ahamed (eds.) Proton Exchange Membrane Fuel Cells: Electrochemical Methods and Computational Fluid Dynamics, (111–136)

environment-friendly, simple structure, flexible and portable, and a wide range of fuel sources. As one of the core components of PEMFC, proton exchange membrane (PEM) plays an important role in the performance of PEMFC. The commercial PEM is mainly perfluorosulfonic acid polytetrafluoroethylene copolymer (Nafion) produced by DuPont company. It has the disadvantages of strong dependence on water for proton conductivity, high permeability of methanol, complex synthesis process, and high price, which hinders the further development and application of Nafion membrane in the fuel cell field. In recent years, non-fluorine high-temperature–resistant sulfonated aromatic polymers have been widely studied. It has good thermal stability and mechanical stability, and its structure can be designed freely, but the balance between proton conductivity and mechanical stability restricts its development. Doping inorganic materials into polymer matrix and combining the advantages of inorganic materials with the characteristics of a polymer matrix to prepare inorganic organic composite PEM is an important method for the modification of PEM. This chapter analyzes the characteristics, advantages, design principles, and research progress of organic-inorganic composite PEM.

6.2 Proton Exchange Membrane Fuel Cell

Fuel cell is a new type of energy conversion device, which directly generates current through the electrochemical oxidation of fuel (such as H_2 and CH_3OH) and the electrochemical reduction of oxidant (such as O_2). Different from ordinary primary cells and batteries, the electrode reactive substances required by fuel cells are not stored in the cell but supplied from the outside of the cell. Theoretically, as long as the external fuel is continuously supplied, the fuel cell can continuously output electric energy [1–7]. Since fuel cell power generation does not go through the combustion process (i.e., fuel cell is not a heat engine) and is not limited by Carnot cycle, it not only has high-energy conversion efficiency (theoretically greater than 80%) but also the power-generation process will not cause environmental pollution. Therefore, fuel cell technology is an efficient and clean energy utilization technology. All countries are vigorously developing relevant technologies. Some achievements have been made [1–25].

According to the different electrolytes used, fuel cells can be divided into alkaline fuel cells, molten carbonate fuel cells, solid oxide fuel cells, phosphoric acid fuel cells, and PEM fuel cells. The performance of various fuel cells [1, 6, 7] is shown in Table 6.1.

Table 6.1 Different types of fuel cells [1, 6, 7].

Fuel cell type	Electrolyte	Fuel/oxidant (Operation temp, °C)	Positives	Negatives	Efficiency %	Power kW	Applications
Alkaline Fuel Cell (AFC)	Potassium Hydroxide Solution	H_2/O_2 (50–200)	30 years of experience, highly efficient	Aging technology, expensive	45–60	≤20	Military, space
Polymer electrolyte Membrane Fuel Cell (PEMFC)	Solid ion Exchange Membrane	H_2/O_2 (50–200)	Small, lightweight potentially low cost	H_2 storage questions	40–60	~100	Automobiles, potable power
Direct Methanol Fuel Cell (DMFC)	Solid ion Exchange Membrane	Methanol/O_2 (50–140)	Small, light, attractive fuel	Technical problems (catalyst and membrane)	40	<1	Potable power, automobiles, military
Solid Oxide Fuel Cell (SOFC)	Mixture of Lithium, Potassium Carbonate	H_2 or CH_4/O_2 (650)	Produce reusable heat and steam	High capital cost, runs very hot, large	50–65	>200	Industrial applications
Molten Carbonate Fuel Cell (MCFC)	Yittria Stabilized Zirconia	H_2 or CH_4/O_2 (950–1,050)	Produce reusable heat, high efficiency	Large units, runs very hot, for large scale only	45–60	>1,000	Offices, industrial applications

Among various fuel cells, PEMFC is recognized as the preferred power supply for electric vehicles, fixed power stations, and spacecraft because of its advantages of low working temperature, fast start-up, high specific power, simple structure, and convenient operation. PEM fuel cell mainly uses hydrogen or reformed hydrogen as fuel, also known as hydrogen oxygen fuel cell. In addition, researchers have also developed a PEM fuel cell, called direct fuel cell, which uses a series of widely sourced and cheap chemicals such as methanol, formic acid, and ether as direct fuel. Direct methanol fuel cell (DMFC) has attracted researchers' attention because of its prospect in portable devices and is often listed as a class of fuel cell [6, 7].

PEMFC is mainly composed of anode, cathode, and electrolyte. It uses solid PEM as electrolyte, Pt/C or Pt-Ru/C as electro-catalyst, hydrogen or purified reforming hydrogen as fuel, air or pure oxygen as the oxidant, and graphite or surface at modified metal plate with gas flow channel as bipolar plate. Figure 6.1 briefly shows the structure and working principle of PEMFC. Taking hydrogen fuel as an example, the fuel on the anode is oxidized to produce e^- and H^+, e^- passes through the load, and H passes through the PEM, and both reach the cathode of the battery at the same

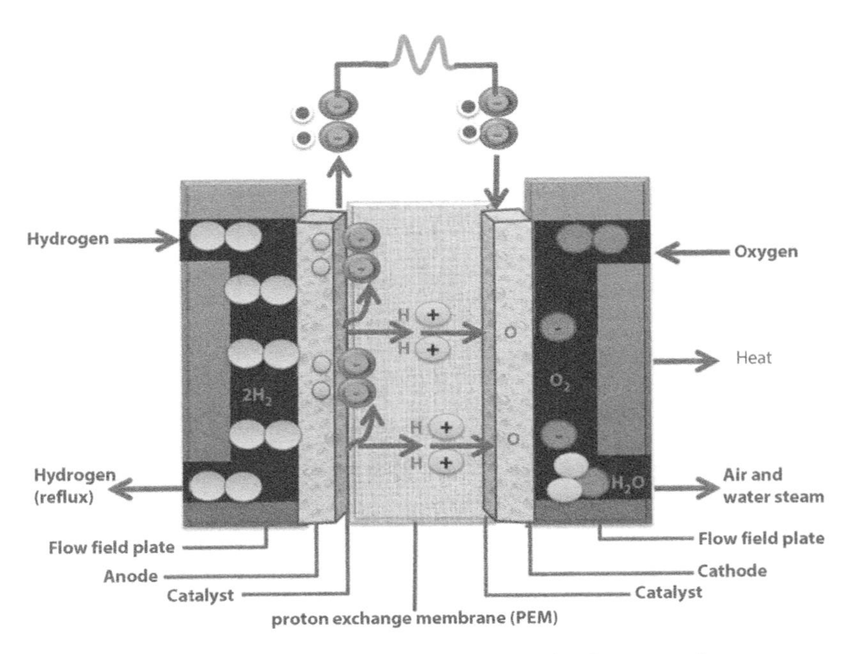

Figure 6.1 Schematic representation and working principle of proton exchange membrane fuel cell (PEMFC).

time and react with O_2 to produce direct current. A plurality of battery cells is connected in series or in parallel as required to form a battery pack (stack) with different power, and output electric energy to the load through appropriate connection. In PEMFC, the fuel is provided by an external high-pressure hydrogen bottle, and the oxidant is also supplied from the outside. The reactions at the membrane electrode are as follows:

$$\text{Anode: } 2H_2 \rightarrow 4H^+ + 4e^- \tag{6.1}$$

$$\text{Cathode: } O_2 + 4H^+ + 4e^- \rightarrow 2H_2O \tag{6.2}$$

$$\text{Total reaction: } 2H_2 \rightarrow O_2 = 2H_2O \tag{6.3}$$

Direct methanol fuels cell (DMFC) is a PEM fuel cell that uses liquid methanol as fuel without methanol reforming; see Figure 6.2. Compared with PEMFC with hydrogen as fuel, it eliminates the potential safety hazards of hydrogen fuel in the process of use, storage, and transportation. It has the advantages of convenient fuel supplement, simple structure, small volume, short start-up time, high specific energy, low noise, convenient carrying, and weak infrared signal; small civil power supply and military portable power supply for mobile electronic equipment have broad application prospects [8]. DMFC and PEMFC have very similar structures, but different fuel substances are used. Because the fuel is liquid, there are different requirements in component material selection, fuel flow field design, and hydrothermal management. The reaction principle is as follows:

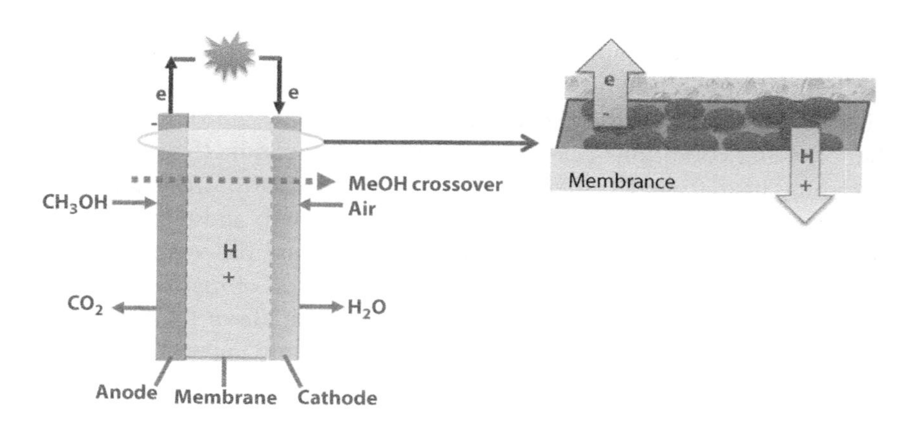

Figure 6.2 Schematic representation of DMFC operating on methanol and O_2.

$$\text{Anode reaction: } CH_3OH + H_2O \rightarrow CO_2\uparrow + 6H^+ + 6e^- \quad (6.4)$$

$$\text{Cathode reaction: } 3/2O_2 + 6H^+ + 6e^- \rightarrow 3H_2O \quad (6.5)$$

$$\text{Total reaction: } CH_3OH + 3/2\ O_2 \rightarrow CO_2\uparrow + H_2O \quad E_{eq}=1.18V \quad (6.6)$$

At present, the factors that affect the comprehensive performance of DMFC and restrict its efficiency and need to be further improved are mainly concentrated in three aspects [9, 10]: first, the activity of anode reaction catalyst is not high, and the intermediate of methanol electrochemical oxidation reaction is easy to poison the catalyst, resulting in the deviation between anode electrode potential and equilibrium electrode potential; second, under the action of concentration difference and electro-osmosis, methanol diffuses from the anode through the PEM to the cathode, which reduces the fuel utilization rate and generates a mixed potential at the cathode, greatly reducing the performance and service life of the battery, that is, the so-called methanol penetration phenomenon; third, the water management and heat management of the battery system are more complex. It can be seen that PEM is the core material of DMFC, and its performance determines the performance of DMFC, which is a bottleneck restricting the commercialization of DMFC.

6.3 Proton Exchange Membrane

PEM is the core component of DMFC and one of the important factors affecting the cost of fuel cell, which determines the cell efficiency, output power, and application performance [12]. As a selective membrane material, PEM has three important functions: insulating electrons, blocking oxidants and fuels, and high H^+ conductivity. In order to obtain fuel cells with low cost, high performance, and long service life, PEMs should have the following conditions [13–15]: (a) high proton conductivity and good electronic insulation; (b) thermal performance, oxidation resistance, and acid-base resistance; (c) excellent mechanical strength and flexibility, stable size, and conducive to large-scale production and processing; (d) low manufacturing cost; (e) try to reduce the penetration of methanol; otherwise, it will affect the service efficiency and service life of the battery. At present, PEM materials that have been studied mainly include the following four categories.

6.3.1 Perfluorosulfonic Acid PEM

Perfluorosulfonic acid PEM was first developed by DuPont, and its representative products are Nafion and series membranes [16, 17]. Its structure is similar to that of polytetrafluoroethylene and consists of two parts: one is the hydrophobic skeleton (fluorocarbon structure), the bond energy of C-F is high, which gives PEM good mechanical strength and oxidation resistance; the other part is the side chain containing SO_3H. Because of the strong electronegativity of fluorine atom, it greatly improves the acidity of SO_3H and enhances the adsorption capacity of water molecules. The two parts form an obvious hydrophilic and hydrophobic water phase separation structure, expand the H^+ conduction channel, and improve the proton conductivity of the membrane. In addition to Nafion® series membranes, there is also Flemion® membrane studied by Asahi Glass Company in Japan, Aciplex® membrane studied by Asahi chemical company and XUS-B204 membrane developed by Dow company in the United States.

Perfluorosulfonic acid PEM has good electrochemical and mechanical properties, but there are still some disadvantages [18]: (a) If the temperature is too high, then the moisture content in the membrane will be too low, resulting in a significant decrease in proton conductivity. (b) The high permeability coefficient of methanol not only causes a large loss of methanol fuel but also reacts after methanol penetrates into the cathode, resulting in poisoning of electrode catalyst, which greatly shortens the service life and efficiency of fuel cell. (c) The synthesis is difficult, the process is complex, and the cost is higher. (d) Monomer synthesis is hard and higher cost, and the waste products are difficult to deal with.

The above shortcomings limit the further commercial development of perfluorosulfonic acid membrane. Therefore, the development of PEM with high proton conductivity and low gas permeability with low cost and good thermodynamic and electrochemical stability has become an urgent problem.

6.3.2 Partial Fluorine-Containing PEM

Because of the shortcomings of perfluoropolymers that cannot be ignored, scientists strive to find substances to replace them, and some fluorine-containing PEMs have been studied and developed. Some fluorine-containing copolymer BAM3G [19] developed by Ballard company has good mechanical and electrochemical properties and excellent battery efficiency. This is because the C-F structure of the polymer skeleton protects the C-C bond, ensures that the polymer has certain mechanical properties, and is less attacked by free radicals. Through the assembly and final test of a membrane electrode of BAM3G, it is found that its continuous operation time can exceed 1–5,000 h. In addition, the cost of BAM3G is also lower than Nafion.

6.3.3 Non-Fluorine PEM

In order to reduce the cost, avoid complex process flow, and reduce the pollution of fluorine to the environment, people have developed non-fluorosulfonic acid PEM. Such electrolyte membranes can be synthesized directly by polymerization between several monomers or modified by polymers. Sulfonated aromatic polymers are easy to prepare; low cost; excellent mechanical, chemical, and thermal properties; and can have proton conductivity by introducing H^+ conducting groups. Therefore, they are often used as materials to replace perfluorosulfonic acid PEM. This includes sulfonated polyphosphones [20, 21] sulfonated polyphthalimides [22–24], sulfonated polyarylene ethers [25–27], phosphorylated or sulfonated polyimilines [28, 29], and other polymers.

6.3.4 Modification of Proton Exchange Membrane

Perfluorosulfonic acid PEM has good proton conductivity and thermal, chemical and mechanical properties, but the process is complex, the use cost is high, and fluorine pollutes the environment; some fluorinated PEMs retain the advantages of perfluorosulfonic acid PEM to a certain extent, but their process, cost, and pollution cannot be ignored; non-fluorinated sulfonated aromatic polymers have attracted extensive attention because of their low cost, relatively easy synthesis process and environmental friendliness. However, it also has corresponding disadvantages. For example, because of the lack of protection of fluorine atoms in its structure, it is easy to be attacked by free radicals and its chemical stability is slightly poor; in addition, the traditional sulfonated aromatic polymers largely rely on the degree of sulfonation to achieve a large number of proton transfers. The increase of the degree of sulfonation will lead to the increase of the water content of the system. However, too much water will aggravate the swelling of the membrane, destroy its dimensional stability and mechanical and chemical stability, increase the penetration of fuel, and reduce the performance and life of the cell. Therefore, it is difficult to meet multiple requirements of PEMFC application on one material. It is necessary to modify the material to improve the comprehensive properties of PEM. There are four common modification methods [30–33].

(1) Block copolymerization: Block copolymerization refers to the method of polycondensation of different polymer segments to form a polymer. In this way, the structure of the polymer can be accurately designed, and its molecular weight and molecular weight distribution can be accurately controlled. In PEM, block copolymerization can be used to significantly increase the phase separation in the membrane and increase the proportion of a hydrophilic region to obtain high H^+ conduction. From the shape of molecular structure, there are star and comb polymers.

(2) Cross-linking modification: Cross-linking is one of the important ways to improve the mechanical strength and thermal and chemical properties of polymers. It is linked between polymer chains in the form of covalent bond. In addition, appropriate cross-linking degree can also reduce the fuel penetration problem of PEM. The commonly used cross-linking methods include thermal initiation cross-linking, UV cross-linking, covalent cross-linking, and ionic cross-linking.

(3) Side chain sulfonation: For polymers containing $-SO_3H$ in the main chain, the benzene ring structure gives them good mechanical and thermal properties. However, such a rigid structure limits the activity of $-SO_3H$ and is not conducive to proton transfer. Generally, the number of $-SO_3H$ should be increased to improve the proton conductivity of PEM. However, too much $-SO_3H$ will cause excessive swelling of PEM but reduce the mechanical properties of PEM.

(4) Organic-inorganic composite: Organic-inorganic composite is to mix inorganic particles and polymers to form new materials and make use of their advantages to obtain better comprehensive properties. Inorganic particles have excellent chemical and thermal properties. When hydrophilic inorganic particles are combined in PEM, they not only can increase the water-holding-capacity of membrane and improve proton conductivity but also can block the penetration of methanol and effectively improve fuel utilization and PEM performance. Common inorganic oxides used to prepare composite PEMs are. Inorganic particles often doped in PEM include heteropoly acids (HPAs), carbon nanotube (CNTs), graphene oxide (GO), silicon dioxide (SiO_2), titanium dioxide (TiO_2), and tin dioxide (SnO_2). HPAs are composed of hydrate cation and complex anion. They have good H^+ storage and transfer ability, such as phosphoplatinum acid, silicotungstic acid, and phosphotungstic acid (PTA). This paper focuses on the research progress of inorganic organic composite PEM.

6.4 Research Progress of Organic-Inorganic Composite PEM

Organic-inorganic composite PEMs have many advantages: (a) reducing fuel permeability of membrane; (b) improving the mechanical, thermal, and oxidation stability of the membrane; (c) enhancing the water absorption and water retention capacity of the membrane; and (d) improving the proton conductivity of the membrane. The design and preparation principles of inorganic organic composite PEM include: (a) the balance between anti-fuel permeability and proton conductivity. The introduction of inorganic filler can effectively reduce the fuel permeability of the membrane. However, the addition of filler has a great dilution effect on the proton exchange clusters in the original polymer matrix, so the inorganic materials need to be modified. At the same time, it meets the requirements of low fuel permeability and high proton conductivity. (b) Appropriate doping proportion of inorganic filler, the maximum density of inorganic filler when reaching the best performance of composite membrane is called "permeability threshold". If it is greater than "permeability threshold", the continuous addition of inorganic filler will reduce the performance of membrane or cause inorganic phase aggregation. Therefore, it is particularly important to add appropriate inorganic filler. (c) The selection of inorganic fillers should consider the hygroscopic capacity, high specific surface area, surface acidity, and good compatibility with polymers. Reasonable selection and design of fillers and determination of the optimal doping ratio of fillers are of great significance to the performance of inorganic organic composite PEM. Organic-inorganic composite has long been a research hotspot, and there are many review works [30–33]. According to the different combination methods, inorganic ions are used in PEM, such as self-assembly, surface modification, sol-gel method, and solution blending method. The research progress of various organic-inorganic PEMs is introduced below.

6.4.1 Inorganic Oxide/Polymer Composite PEM

Inorganic oxides have excellent stability and hygroscopic capacity and are easy to be modified by organic functions. Their introduction into a polymer matrix can improve the water absorption, water retention, oxidation resistance, mechanical stability [34], and fuel penetration resistance [35] of PEM. They are often selected inorganic fillers for PEM modification. However, inorganic oxides are generally non-proton conductors and have poor compatibility with polymer matrix. In order to make up for the loss of proton conductivity of composite membrane, functionalization modification of its surface is a common method [36].

Boutsika *et al.* [37] incorporated silica (SiO_2) nanoparticles modified by phosphonate or sulfonate into Nafion to prepare acidified SiO_2/Nafion composite membrane. The proton conductivity of the composite membrane is about 50 mS/cm at 130°C and 30% RH (relative humidity), and the thermal stability of the membrane is also improved. Wu *et al.* [35] studied the properties of amino acid modified titanium dioxide (TiO_2) particles/SPEEK composite PEM. The results show that, compared with the pure SPEEK membrane, the methanol permeability of the composite membrane is reduced, and the mechanical strength is enhanced after adding the modified TiO_2 particles. At the same time, the composite membrane shows excellent proton conductivity, and the highest proton conductivity of the composite membrane reaches 0.258 S/cm at 60°C. Zhao *et al.* [38] prepared two kinds of phosphorylated SiO_2 microspheres (SiP-I and SiP-II) with short chain and long chain on the surface by silane coupling agent modification and polymer grafting. The two kinds of phosphorylated silica microspheres were mixed into SPEEK matrix to prepare composite membrane. At 60°C and 100% RH, the composite membrane doped with 20% (WT, mass fraction, the same below) SiP-I has the highest proton conductivity of 0.335s/cm. After that, Zhao *et al.* [39] prepared phosphorylated mesoporous hollow SiO_2 microspheres and doped them into chitosan (CS) matrix to prepare phosphorylated mesoporous hollow SiO_2/CS composite membrane. It was found that the proton conductivity of the composite membrane doped with 7.5% phosphorylated mesoporous hollow SiO_2 microspheres was 9.4×10^{-2} s/cm at 110°C and 100% RH, showing excellent proton conductivity at high temperature.

Wang *et al.* [40] introduced amino modified SiO_2(SiO_2-NH_2) into sulfonated polyethersulfone (SPES) nanofibers and compounded with Nafion membrane to prepare Nafion/SPES/SiO_2 composite membrane. Nafion/SPES/SiO_2-3% membrane (3% is the SIO_2-NH_2 content) has low methanol permeability of 7.22×10^{-7} cm^2/s and proton conductivity of 0.23 S/cm at 80°C. Liu *et al.* [41] prepared sulfonated silica–coated polyvinylidene fluoride sulfone (PVDF) nanofibers with high strength and high sulfonate surface concentration and introduced them into CS matrix to prepare composite proton exchange films for DMFC. The proton conductivity of the composite membrane is about 2.8 times higher than that of pure CS membrane, and the methanol permeability is as low as 26% of Nafion 115 membrane.

Zhang *et al.* [42] prepared sulfonated poly(aryl ether sulfone) (SPAES) nanocomposite PEM doped with sulfonated titanium dioxide (S-TiO_2) by solution casting method. The results show that the proton conductivity of SPAES membrane doped with S-TiO_2 is slightly lower than that of Nafion117 membrane (60 mS/cm), but it shows lower methanol permeability (2.1×10^{-7} cm^2/s) and better proton selectivity. Yuan *et al.* [43] synthesized sulfonated polyaryl ether ketone sulfone (SPAEKS) with local high concentration sulfonic acid group and doped different content of TiO_2 into the SPAEKS matrix to prepare SPAEKS/TiO_2 composite membrane. It was found that the proton conductivity of the composite membrane with TiO_2 content of 3% was close to that of Nafion membrane, and the methanol permeability decreased significantly.

Inorganic oxides are designed as nanoparticles with different topological structures (such as mesoporous and hollow) [39], which increases the specific surface area and volume ratio of fillers, thus increasing the number of surface modified proton transfer groups. In addition, the proton conductivity of the membrane can be greatly improved by controlling the nanophase separation behavior of the composite membrane and making it spontaneously assembled into ion nanochannels through the shape and functional design of the filler [38, 39, 44], but the main reasons restricting its development are the dispersion state of the filler in the matrix and its compatibility with the polymer matrix.

6.4.2 Two-Dimensional Inorganic Material/Polymer Composite PEM

Two-dimensional materials, such as montmorillonite [45, 46], GO [47, 48], and transition metal carbide/carbon nitride (MXens) [49, 50], have high aspect ratio and monatomic thickness. Their special two-dimensional sheet shape makes them have excellent electrochemical activity and mechanical properties.

Lin *et al.* [45] prepared a new functional montmorillonite with a pop skeleton quaternary ammonium salt containing sulfonic acid group (SO_3H) and directly blended it with Nafion solution to prepare composite membrane. It was found that the alcohol resistance and proton conductivity of the composite membrane were improved. Wu *et al.* [46] mixed different kinds of montmorillonite into Nafion membrane, which well reduced the ethanol permeability of the membrane.

GO, like graphene, has a single-layer two-dimensional structure and is widely used in the study of alcohol resistance of PEMFC [51].

Its oxygen-containing group makes it have excellent electrophilicity. As a membrane material, it can improve the proton conductivity of PEMFC [51]. At the same time, the large specific surface area of go can also shield the movement of methanol molecules [52], so as to inhibit the penetration of methanol molecules and improve the selectivity of the membrane. It is reported in the literature [53] that the methanol permeability of a group of nanocomposite proton exchange membranes based on PVA/sulfosuccinic acid (SSA)/GO prepared by nanotechnology is significantly reduced by the presence of GO.

He *et al.* [54] prepared go with different statistical average radius and doped it into sulfonated polyimide (SPI) to prepare GO/SPI composite film. It is found that the incorporation of small-size graphene makes the sulfonic acid ion clusters more dispersed, the volume decreases, the number increases, and the continuity of proton transfer channels are improved. Because of the increase of the number of ion clusters, the barrier sites for a methanol penetration increase, which makes the methanol penetration path more tortuous, and the alcohol resistance of the composite membrane is improved. He *et al.* [55] prepared sulfonated GO (S-GO)/sulfonated polyether ether ketone (SPEEK) composite membrane. The introduction of S-GO increased the number of sulfonic acid groups in the composite membrane, the dispersion of sulfonic acid ion clusters was more uniform, a more continuous proton transport network was formed in the membrane, the proton conductivity of the composite membrane reached 8.41×10^{-3} S/cm, and the methanol permeability coefficient is reduced to 2.6388×10^{-7} cm^2/s.

Wang *et al.* [56] designed a high-performance PEM (SS DNA@ GO/Nafion) based on single-stranded deoxyribonucleic acid (DNA)/GO/Nafion. It maintains higher proton conductivity (351.8 ms/cm) and lower methanol permeability (1.63×10^{-7} cm^2/s) than Nafion membrane at 800°C and 100% RH. Zhong *et al.* [57] obtained self–cross-linked SPEEK (SCSPEEK) by self–cross-linking propylene containing SPEEK, and then combined it with S-GO. The composite membrane (SCSP/SF) was successfully constructed. Compared with the original membrane, when the S-GO content of SCSP/SF membrane is 3%, the methanol diffusion coefficient is the smallest (about 1.5×10^{-8} cm^2/s), and the selectivity is about 6.8 and 19.3 times higher than that of pure speek membrane and Nafion 117 membrane, respectively.

Liu *et al.* [49, 50] studied the properties of two-dimensional layered transition metal carbide (Ti_3C_2Tx) nanosheet composite PEM. It is found that (Ti_3C_2Tx) nanosheets connect the unconnected proton transfer channels in the membrane and form a remote interface path for proton transfer in the membrane. At the same time, the –OH on (Ti_3C_2Tx) enriches the acidic/basic groups of the polymer matrix at the nano-sheets interface, provides effective proton jump sites, and contributes to the improvement of proton conductivity.

A large number of studies show that the special lamellar structure and high aspect ratio of two-dimensional materials can effectively improve the gas-liquid barrier properties of polymer membranes, significantly interfere with the movement of polymer macromolecular chains, and improve the mechanical stability and solvent resistance of composite membranes. After functionalization, these materials establish a remote interface path in the composite membrane with the help of huge specific surface area and aspect ratio, so as to reduce the energy barrier of proton migration, so as to improve the performance of the composite membrane.

6.4.3 Carbon Nanotube/Polymer Composite PEM

CNTs are commonly used nanocarbon materials for PEM modification Doping CNTs can significantly improve the mechanical properties of a polymer matrix, but the intrinsic conductivity of CNTs is easy to cause short circuit of a membrane electrodes. The research shows that controlling the doping amount of CNTs below the penetration threshold will avoid short circuit [58]. The organic functionalization modification of CNTs and the control of a reasonable doping amount provide an important idea for the preparation of composite films with CNTs as inorganic filler. Thomassin *et al.* [59] prepared carboxyl modified multi-wall CNTs (MWCNTs)/Nafion composite membrane. The research showed that the methanol permeability of the composite membrane decreased by about 60%, while the proton conductivity did not decrease. Compared with pure Nafion, the young's modulus of the composite membrane containing 2% MWCNTs is increased by 160%. Liu *et al.* [58] prepared Nafion functionalized MWCNT/Nafion composite films. Compared with the original Nafion membrane, the mechanical strength of the composite membrane with MWCNT Nafion loading of 0.05% increased by 1.5 times and the proton conductivity increased by five times. Because CNTs have a very large aspect ratio, its addition increases the cross resistance of the composite membrane to the fuel, thus reducing the fuel permeability of the composite membrane. However, the proton conduction mechanism of CNTs containing composite membranes has not been fully studied, and it is difficult to explain the relationship between the distribution of CNTs in polymer matrix and proton conductivity.

6.4.4 Inorganic Acid–Doped Composite Film

Inorganic acid doped composite membrane still maintains certain humidity at a high temperature due to the restriction of water molecules by strong acidic inorganic components in the membrane. Inorganic liquid acid has a high boiling point. Even under the conditions of high temperature and low humidity, the liquid acid can maintain fluidity, so that the membrane has high proton conductivity. Xing *et al.* [60] showed that, under the specific doping conditions of high concentration, the conductivity of inorganic acid is as follows: sulfuric acid (H_2SO_4) > phosphoric acid (H_3PO_4) > perchloric acid ($HClO_4$) > nitric acid (HNO_3) > hydrochloric acid (HCl). For PBI doped with H_2SO_4 (16 mol/L), the highest conductivity was 0.0601 S/cm, which was as good as that of Nafion 117. Although sulfuric acid has high proton transfer ability, it depends on water. Phosphoric acid has higher thermal stability, and phosphoric acid is an amphoteric acid. It is not only a proton donor but also a proton receptor. It has high proton self-dissociation ability. Through the formation and fracture of intermolecular dynamic hydrogen bonds, protons can jump and transfer between phosphoric acid molecules. Che *et al.* [61] prepared SPEEK/polyurethane (PU) composite membrane and doped phosphoric acid into the composite membrane. The decomposition temperature of the composite membrane was as high as 180°C, and the maximum proton conductivity of the composite membrane was 3.0×10^{-2} S/cm at 160°C without water. Yue *et al.* [62] synthesized a series of covalently cross-linked sulfonated poly(imide benzimidazole) (CBrSPIBI) PEMs and then soaked them in phosphoric acid to prepare (PA CBrSPIBI) membranes. The research shows that the highest proton conductivity of the composite membrane is 0.042 S/cm at 130°C and 30% RH, which is about one to two orders of magnitude higher than CBrSPIBI and Nafion.

Compared with acidic polymers, basic polymers have a better binding ability with inorganic acid molecules. Polybenzimidazole (PBI) has excellent mechanical and thermal stability, and its own conductivity reaches 10^{-10} to 10^{-9} S/cm [60]. In addition, the –N=N– nitrogen atom on the imidazole ring in PBI has strong alkalinity and is easy to be protonated, so it is easy to bind to strong acids. Phosphoric acid–doped PBI composite membrane has quite high proton conductivity of 0.3 S/cm [63] under high temperature and low humidity working conditions, and even 0.14 S/cm [64] under anhydrous conditions. Therefore, such membranes can achieve high proton conductivity under completely dry conditions. However, relevant research results show that RH has a certain impact on the proton conductivity of such membranes. With the increase of RH, the proton conductivity gradually increases, but the dependence of the proton conductivity of such membranes on RH is far less strong than that of sulfonated polymer membranes [65].

Although doping small molecular inorganic acids can significantly improve the proton conductivity of the membrane, because small molecular inorganic acids are soluble in water and lack binding force on small molecular acids in the composite membrane, the doped acids are easy to be lost during the use of the membrane, resulting in the decline of membrane performance. How to better increase the binding of a polymer matrix to inorganic acid and reduce the leakage of inorganic acid has become the research focus of this kind of membrane.

6.4.5 Heteropoly Acid–Doped Composite PEM

HPA is a solid crystalline material with polyoxometalate inorganic cage structure, in which the proton conductivity of PTA reaches 0.18 S/cm [66]. Therefore, HPA is an excellent filler for preparing inorganic organic composite PEM.

Zhao *et al.* [67] constructed polycationic chitosan (CTS) and negatively charged inorganic particle PTA on the surface of carboxyl containing sulfonated poly aryl ether ketone (SPAEK-COOH) film by layer by layer self-assembly method, and prepared (SPAEK-COOH)–$(CTS/PTA)_n$ multilayer composite film. The proton conductivity of the composite membrane is as high as 0.086 S/cm at 25°C and 0.24 S/cm at 80°C, and shows good alcohol resistance. Yu *et al.* [68] prepared HPA/sulfonated poly(aryl ether sulfone) composite membrane. The results show that the introduction of HPA into sulfonated poly(aryl ether sulfone) significantly reduces the swelling rate of the membrane without affecting the proton conductivity at room temperature. The proton conductivity of the composite membrane is as high as 0.15 S/cm at 100°C–130°C. In the space between the acidic groups of the polymer, because of the existence of HPA anion, more water can be maintained, so as to improve the water retention of the composite membrane. At the same time, the introduction of HPA increased the density of proton transfer sites in the composite membrane and improved the proton conductivity. However, HPA itself is water-soluble, so the fuel cell in a humid environment will lead to the instability of the composite membrane. Although researchers are committed to immobilizing HPA in other inorganic fillers or polymer matrix, it will affect the continuity of HPA distribution in polymer matrix, lead to the blocking of proton transfer channel, and then affect the proton conductivity. Therefore, how to retain more

HPA in the polymer matrix has become one of the challenges of this kind of membrane.

6.4.6 Zirconium Phosphate–Doped Composite PEM

Zirconium phosphate (ZRP) inorganic salt is a kind of inorganic proton conductor, which has been studied by researchers. α-ZRP still has high ion exchange capacity at 300°C, so ZRP inorganic salts have become common inorganic fillers for the preparation of inorganic organic composite PEMs.

Zhang *et al.* [69] synthesized PBI/ZRP composite PEM (PBIZrP) by nucleation crystallization isolation (SNAs). It was found that the introduction of ZRP improved the thermal stability and mechanical strength of the composite membrane. The tensile strength of synthesized PBI/ZRP composite film with adding 20% α-ZrP can reach 67.4 MPa. The proton conductivity of synthesized PBI/ZRP composite membrane with adding 10% α-ZrP can reach 0.192 S/cm at 160°C without water. Qian *et al.* [70] prepared sulfonated PBI (SPBI-10) with a 10% sulfonation degree and directly physically blended ZRP and SPBI-10 to prepare inorganic/organic composite PEM. The composite membrane maintained good swelling resistance and mechanical properties. The highest proton conductivity of the composite membrane is 0.08 S/cm at 180°C, which is about 20% higher than that of pure SPBI-10 membrane. Wei *et al.* [71] prepared sulfonated poly(2,3-azanaphthone ether ketone) (SPPEK)/ZRP composite PEM (SPPEK/ZRP) membrane by impregnation and direct blending. It was found that the highest proton conductivity of the composite membrane could reach 0.074 S/cm at 80°C and 100% RH, and the methanol permeability at room temperature was more than 10 times lower than Nafion 117. ZRP inorganic organic composite PEM has high proton conductivity and stability at a high temperature. However, how to fix ZRP on the polymer matrix and enhance its stability in the composite membrane is a problem to be considered in the design and preparation of this kind of membrane.

6.4.7 Polyvinyl Alcohol/Inorganic Composite Membrane

PVA membrane is a kind of fat permeable total evaporation membrane. It has excellent methanol barrier performance, but its proton conductivity is very weak. It can be used for DMFC after adding functional inorganic substances or blending with other proton conductive polymers.

Matsuda *et al.* [72] found that, after mixing PVA with perchloric acid, the conductivity of 5×10^{-2} S/cm containing silica gel could reach, and the thermal stability and water retention capacity were not mentioned. Nakane [73] prepared PVA/SiO_2 composite membrane. X-ray diffraction analysis shows that the SiO_2 network structure can limit the growth of PVA crystal, reduce the crystallinity of PVA, improve the activity of polymer chain, and facilitate proton conduction. Nano-SiO_2 was mixed into SSA (PVA) membrane with sol-gel method by Kim *et al.* [74], and cross-linked hybrid PEM was prepared. DSC study showed that when the content of SSA was lower than and higher than 20%, the contents of free water and adsorbed water in the composite membrane gradually decreased and increased, respectively, and the same change trend was obtained in the study of proton conductivity and methanol permeability. The research group also prepared cross-linked PVA/PAA (polyacrylic acid)/SiO_2 composite membrane [75] by a similar method to improve the alcohol resistance of the membrane without sacrificing proton conductivity. SiO_2 can keep water at high temperature and reduce methanol penetration.

6.5 Conclusion and Prospection

PEMFC and DMFC have become a kind of fuel cell with rapid development in recent years because of their advantages such as high-energy conversion rate, high-power density, environment-friendly, and fast start-up. Proton exchange membrane (PEM) is one of the core components of PEMFC and DMFC. At present, Nafion membrane is the most widely used commercially, but its high cost, low permeability of fuel (methanol, ethanol, etc.) and significant reduction of proton conductivity under high temperature and low humidity limits its large-scale application. Therefore, the development of PEM with high-price ratio and excellent comprehensive performance has become one of the hotspots of fuel cell research. Organic/inorganic composite PEMs are one of the important membranes for PEMFC, which generally have low methanol permeability and good mechanical stability. However, the compatibility of inorganic fillers in a polymer matrix is still an urgent problem to be solved. The leakage of a small-molecule proton conductor in inorganic organic composite membrane also restricts the further improvement of the performance of this kind of composite membrane. Water is still the focus of PEM research, but water loss from the membrane under high-temperature environment is an unavoidable problem, and the complex hydrothermal management system also restricts the cost of fuel cell.

Structure determines performance. In order to develop PEM with similar performance to Nafion membrane, we can modify the existing materials by imitating its chemical structure. Starting from the mechanism, finding the universal law is also one of the research directions in the future. Molecular simulation can provide strong support for the study of proton transfer mechanism and the construction of fuel permeation channel, better guide the preparation of PEM with superior performance, and promote the further research of PEM. The proton transfer mechanism of inorganic organic composite PEM and the construction of a proton transfer model still need more in-depth research. How to design composite membrane with reasonable and efficient proton transfer channel from the perspective of proton transfer mechanism is also the focus of future research.

With the application and development of PEMFC, the working conditions of PEM are becoming harsher. It is very necessary to prolong its service life while maintaining its excellent performance. High-performance and special composite PEMs such as self-humidification and self-repair are main research direction. These special composite films not only maintain the self-height performance but also have certain intelligence that can change with the environment. A composite PEM with excellent water retention was prepared, and its filler could act as a water reservoir to inhibit water diffusion and enhance water retention. Manufacturing self-humidifying membrane, improving proton conductivity of PEM under low RH, and developing non-water subcarrier composite PEMs with good comprehensive performance and durability are one of the research directions in the future. With the deepening of the research on composite PEM, the new composite PEM will develop toward high performance and intelligence, so as to meet the various requirements of fuel cell and improve its performance more effectively. Therefore, non-fluorination, high performance, and intelligence have become an important direction for the future development of composite PEM, which is worthy of the attention of researchers and industry.

In addition, the various composite PEMs mentioned above are the research results of the laboratory, rarely considering the problems of raw materials, processes, and equipment in the industrial preparation process, and the research on further industrialization has not been involved. From laboratory to industrial application, more manpower and resources need to be invested for in-depth research.

Acknowledgments

The author would like to thank the financial support by the National Natural Science Foundation of China (51774158 and 51264021) and Back-up Personnel Foundation of Academic and Technology Leaders of Yunnan Province (2011HR013).

Conflict of Interest

The authors declare no conflict of interest.

References

1. Jin, H., *Modification of sulfonated peek proton exchange membrane for DMFC*, pp. 1–18, Beijing University of Chemical Technology, Beijing, China, 2007 (In Chinese).
2. Vielstich, W., Lamm, A., Gasteiger, H., *Handbook of Fuel Cells, Fundamentals Technology and Applications*, pp. 100–120, John Wiley & Sons Ltd, Chichester, UK, 2003.
3. Bruijn, E., The current status of fuel cell technology for mobile and stationary applications. *Green Chem.*, 7, 132, 2005.
4. Kerres, J., Development of ionomer membranes for fuel cells. *J. Membr. Sci.*, 185, 3, 2001.
5. Rikukawa, M. and Sanui, K., Proton-conducting polymer electrolyte membranes based on hydrocarbon polymers. *Prog. Polym. Sci.*, 25, 1463, 2000.
6. Xie, X.F. and Fan, X.H., *Fuel Cell Technology*, pp. 1–16, Chemical Industry Press, Beijing, China, 2004.
7. Mao, Z.Q., *Fuel Cell*, pp. 1–21, Chemical Industry Press, Beijing, China, 2005.
8. Deluca, N.W. and Elabd, Y.A., Polymer electrolyte membranes for the direct methanol fuel cell: A review. *J. Polym. Sci. B Polym. Phys.*, 44, 2201, 2006.
9. Wang, G., Sun, G., Gong, G., Direct methanol fuel cells. *Physics*, 33, 4, 165, 2004 (In Chinese).
10. Liu, J., Yi, B., Wei, Z., Principle, progress and main technical problems of direct methanol fuel cell. *Power Supply Technol.*, 25, 363, 2001.
11. Steele, B.C.H. and Heinzel, A., Materials for fuel-cell technologies, in: *Materials for sustainable energy: a collection of peer-reviewed research and review articles from nature publishing group*, V. Dusastre, (Ed.), pp. 224–231, World Scientific, Europe, 2010.

12. Shin, D.W., Lee, S.Y., Kang, N.R., Lee, H., Guiver, M.D., Lee, Y.M., Durable sulfonated poly (arylene sulfide sulfone nitrile) s containing naphthalene units for direct methanol fuel cells (DMFCs). *Macromolecules*, 46, 9, 3452–3460, 2013.
13. Mondal, A.N., He, Y., Ge, L., Preparation and characterization of click-driven N-vinylcarbazole-based anion exchange membranes with improved water uptake for fuel cells. *RSC Adv.*, 7, 29794, 2017.
14. Sutradhar, S.C., Jang, H., Banik, N., Yoo, J., Ryu, T., Yang, H., Yoon, S., Kim, W., Synthesis and characterization of proton exchange poly (phenylenebenzophenone) s membranes grafted with propane sulfonic acid on pendant phenyl groups. *Int. J. Hydrog. Energy*, 42, 12749, 2017.
15. Wang, G. and Guiver, M.D., Proton exchange membranes derived from sulfonated polybenzothiazoles containing naphthalene units. *J. Membr. Sci.*, 542, 159, 2017.
16. Pang, J., Feng, S., Yu, Y., Zhang, H., Jiang, Z., Poly(aryl ether ketone) containing flexible tetra-sulfonated side chains as proton exchange membranes. *Polym. Chem.*, 5, 1477, 2014.
17. Lai, A.N., Zhuo, Y.Z., Lin, C.X., Zhang, Q.G., Zhu, A.M., Ye, M.L., Liu, Q.L., Side-chain-type phenolphthalein-based poly(arylene ether sulfone nitriles anion exchange membrane for fuel cells. *J. Membr. Sci.*, 502, 94, 2016.
18. Roziere, J. and Jones, D.J., Non-fluorinated polymer materials for proton exchange membrane fuel cells. *Annu. Rev. Mater. Res.*, 33, 503, 2003.
19. Basura, V.I., Beanie, P.D., Holdcroft, S., Solid-state electrochemical oxygen reduction at Pt/Nafion. 117 and Pt/BAM3GTM 407 interfaces. *J. Electroanal. Chem.*, 458, 1, 1998.
20. Luo, T., Zhang, Y., Xu, H., Zhang, Z., Fu, F., Gao, S., Ouadah, A., Dong, Y., Wang, S., Zhu, C., Highly conductive proton exchange membranes from sulfonated polyphosphazene-graft-copolystyrenes doped with sulfonated single-walled carbon nanotubes. *J. Membr. Sci.*, 514, 527, 2016.
21. He, M.L., Zhu, C.J., Jing, C.J., Sulfonated polyphosphazene-montmorillonite hybrid composite membranes for fuel cells. *Adv. Mater. Res.*, 724, 744, 2013.
22. Yao, H., Shi, K., Song, N., Zhang, N., Huo, P., Zhu, S., Zhang, Y., Guan, S., Polymer electrolyte membranes based on cross-linked highly sulfonated co-polyimides. *Polymers*, 103, 171, 2016.
23. Zhang, B., Ni, J., Xiang, X., Wang, L., Chen, Y., Synthesis and properties of reprocessable sulfonated polyimides cross-linked via acid stimulation for use as proton exchange membranes. *J. Power Sources*, 337, 110, 2017.
24. Wang, C., Cao, S., Chen, W., Xu, C., Zhao, X., Li, J., Ren, Q., Synthesis and properties of fluorinated polyimides with multi-bulky pendant groups. *RSC Adv.*, 7, 26420, 2017.
25. Li, C., Huang, N., Jiang, Z., Tian, X., Zhao, X., Xu, Z.L., Yang, H., Jiang, Z.J., Sulfonated holey graphene oxide paper with SPEEK membranes on its both sides: A sandwiched membrane with high performance for semi-passive direct methanol fuel cells. *Electrochim. Acta*, 250, 68, 2017.

26. Lee, K.H., Chu, J.Y., Kim, A.R., Yoo, D.J., Enhanced performance of a sulfonated poly(arylene-ether-ketone) block copolymer bearing pendant sulfonic acid groups for polymer electrolyte membrane fuel cells operating at 80% relative humidity. *ACS Appl. Mater. Interfaces*, 10, 20835, 2018.
27. Oh, K., Son, B., Sanetuntikul, J., Shanmugam, S., Polyoxometalate decorated graphene oxide/sulfonated poly (arylene ether ketone) block copolymer composite membrane for proton exchange membrane fuel cell operating under low relative humidity. *J. Membr. Sci.*, 541, 386, 2017.
28. Singha, S., Jana, T., Modestra, J.A., Naresh, A., Kumar, S.V., Highly efficient sulfonated polybenz- imidazole as a proton exchange membrane for microbial fuel cells. *J. Power Sources*, 317, 143, 2016.
29. Xia, Z., Ying, L., Fang, J., Du, Y.Y., Zhang, W.M., Guo, X., Yin, J., Preparation of covalently cross-linked sulfonated polybenzimidazole membranes for vanadium redox flow battery applications. *J. Membr. Sci.*, 525, 229, 2017.
30. Wu, Y.H., Kang, F.F., Bing, K.L., Tong, M., Research progress in inorganic-organic composite proton exchange membrane. *Chin. J. Power Sources*, 30, 169, 2006 (In Chinese).
31. Chen, Y., *Research on Preparation of organic/inorganic composite proton exchange membrane for DMFC*, Central South University, Changsha, 2008 (In Chinese).
32. Hou, J.H., Liu, S.S., Xiao, Z.Y., Cai, Y.X., Ding, H.L., Research progress of inorganic-organic composite proton exchange membrane for fuel cells. *New Chem. Mater.*, 46, 44, 2018.
33. Xu, H.T. and Jiang, J.Y., Research progress of organic-inorganic hybrid proton exchange membranes for fuel cells applications. *Guangzhou Chem. Eng.*, 38, 40, 2010 (In Chinese).
34. Taghizadeh, M. and Vatanparast, M., Ultrasonic-assisted synthesis of ZrO_2 nanoparticles and their application to improve the chemical stability of Nafion membrane in proton exchange membrane (PEM) fuel cells. *J. Colloid Interface Sci.*, 483, 1, 2016.
35. Wu, H., Shen, X., Xu, T., Sulfonated poly(ether ether ketone)/amino-acid functionalized titania hybrid proton conductive membranes. *J. Power Sources*, 213, 83, 2012.
36. Wu, H., Hou, W., Wang, J., Preparation and properties of hybrid direct methanol fuel cell membranes by embedding or ganophosphorylated titania submicrospheres into a chitosan polymer matrix. *J. Power Sources*, 195, 4104, 2010.
37. Boutsika, L., Enotiadis, A., Nicotera, I., Nafion nanocomposite membranes with enhanced properties at high temperature and low humidity environments. *Int. J. Hydrog. Energy*, 41, 22406, 2016.
38. Zhao, Y., Jiang, Z., Lin, D., Enhanced proton conductivity of the proton exchange membranes by the phosphorylated silica submicrospheres. *J. Power Sources*, 224, 28, 2013.

39. Zhao, Y., Yang, H., Wu, H., Enhanced proton conductivity of hybrid membranes by incorporating phosphorylated hollow mesoporous silica submicrospheres. *J. Membr. Sci.*, 469, 418, 2014.
40. Wang, H., Wang, X., Fan., T., Fabrication of electrospun sulfonated poly(ether sulfone) nanofibers with amino modified SiO_2 nanosphere for optimization of nanochannels in proton exchange membrane. *Solid State Ionics*, 349, 115300, 2020.
41. Liu, G., Tsen, W.C., Wen, S., Sulfonated silica coated polyvinylidene fluoride electrospun nanofiber-based composite membranes for direct methanol fuel cells. *Mater. Des.*, 193, 108806, 2020.
42. Zhang, X., Xia, Y., Gong, X., Preparation of sulfonated polysulfone/sulfonated titanium dioxide hybrid membranes for DMFC applications. *J. Appl. Polym. Sci.*, 137, 48938, 2020.
43. Yuan, C. and Wang, Y., The preparation of novel sulfonated poly(aryl ether ketone sulfone)/TiO_2 composite membranes with low methanol permeability for direct methanol fuel cells. *High Perform. Polym.*, 33, 326, 2021.
44. He, G., Chang, C., Xu, M., Tunable nanochannels along graphene oxide/polymer core-shell nanosheets to enhance proton conductivity. *Adv. Funct. Mater.*, 25, 7502, 2016.
45. Lin, Y., Yen, C., Hung, C., A novel composite membranes based on sulfonated montmorillonite modified nafion® for DMFCs. *J. Power Sources*, 168, 162, 2007.
46. Wu, X., Wu, N., Shi, C., Proton conductive montmorillonite-nafion composite membranes for direct ethanol fuel cells. *Appl. Surf. Sci.*, 388, 239, 2016.
47. Beydaghi, H., Javanbakht, M., Kowsari, E., Preparation and physic-chemical performance study of proton exchange membranes based on phenyl sulfonated graphene oxide nanosheets decorated with iron titanate nanoparticles. *Polymers*, 87, 26, 2016.
48. Yin, Y., Wang, H., Cao, L., Sulfonated poly(ether ether ketone)-based hybrid membranes containing graphene oxide with acid-base pairs for direct methanol fuel cells. *Electrochim. Acta*, 203, 178, 2016.
49. Liu, Y., Zhang, J., Zhang, X., Ti_3C_2Tx filler effect on the proton conduction property of polymer electrolyte membrane. *ACS Appl. Mater. Interfaces*, 8, 20352, 2016.
50. Wu, X., Hao, L., Zhang, J., Polymer-Ti_3C_2Tx composite membranes to overcome the trade-off in solvent resistant nanofiltration for alcohol-based system. *J. Membr. Sci.*, 515, 175, 2016.
51. Fu, F.Y., Zhang, J., Cheng, J.Q., Application of graphene oxide in fuel cell proton exchange membrane. *Chem. Prog.*, 38, 2233, 2019.
52. Cheng, T., *Preparation and properties of functional fossil grapheme/sulfonated polyarylethene eye proton exchange membrane*, pp. 1–25, University of Electronic Science and Technology, Chengdu, 2019 In Chinese.

53. Gil-Castell, O., Santiago, O., Pascual-Jose., B., Performance of sulfonated poly(vinyl alcohol)/graphene oxide polyelectrolytes for direct methanol fuel cells. *Energy Technol.*, 8, 2000124, 2020.
54. He, Y., Tong, C., Geng, L., Enhanced performance of the sulfonated polyimide proton exchange membranes by graphene oxide:size effect of graphene oxide. *J. Membr. Sci.*, 458, 36, 2014.
55. Heo, Y., Im, H., Kim, J., The effect of sulfonated graphene oxide on sulfonated poly (ether ether keto.ne) membrane for direct methanol fuel cells. *J. Membr. Sci.*, 425–426, 11, 2013.
56. Wang, H., Sun, N., Zhang, L., Ordered proton channels constructed from deoxyribonucleic acid- functionalized grapheme oxide for proton exchange membranes via electrostatic layer-by-layer deposition. *Int. J. Hydrog. Energy*, 45, 27772, 2020.
57. Zhong, S., Ding, C., Gao, Y., Sulfonic group-functionalized graphene oxide-filled self-cross-linked sulfonated poly(etherether ketone) membranes with excellent mechanical property and selectivity. *Energ. Fuel*, 34, 11429, 2020.
58. Liu, Y., Su, Y., Chang, C., Preparation and applications of nafion-functionalized multiwalled carbon nanotubes for proton exchange membrane fuel cells. *J. Mater. Chem.*, 20, 4409, 2010.
59. Thomassin, J., Kollar, J., Caldarella, G., Beneficial effect of carbon nanotubes on the performances of nafion membranes in fuel cell applications. *J. Membr. Sci.*, 303, 252, 2007.
60. Xing, B. and Savadogo, O., The effect of acid doping on the conductivity of polybenzimidazole (PBI). *J. New Mater. Electrochem. Syst.*, 2, 95, 1999.
61. Che, Q., Chen, N., Yu, J., Sulfonated poly(ether ether) ketone/polyurethane composites doped with phosphoric acids for proton exchange membranes. *Solid State Ionics*, 289, 199, 2016.
62. Yue, Z., Cai, Y., Xu, S., Phosphoric acid-doped cross-linked sulfonated poly(imide-benzimidazole) for proton exchange membrane fuel cell applications. *J. Membr. Sci.*, 501, 220, 2016.
63. Li, M. and Scott, K., A polymer electrolyte membrane for high temperature fuel cells to fit vehicle applications. *Electrochim. Acta*, 55, 2123, 2010.
64. Yang, J., Cleemann, L., Steenberg, T., High molecular weight polybenzimidazole membranes for high temperature PEMFC. *Fuel Cells*, 14, 7, 2014.
65. Li, Q., He, R., Jensen, J.O., PBI-Based polymer membranes for high temperature fuel cells-preparation, characterization and fuel cell Demonstration. *Fuel Cells*, 4, 147, 2004.
66. Malers, J.L., Sweikart, M.A., Horan, J.L., Studies of heteropoly acid/polyvinylidenedi-fluoride- hexafluoroproylene composite membranes and implication for the use of heteropoly acids as the proton conducting component in a fuel cell membrane. *J. Power Sources*, 172, 83, 2007.
67. Zhao, C., Lin, H., Cui, Z., Highly conductive, methanol resistant fuel cell membranes fabricated by layer-by-layer self-assembly of inorganic heteropolyacid. *J. Power Sources*, 194, 168, 2009.

68. Yu, S.K., Feng, W., Hickner, M., Fabrication and characterization of heteropolyacid ($H_3PW_{12}O_{40}$)/directly polymerized sulfonated poly(arylene ether sulfone) copolymer composite membranes for higher temperature fuel cell applications. *J. Membr. Sci.*, 212, 263, 2003.
69. Zhang, Q., Liu, H., Li, X., Synthesis and characterization of polybenzimidazole/α-zirconium phosphate composites as proton exchange membrane. *Polym. Eng. Sci.*, 56, 622, 2016.
70. Qian, W., Shang, Y., Fang, M., Sulfonated polybenzimidazole/zirconium phosphate composite membranes for high temperature applications. *Int. J. Hydrog. Energy*, 37, 12919, 2012.
71. Wei, Z.H., Du, C.H., Yi, X.Y., Hybrid proton exchange membranes based on sulfonated poly(phthalazinone ether ketone) and zirconium hydrogen phosphate. *Polym. Adv. Technol.*, 18, 373, 2010.
72. Matsuda, A., Honjo, H., Hiram, K., Electric double-layer capacitor using composites composed of phosphoric acid doped silica gel and styrene ethylene butylene styrene elastomer as a solid electrolyte. *J. Power Sources*, 77, 12, 1999.
73. Nakane, K., Yamashita, T., Iwakura, K., Properties and structure of poly(vinyl alcohol)/silica composites. *J. Appl. Polym Sci.*, 74, 133, 1999.
74. Kim, D.S., Park, H.B., Rhim, J.W., Preparation and characterization of crosslinked PVA/SiO_2 hybrid membranes containing sulfonic acid groups for direct methanol fuel cell applications. *J. Memb. Sci.*, 240, 37, 2004.
75. Kim, D.S., Park, H.B., Rhim, J.W., Proton conductivity and methanol transport behavior of cross—linked PVA/PAA/Silica hybrid membranes. *Solid State Ionics*, 176, 117, 2005.

7

Thermoset-Based Composite Bipolar Plates in Proton Exchange Membrane Fuel Cell: Recent Developments and Challenges

Salah M.S. Al-Mufti and S.J.A. Rizvi*

Department of Petroleum Studies, Aligarh Muslim University, Aligarh, India

Abstract

Bipolar plates (BPs) are one of the primary components that taking responsibility for multi-functions of vital importance to the long-term performance of proton exchange membrane (PEM) fuel cells. BPs provide conducting networks for electrons between cells, homogeneous distribution of fuel gases, water and heat management, and structural support for the thin membrane. BPs primarily have been manufactured from high-density graphite and metals. Recently, novel characteristics of composite materials have captured the great attention of researchers for BP applications. Two categories of composites have aroused extensive interest: thermoplastic and thermoset. This chapter overviews the present research studies being performed on thermoset composites for BP applications. In this research, types of the polymer matrix (thermoset resins) and carbon fillers, as well as manufacturing methods, are investigated. Characteristics of polymer matrix and conductive fillers play a major role in the electrical, mechanical, and thermal properties of thermoset-based composite BPs. Herein, the effect of different conductive fillers and the influence of parameters of manufacturing methods on the properties of thermoset-based composite BPs have been discussed in detail. This chapter also includes theories of electrical conductivity in polymer composites and the testing and characterization of polymer composite-based BPs.

Keywords: Bipolar plate, fuel cell, thermoset composite, conductive fillers, percolation theory, compression molding

**Corresponding author*: javedrizvi1976@gmail.com

Inamuddin, Omid Moradi and Mohd Imran Ahamed (eds.) Proton Exchange Membrane Fuel Cells: Electrochemical Methods and Computational Fluid Dynamics, (137–212) © 2023 Scrivener Publishing LLC

7.1 Introduction

Recently, renewable and alternative energy sources such as hydrogen energy, solar energy, biomass energy, geothermal energy, wind energy, and fuel cell (FC) technology have attracted attention from countries, companies, and researchers due to the negative climate change, limited energy resources, and increasing energy requirement [1–5]. As a result of air pollution, each year around 7–8.8 million people have died [6–8]. Most of these problems might be solved by using FCs as a new technology for producing clean energy [9–11]. FCs have numerous advantages such as low/non-polluting, low-temperature operation, high efficiency, silent operation, size reduction, fast start-up, and renewable fuel sources [12]. A FC is a device that converts chemical energy (H_2 and O_2) into electricity via an electrochemical reaction between hydrogen and oxygen [13]. FCs are not widely used commercially according to cost, endurance, and mass. A practical example, in relation to 1.1 billion diesel/petrol cars, in the world in 2018, there were only 6,500 FC vehicles (FCVs) due to the high cost of FCVs, whereas the average cost of FCV is between 60,000 and 70,000 $ [14–16]. Lower fabrication costs and inexpensive materials are needed to be cost-competitive. The PEMFC is composed of a membrane electrode assembly (MEA), end plates, gas diffusion layer (GDL), and bipolar plates (BPs). A brief illustration of the proton-exchange membrane FCs is shown

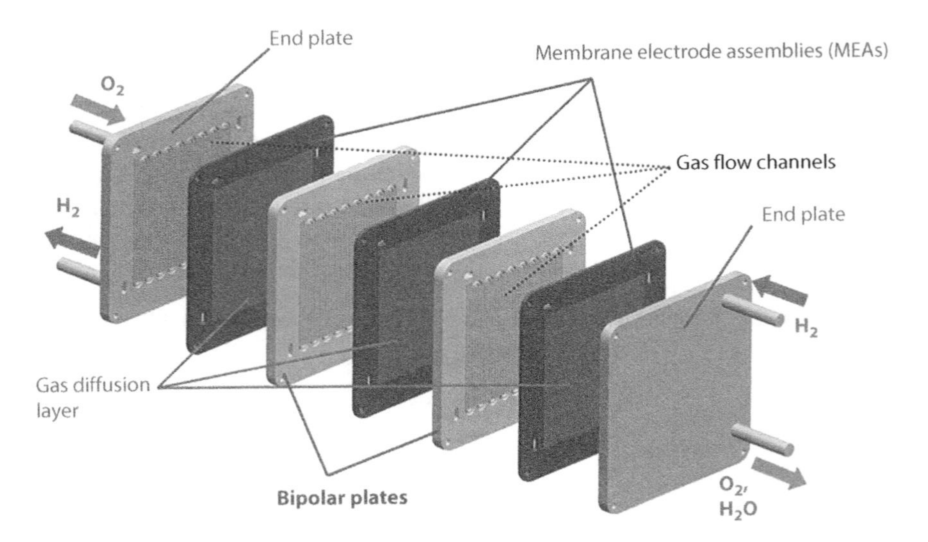

Figure 7.1 Components of the PEM fuel cell. Reprinted with permission from Ref. [42], copyright of John Wiley and Sons.

in Figure 7.1. In a FC stack, BPs represent a major part of this device and play an important role in the weight, volume, cost, and efficiency of the FCs. The BP consumes around 60%–80% of the weight, constitutes over 30%–45% of the total cost, and contributes about 70%–80% of the volume of FC stacks [17–19]. BPs have a number of functions as follows: electrical performance (collecting and transport electrons within the FC), provide structural support (mechanical strength) for the thin membrane, feeding the reactive gases to anode and cathode, separate the fuel and oxidant gases (hydrogen and oxygen), help in heat and water management, and separate individual cells in the stack [20–23]. According to the important role of BPs in PEMFC, the BPs must have the following properties: high electrical conductivity, lightweight materials, high mechanical strength, low-cost, chemical stability in the PEMFC, low gas permeability, and high corrosion resistance [24–28]. The US Department of Energy (US DOE) has been given specific measurements and technical targets for the BP as tabulated in Table 7.1. Coated metals, non-coated metals, non-metals, and composites are the calcification of BP materials. Metallic BPs have high electrical conductivity, mechanical strength, thermal conductivity, ease of manufacturing, and high gas impermeability [29, 30]. On the other hand, metallic BPs have many problems due to low corrosion resistance and high cost [31, 32]. The electro-corrosion generates cations, which may act as a membrane contaminant [33, 34]. This leading to decreasing the cell performance by disturbing the conductivity of the membrane [35]. Graphite plates have excellent corrosion resistance and good electrical conductivity. Graphite BPs also have many problems such as high processing cost, high weight and volume, and high porosity [36]. Recently, the carbon-based composite materials such as thermoplastic and thermoset composites have attracted research attention from researchers due to good processing, high resistance to corrosion, low cost, low weight, and diverse functional application. Table 7.2 shows the comparison between properties of different types of bipolar plate materials.

In the fabrication of BP, thermoset resins are used more than thermoplastic resins for diverse reasons as follows [37, 38]:

i. Have higher strength
ii. Have lower toughness
iii. Have a creep resistance
iv. Have a shorter cure time
v. Can be loaded with a higher level of conductive fillers due to lower viscosity than thermoplastic resins
vi. Have thermal stability better than thermoplastics resins.

Table 7.1 US DOE technical targets for composite bipolar plates [47, 175, 200, 201].

Characteristic	Units	2015 status	2020 status	2025 target
Electrical conductivity	S cm^{-1}	>100	100	>100
Areal specific resistance	Ωcm^2	0.006	0.01	<0.01
Flexural strength	MPa	>34 (carbon plate)	25	>40
Cost	\$kW^{-1}	7	3	2
Plate weight	kg/kW	<0.4	0.4	0.18
Corrosion, anode	μA cm^{-2}	No active peak	1 and no active peak	<1 and no active peak
Corrosion, cathode	μA cm^{-2}	<0.1	1	<1
H_2 permeability	cm^3 s^{-1}cm^{-2}	0	1.3×10^{-14}	2×10^{-6}
Thermal conductivity	Wm^{-1} K^{-1}	>20	>20	/
Forming elongation	%	20-40	40	40

Lots of review papers about BPs for PEMFC have been published.

Antunes *et al.* [39], many years ago (2011), have been studied the carbon materials in composite BPs for PEMFC and discussed the topics which are related to the structural aspects of carbon-based fillers. Furthermore, the carbon-based fillers effect on the final effectiveness of electrical conductivity of composite BPs was debated. Planes *et al.* [40], a few years ago (2012), summarized the current improvement, manufacturing, and structure property of polymer composites designed for applications of BPs. In addition, they are investigated and considered thermoplastics and thermosets resins as polymer matrices. They reported that matrix combined with

Table 7.2 Comparative study between the properties of various categories for bipolar plates. Reproduced with permission from Ref. [202], copyright of Elsevier.

Property	Coated metallic BP	Graphite BP	Polymer composite (thermoset/ thermoplastic) BP
Electrical Conductivity	Excellent	Weak	Good
Strength	Excellent	Weak	Good
Corrosion resistance	Weak	Excellent	Excellent
Cost	High	Very high	Very low
Gas permeation	Excellent	Weak	Good
Density	More than (6 gr/cm^3)	Around (2 gr/ cm^3)	Less than (2 gr/cm^3)
Needed thickness	<2 mm	>4 mm	2–3 mm

the different fillers such as carbon nanotubes (CNTs), carbon fiber, carbon black (CB), and graphite. Karimi *et al.* [41] in (2012) have published a review about the materials of metallic BPs and their fabrication methods. San *et al.* [42] in (2013) have been covered using thermoplastic composites in different applications of BPs. In the same review paper, they offered in detail the many effects on the performance of the BPs like filler type, polymer types, post-processing, blending additives, filler form, filler synergy, and orientation. Taherian [43] (2014) analyzed the materials, materials selection, and fabrication methods of composite and metallic BPs. Furthermore, he evaluated the properties of composite BPs completely as electrical conductivity and mechanical. In addition, he investigated the materials, properties, coatings, coating method, and ionic contaminations of metallic BPs.

Kausar *et al.* [44] (2016) reported that one of the important applications of Epoxy/CNT composite is composite BP. Wilberforce *et al.* [45] (2018) studied the geometry design of the BP, which is relevant to the FC function. They debated the specific parameters (flow direction and channel length), which can be improved the fuel stack. The influence geometry design of BP correlated to heat transfer and mass transport, in addition to electrical conductivity, has been investigated. They reported that the effective design of BP has the ability to increase the overall performance of FCs by nearly 50%. Yi *et al.* [46] (2019) presented the recent development on carbon-based coating, which is used in polymer electrode membrane FCs for metallic BP applications. Recently, Song *et al.* [47] (2019) debated comprehensive research studies about current materials, fabrication methods, and covering the applications of the BP for PEMFCs. They reported that, at the time of operating the PEMFC, the flow field design can help BP to work well, especially for removing the water from the channels of BP.

7.2 Theories of Electrical Conductivity in Polymer Composites

Polymer matrix is electrically insulating materials, where their electrical conductivity values are 10^{-14} to 10^{-17} S/cm (see Table 7.4). Typical conductive values for carbon fillers are approximately 10^2 to 10^5 S/cm (see Table 7.5). By mixing insulating polymer matrix and fillers with high electrical conductivity as graphite, CB, CNT, carbon fiber, expanded graphite (EG), and graphene, the conductive polymer composites are obtained.

Numerous factors play a major role in the electrical conductivity of polymer composites such as filler distribution, aspect ratio, shape and size, polymer matrix nature, orientation, wettability, surface energy, filler conductivity, matrix interaction, and processing technique. In general, the classical percolation theory explained the composite conductivity is carried out when the filler volume fraction is above the percolation threshold phenomena. To predict the electrical conductivity of the conductive polymer composite, different models have been developed. Lux [48] explained in detail the statistical, thermodynamic, structure-oriented, and geometrical models, which are the main classes of conductivity models. These four models have been modified by a number of researchers; some of the models are explained below. Some of the common electrical conductivity models for polymer composite is clarified in Table 7.3.

Table 7.3 Description of some of the common electrical conductivity models for polymer composite.

Developer	Model equation	References
Bruggeman	$\left(\frac{\sigma_d-\sigma}{\sigma_d+2\sigma}\right)\phi+\left(\frac{\sigma_m-\sigma}{\sigma_m+2\sigma}\right)(1-\phi)=0$	[203]
Reuss	$\frac{1}{\sigma}=\left[\frac{\phi}{\sigma_d}\right]+\left[\frac{(1-\phi)}{\sigma_m}\right]$	[203]
Maxwell	$\sigma_r=\frac{\sigma}{\sigma_m}=\left[\frac{1+2\phi\left(\frac{\lambda-1}{\lambda+2}\right)}{1-\phi\left(\frac{\lambda-1}{\lambda+2}\right)}\right]$	[204]
Pal	$\frac{\sigma}{\sigma_m}=1+3\alpha\left(\frac{\sigma_d-\sigma_m}{\sigma_d+2\sigma_m}\right)\phi$	[205]
Clingerman	$\log\sigma_m=\log\sigma_p\text{, for }\phi\le\phi_c$ $\log\sigma_m=\log\sigma_p+\text{D}\log\sigma_f(\phi-\phi_c)^q{}_c$ $+\text{h(a)}\cos\theta-\text{C}\gamma\text{pf, for }\phi>\phi\text{q}=\frac{\text{B}\phi_c}{(\phi-\phi_c)^N}$ $\text{h(a)}=\text{A}^2[1-0.5(\text{A}-1/\text{A})$ $\times\ln[(\text{A}+1)/(\text{A}-1)]]\text{A}^2=\frac{\text{a}^2}{\text{a}^2-1}$	[206]
Keith	$\log\sigma=\log\sigma_p+\text{H}(\phi-\phi_c)^{\frac{G}{(\phi-\phi_c)^n}}+\text{E}$	[207]
McCullough	$P_i=\phi_f P_f+\phi_m P_p-\left[\frac{\lambda_i\phi_f\phi_m(P_f-P_p)^2}{V_{fi}P_f+V_{mi}P_p}\right]$ $V_{fi}=(1-\lambda_i)\phi_f+\lambda_i\phi_m$ $V_{mi}=(1-\lambda_i)\phi_m+\lambda_i\phi_f$	[208]
Taherian	$\sigma_{\text{composite}}=\frac{a_0\cdot(AR)(\sigma_f)}{c+\exp\left[-\frac{(x-\text{ roundness})}{\cos\theta}\right]}$ $\cos(\theta)=\frac{\gamma_S-\gamma_{SL}}{\gamma_L}$ $\gamma_{SL}=\gamma_S+\gamma_L-2(\gamma_S\cdot\gamma_L)^{0.5}$	[209]

(*Continued*)

Table 7.3 Description of some of the common electrical conductivity models for polymer composite. (*Continued*)

Developer	Model equation	References
Weber	$\sigma_c = \sigma_m + \left[4/\pi\left(d_c \cdot \frac{l}{d^2\cos^2\phi}\right)(v_p\sigma_f)X\right.$ $X = 1/(0.59+0.15\text{m})$	[210]
Nielson	$\sigma_c = \sigma_{\text{poly}} \frac{1+AB\phi_f}{1-B\Psi\phi_f}$ $B = \frac{\sigma_f/\sigma_{\text{poly}}-1}{\sigma_f/\sigma_{\text{poly}}+A}, \quad \Psi \approx 1+\left(\frac{1-\phi_m}{\phi_m^2}\right)\phi_f$	[60, 211]
Voet	$\ln\delta_c = \text{KV}_\text{f}^{1/3}$	[212, 213]
Kirkpatrick and Zallen	$\sigma_m = \sigma_h(\phi - \phi_c)^t$	[63]
Tiusanen	$\sigma_{\text{DC}} = \sigma_0\left(\frac{\text{P}_\text{A}-\text{P}_\text{c}}{1-\text{P}_\text{c}}\right)^\text{x}$	[214]
Zare	$\sigma = \left(\frac{\phi_{eff}^{1/3}-\phi_p^{1/3}}{1-\phi_p^{1/3}}\right)\phi_f\left(1+\frac{t}{R}\right)^2$ $\left[\frac{1+2\alpha\left(\frac{\phi_{eff}^{1/3}-\phi_p^{1/3}}{1-\phi_p^{1/3}}\right)\phi_f\left(1+\frac{t}{R}\right)^2}{1-\left(\frac{\phi_{eff}^{1/3}-\phi_p^{1/3}}{1-\phi_p^{1/3}}\right)\phi_f\left(1+\frac{t}{R}\right)^2}\right]\left(\frac{d}{\lambda}\right)^2$	[215]

σ, σ_m: electrical conductivity of composite; σ_c: composite conductivity in percolation threshold; σ_F: electrical conductivity in maximum volume fraction; σ_f, σ_h, σ_0: electrical conductivity of filler; σ_p: electrical conductivity of polymer; ϕ, V_f, P_A, V: volume fraction of filler; ϕ_c, P_c, V_C: volume fraction of filler in percolation threshold; v_p: effective volume fraction; μ: volume fraction value of filler at the pick of dσ/dx – x curve; ρ: the resistivity of composite; ρ_m: resistivity of polymer; ρ_f: resistivity of filler; t, s: critical exponent; f_c: the percolation threshold; f: volume fraction; F: compact factor; θ: wetting angle; ϕ: angle of filler orientation in the matrix; : aspect ratio; m: contact number between fillers; γ_{pf}: interface surface energy; d: filler diameter; *l*: length of filler; d_c: critical filler diameter.

7.2.1 Percolation Theory

In general, the electrical conductivity of a conductive composite depends on the content of conductive fillers. Figure 7.2 shows the filler distribution in thermoset composites at low content of filler and conductive network at sufficient filler content.

In which, typical conductive values for carbon fillers are approximately 10^2 to 10^5 S/cm. For making the free electrons travel easily in the polymer composite, the conductive network must be formed, which leads to an increase in electrical conductivity. The conductive path formation is built on the principles of percolation theory. The concept of percolation is a substantial theory to understand the conductivity within conductive polymer composite, particularly when the polymer matrix and carbon filler possess different properties. The percolation theory clearly demonstrated the polymer composite electrical conductivity close to the value of the percolation threshold [49–51]. The percolation threshold in polymer composite is based on various factors such as aspect ratio, surface treatment, dispersion, orientation, distribution, filler agglomeration, type of polymer, mixing method, surface energies, and phase structure [52–57]. Figure 7.3 clearly shows the dependence of the electrical conductivity on the volume fraction of filler, where in region 1 (at low filler loading), the value of electrical conductivity is close to the electrical conductivity value of polymer matrix due to the unavailable path for electron transport. Region 2 (at some critical filler loading) is called the percolation threshold. In this region, with a very small increase in the filler amount, the conductivity significantly increases. In region 3 and after increasing the conductivity drastically, the

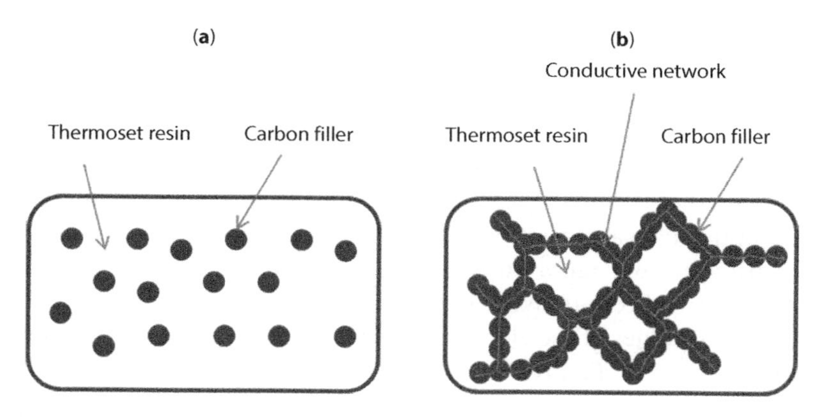

Figure 7.2 Filler distribution in thermoset composite, (a) at low filler content, (b) conductive network at high filler content.

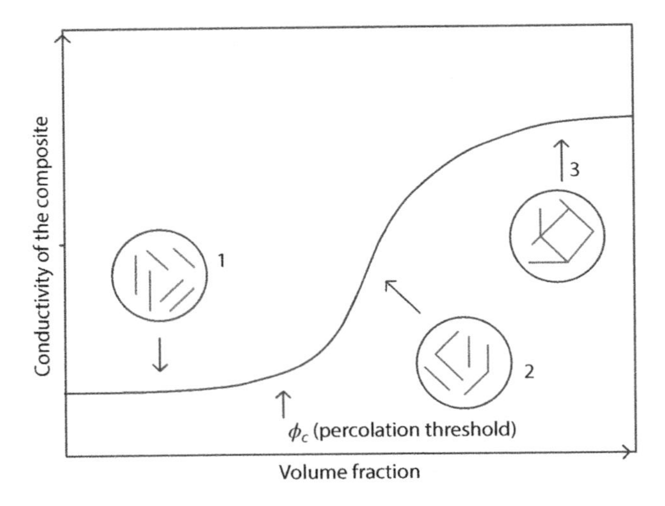

Figure 7.3 The percolation zone of conductive fillers reinforced polymers composites [194].

conductivity started to increase slowly and approaches that of the filler material value. Ultimately, the maximum percolation threshold is reached, in which any more addition of filler will not increase the conductivity of the composite.

7.2.2 General Effective Media Model

McLachlan *et al.* [58, 59] developed the general effective media (GEM) model for predicting the polymer composite electrical conductivity. They combined percolation theories and Bruggeman together for producing the GEM model. By many researchers, the GEM model has been investigated [60, 61]. The GEM model is appropriate for single or multiple filler composites. Zakaria *et al.* [23] found that the GEM model successfully was used for predicting the in-plane and through-plane electrical conductivities for both single and multiple filler polymer composites.

Kakati *et al.* [62] reported that the GEM equation is suitable to predict the electrical conductivity of the BP and the model well predicted the experimental values of the electrical conductivity of the polymer composites BPs as shown in Figure 7.4.

This model is given by

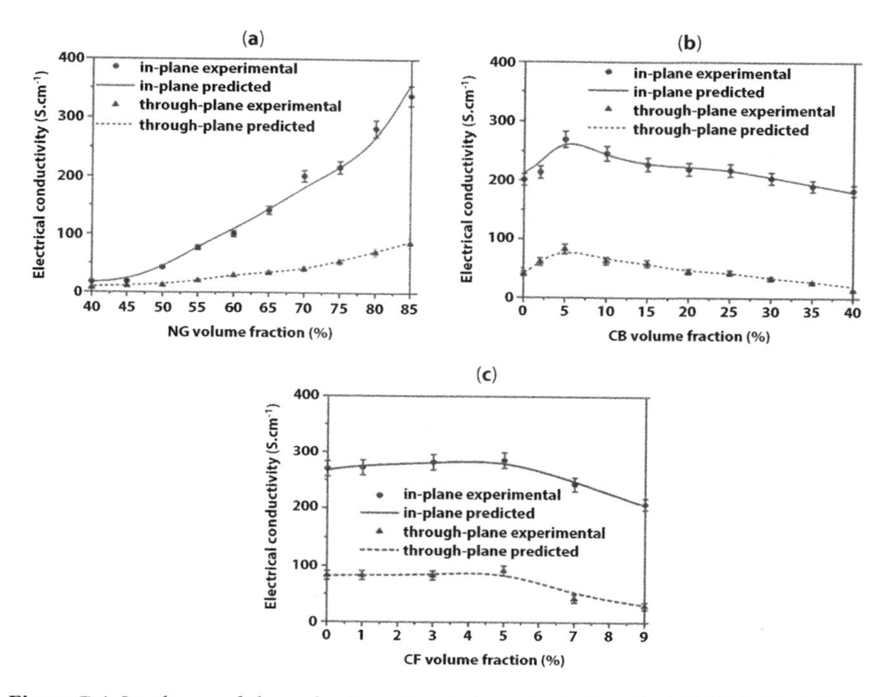

Figure 7.4 In-plane and through-plane electrical conductivities of (a) NG/PF, (b) (NG + CB)/PF 70/30 vol.%, and (c) PF/(NG+ CF)/CB 30/65/5 vol.% composite BPs. Reprinted with permission from Ref. [62], copyright of Elsevier.

$$\frac{(1-\varphi)\left(\sigma_l^{\frac{1}{t}}-\sigma_m^{\frac{1}{t}}\right)}{\sigma_l^{\frac{1}{t}}+A\sigma_m^{\frac{1}{t}}}+\frac{\varphi\left(\sigma_h^{\frac{1}{t}}-\sigma_m^{\frac{1}{t}}\right)}{\sigma_h^{\frac{1}{t}}+A\sigma_m^{\frac{1}{t}}}=0 \tag{7.1}$$

where σ_m is the composite conductivity, σ_h and σ_l are the conductivity of the high and low conductivity phases, φ is the volume fraction, $A = ((1 - \phi_c)/\phi_c)$ where ϕ_c is the percolation threshold of the high conductivity phase, and t is the GEM exponent.

7.2.3 McLachlan Model

This model fits into the statistical percolation model category. Typically, statistical percolation models such as Kirkpatrick [63], Zallen [64], and

McLachlan models predict that the electrical conductivity depends on the particle contacts probability into the polymer composite *c*. In this model, the critical exponential *t* is required, which is calculated from the experimental data. The McLachlan model is given by

$$\frac{(1-\phi)\left(\rho_m^{\frac{1}{t}}-\rho_h^{\frac{1}{t}}\right)}{\rho_m^{\frac{1}{t}}+((1-\phi_c)/\phi_c)\rho_h^{\frac{1}{t}}}+\frac{\phi\left(\rho_m^{\frac{1}{t}}-\rho_l^{\frac{1}{t}}\right)}{\rho_m^{1/t}+((1-\phi_c)/\phi_c)\rho_l^{\frac{1}{t}}}=0 \tag{7.2}$$

where ρ_m is the resistivity of the composite, ρ_h and ρ_l are the resistivity of the component with high and low resistivity, t is the critical exponent, ϕ is the volume fraction, and ϕ_c is the percolation threshold.

7.2.4 Mamunya Model

Mamunya model agrees with the thermodynamic model category. Mamunya *et al.* [65, 66] have evaluated the effect of different factors, including polymer filler, polymer melt viscosity, and polymer surface energies on the electrical conductivity, through study the composite conductivity based on the concentration of filler for various polymers. This model displayed that the behavior of percolation is based on the interaction of polymer filler as well as the size and amount of the conductive filler. Beyond the percolation threshold, the electrical conductivity of the polymer composite is given by

$$\log\sigma=\log\sigma_c+(\log\sigma_m-\log\sigma_c)\left(\frac{\phi-\phi_c}{F-\phi_c}\right)^k \tag{7.3}$$

$$k=\frac{K\phi_c}{(\phi-\phi_c)^{0.75}},\ K=A-B\gamma_{\text{pf}} \tag{7.4}$$

where σ is the composite electrical conductivity, σ_c is the composite conductivity at ϕ_c, F is the packing fraction at maximum value, σ_m is the electrical conductivity at F, ϕ_c is the percolation threshold, ϕ is the volume fraction, A and B are constants, γ_{pf} is the interfacial surface tension, and the value k is based on the volume fraction of filler, surface energy, and percolation threshold.

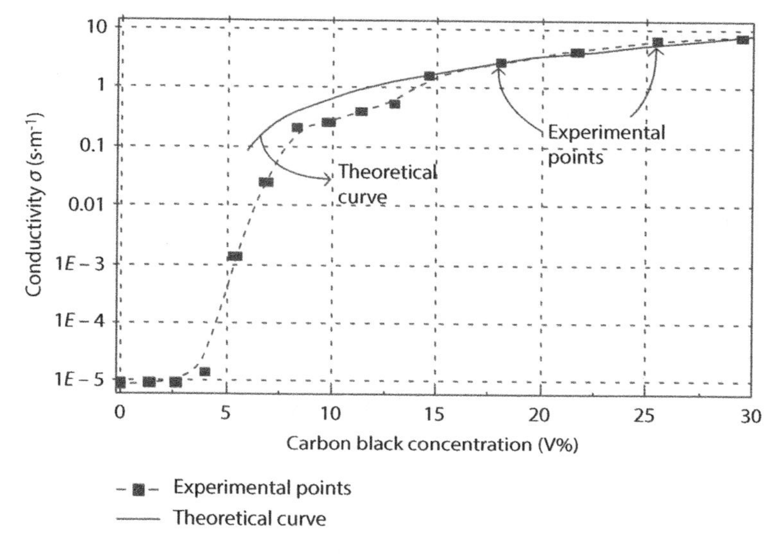

Figure 7.5 Comparison of Mamunya model with experimental conductivity data as a function of CB content in the PP-based composites [67].

Merzouki *et al.* [67] reported that the Mamunya model shows good accordance between the results of experimental and calculated for CB in various polymers, but this model is not righteous for other types of fillers. In addition, They confirmed that for the low concentrations of filler comparable to the filler concentrations of the percolation, the Mamunya model could not be used as shown in Figure 7.5.

7.2.5 Taherian Model

Taherian *et al.* [68] have been developed a statistical-thermodynamic formula for predicting the polymer composite BPs electrical conductivity. This model depends on the phenomenon of percolation threshold, which is supposed that the relevance between polymer composite electrical conductivity and volume fraction of filler subjected to a sigmoidal equation. There are four effective factors efficient on the electrical conductivity of composite: filler aspect ratio, the electrical conductivity of filler, interface contact resistance, and wettability. These factors can be used instead of constant parameters for the sigmoidal function. This model has well predicted the electrical conductivity of some single-filler composites such as PF/G, PF/EG, and PF/CF, as shown in Figure 7.6. This model is given by

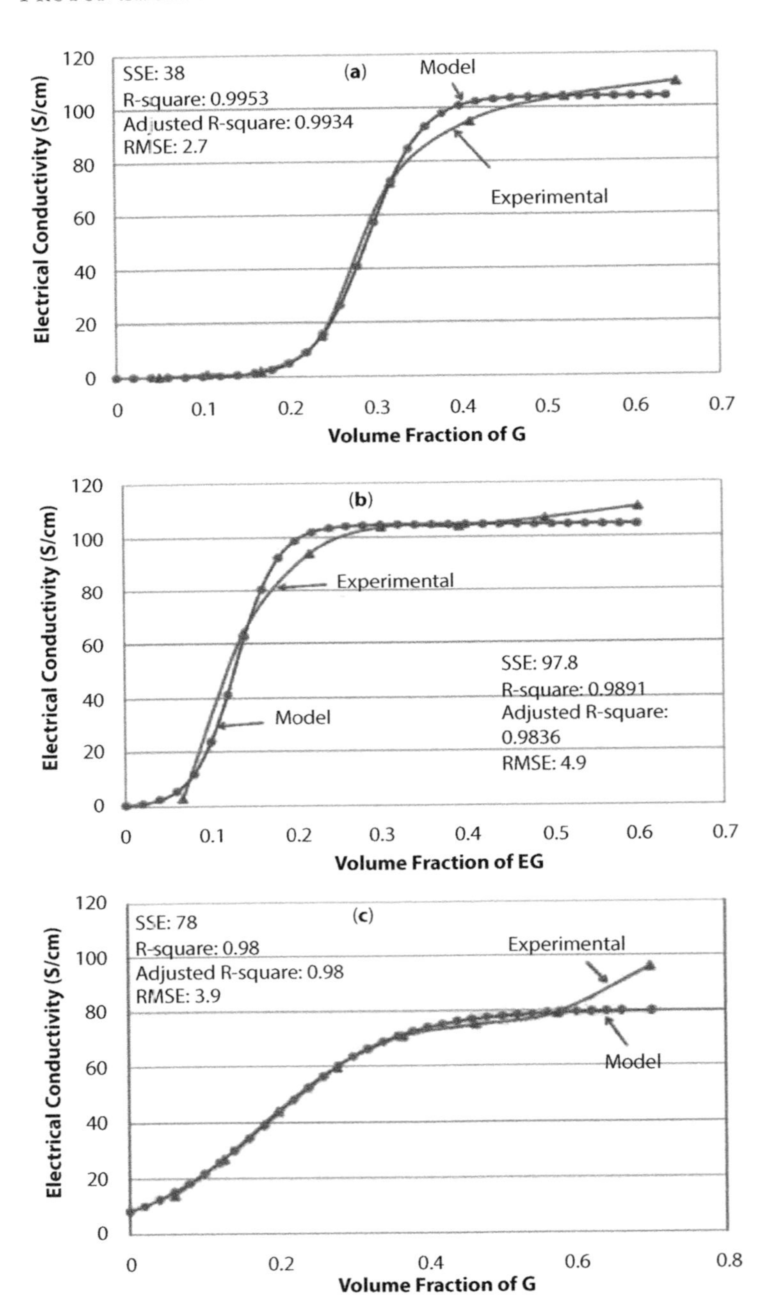

Figure 7.6 Comparison experimental data with Taherian model of polymer composites BPs: (a) P/G, (b) P/EG, and (c) P/CF. Reprinted with permission from Ref. [68], copyright of John Wiley and Sons.

$$\sigma_{\text{composite}} = \frac{a_0 \times \dfrac{\text{AR}}{\text{CR}} \sigma_f}{0.01089 + \exp(-b_0 \times \text{AR} \cdot x + d_0 \times \cos\theta)}$$

$$\cos(\theta) = \frac{\gamma_S - \gamma_{\text{SL}}}{\gamma_L} \tag{7.5}$$

$$\gamma_{\text{SL}} = \gamma_S + \gamma_L - 2(\gamma_S \cdot \gamma_L)^{0.5}$$

where σ_f is the electrical conductivity of filler; x is the volume fraction of filler; AR is the filler aspect ratio; cos (θ) is surface energy; CR is the interface contact resistance among fillers; and a_0, b_0, and d_0 are constant parameters, which are adjustable parameters and can be calculated by fitting the model on experimental data. γ_L, γ_S, γ_{SL}, and θ are the surface energies of polymer, filler, filler/polymer, and wetting angle, respectively.

7.3 Matrix and Fillers

Thermoset-based composite BP materials are divided into thermoset resin as a binder and fillers as conductive materials. In this section, thermoset resins and fillers for the fabrication of thermoset-based composite BPs are introduced.

7.3.1 Thermoset Resins

In the fabrication process of BP, the polymer affects BP performance, such as electrical, thermal conductivity, and flexural strength, so polymer not only plays a matrix role [69]. Thermoset polymers are the most popular type of matrix system (binder) for producing composite BPs. They are cross-linked [70, 71]. By strong covalently bonded atoms, the molecular chains of thermoset polymer are linked to composing a continuous three-dimensional network [72]. Thermoset polymers have numerous advantages, like low viscosity, which leads to having the capability for loading with a high quantity of conductive fillers. Furthermore, thermoset polymers have good corrosion resistance, low density, high mechanical strength, low toughness, short cure time, low shrinkage/zero shrinkage, long-term thermal and mechanical stability, and low cost [73]. Another benefit of thermosets is that during the fabrication of thermoset-based composite BP, thermoset polymers do not require any cooling. In additon, the thermoset resins during the compression molding process allow for the molding of intricate details. The main thermosetting resins are unsaturated polyester, epoxy, phenolic, vinyl esters (VEs), bismaleimides, polyimides, furan, and benzoxazine [74–76]. In the literature for the fabrication of composite-based BPs for PEM FCs, numerous types of thermoset resin have been investigated, such as epoxy [73, 77–80], VE [81–88], phenolic [89–95], and unsaturated polyester and benzoxazine [35, 96–98]. The physical, electrical, and mechanical properties of thermoset resins are presented in Table 7.4.

7.3.1.1 Epoxy [99–103]

Epoxy (EP) resins are low–molecular weight organics, usually given from the reaction Bisphenol-A or Bisphenol-F and epichlorohydrin. Epoxy resins contain a resin (binder) and catalyst (hardener) in both cases of epoxy resin's shape (solid powder or liquid). Epoxy resins have many benefits, such as excellent electrical, mechanical, and thermal properties, high adhesion strength, high corrosive resistance, low shrinkage during cure, good processing ability, and low cost [104–107]. Furthermore, epoxy resins have the capability for loading with a high quantity of fillers for improving their final properties [108].

7.3.1.2 Unsaturated Polyester Resin [109–113]

Unsaturated polyester resins (UPRs) have been used in very varied technology fields, and it is most commonly used as a matrix (binder) for composite materials. By the condensation of a diol with a blend of saturated anhydrides and unsaturated anhydrides, UPRs are produced. Cross-linking of the condensation products with reactive vinyl monomers (like styrene) can form very durable structures. It has low cost, low viscosity, and good thermal stability as well as can be cured at room temperature. On the other hand, UPR has large shrinkage (4%–6%) on curing, leading to causing sink marks on the product surface. Because of this disadvantage, UPR is not widely used in the matrix for composite BPs.

7.3.1.3 Vinyl Ester Resins [114–118]

VEs are the binary resin produced by the reaction (esterification) between an epoxy resin and an unsaturated monocarboxylic acid (methacrylic acids or acrylic). The vinyl groups are indicated to ester substituents that prone to polymerize. A cured VE resin has a higher fracture toughness and is more flexible than a cured polyester resin. One of the unique properties of a VE molecule is that it consists of hydroxyl (-OH) groups along its molecular chain. Hydrogen bonds with similar groups can be formed by OH groups on a glass-fiber surface, leading to a good wet-out and adhesion with glass fibers. The viscosity of VERs (as UPRs) is reduced by dissolved in styrene monomer. VEs have a good chemical resistance of epoxy and good mechanical properties (like the excellent corrosion-fatigue, which correlated to a high specific strength), as well as the benefits of UPRs, such as fast curing and low viscosity. However, there are some drawbacks for this resin, as the volumetric shrinkage of VERs is somewhere between 5% and 10%, which is higher than EP resins, and also exhibits only moderate adhesive strengths with comparing to EP resins.

7.3.1.4 Phenolic Resins [119–123]

Currently, phenolic resins (PRs) are a more fascinated polymer matrix to be used for producing composite BPs, because it has low cost, high mechanical strength, good heat resistance, and stable performance. PRs are produced by the condensation reaction of formaldehyde and phenol. Resole and Novalac are the compositions of PRs, in which resole is generated via the reaction of a phenol with an excess amount of molar formaldehyde with an alkaline catalyst. While, Novolac is produced through the acid-catalyzed reaction of phenol and formaldehyde with an excess of phenol.

7.3.1.5 Polybenzoxazine Resins [124–128]

Recently, polybenzoxazine (PB-a) resins as a new thermoset resin have been used for fabricating composite BPs as a polymer matrix. It is a novel type of PRs. Benzoxazine resins are essentially produced from formaldehyde, phenoxide group, and amine functional group. PB-a–based resin has intrinsic characteristics such as near-zero shrinkage, water resistance, high mechanical properties, electrical insulation properties, and dimensional stability (see Table 7.4).

7.3.2 Fillers

The type, shape, and size of the fillers play an important role in specifying the properties of the composite BPs. Polymers are usually electrical insulators; thus, by loading conductive fillers, the conductivity of composite can be improved. In the literature, the most widespread carbon fillers used for fabricating thermoset and thermoplastic composite BPs are graphite,

Table 7.4 Physical, mechanical, and electrical properties of thermoset resins [216–218].

Property	Epoxy	Phenolics	Polyester resin [217]	Vinylster [219, 220]	Polybenzoxazine resin
Density (g/cc)	1.2–1.25	1.24–1.32	1.10–1.46	1.2–1.4	1.19
Tensile strength (MPa)	55–130	24–45	42–71	69–75 [221]	100–125
Electrical conductivity (S/cm)	10^{-14} [222]	10^{-15} [222]	10^{-14} [222]	10^{-16} [222]	-
Flexural strength (MPa)	~130	78–120	60–120	124–131	-
Compressive strength (MPa)	~11	88–110	92–190	100	-
Elongation at break (%)	3–4.3	0.3	5	4–7	2.3–2.9
Tensile elastic modulus (GPa)	3.1–3.8	3–5	2.1–4.5	3.00–3.35	3.8–4.5
Cure Temperature (°C)	RT-180	150–190	-	-	160–220
Max use temperature (°C)	180	200	-	-	130–280
Tg (°C)	150–220	170	100 [223]	110 [223]	170–340
Shrinkage rate (%)	1–2	0.002	4-6	2–7 [224]	~0

Table 7.5 Properties of different conductive filler for bipolar plate application. Reproduced with permission from Ref. [202], copyright of Elsevier [40, 202, 225–227].

Properties	Unit	Graphite (G)	Carbon fibers (CF)	Carbon black (CB)	Expanded graphite (EG)	Carbon nanotubes (CNT)	Graphene (GP)
Density	(g/cm^3)	2–2.25	1.79–1.99	1.7–1.9	1.7	2	2.2
Particle size	μm	6–100	L: 10–100 D:4–10	14–250	100–150 sheet thickness diameter 1 μm^2	L:10–100 D: few nm	5–25
Specific surface area	(m^2/g)	6.5–20	0.27–0.98	7–560	100	200–250	2630
Aspect ratio		Close to 1	6–30	Close to 1	≈ 100	1,000–50,000	~1,000
Volume	$cm^3(100\ g)^{-1}$	–	nil	480–510	2–10	–	–
Electrical conductivity	(S/cm)	400–1250	598	2.5–20	a-axes:2.5×10^4 c-axes: 8.3	10^{-4}–10^2	>6,000
Tensile strength	MPa	Flexural: 7–10	1,000–4,000	Nil	Nil	60–150 GPa [228, 229]	130 GPa
Carbon content	(wt.%)	99.91	94–96	99.5	99.5	95	–

CB [78, 79, 94, 129–133], CNT [57, 134–140], carbon fiber [20, 36, 141–146], and EG and graphene. The physical and mechanical specifications of graphite and other carbon fillers are listed in Table 7.5.

7.3.2.1 Graphite

Graphite is the most common filler used for the fabrication of BPs. This material is a crystalline form of the element carbon, which contains 99.91 wt.% of carbon [147]. Graphite consists of many layers of hexagonal structure, which is made from carbon rings with arranging atoms. Figure 7.7 shows the structure of graphite. This filler has micro-size particles with less specific surface and also a low aspect ratio of one. Hence, there is a possibility of increasing the mechanical property of graphite. It has unique properties such as good electrical and thermal conductivities as well as excellent resistance to corrosion and high chemical resistance. The graphite is having

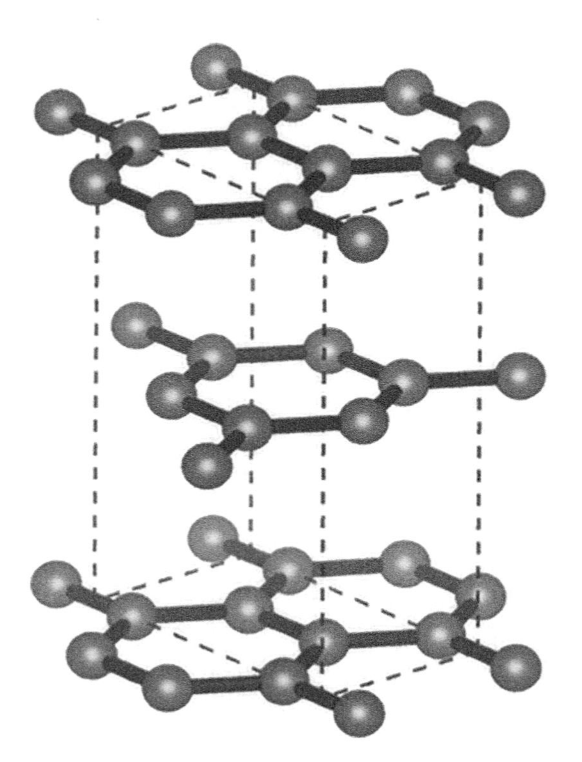

Figure 7.7 Graphite structure.

rather high conductivity with comparing to others fillers. The electrical conductivity depends on the crystallographic direction, in which the existence of delocalized electrons in the crystallite leading to generate a high in-plane electrical conductivity. Therefore, graphite is used to enhance the electrical, thermal, and mechanical properties of composite BPs.

7.3.2.2 Graphene

Graphene is basically one single layer of graphite that contains carbon atoms arranged in a hexagonal lattice. Figure 7.8 shows the structure of a single graphene sheet. Graphene is considered a suitable filler for the fabrication of BPs, in which at a low quantity of this filler can be enhanced the properties of BPs. It provides very high thermal conductivity (3,000–5,000 W/mK) [148, 149], electrical conductivity (>6,000 S/cm) [150, 151], very light (1 m^2 sheet weighing 0.77 mg), surface area (2630 m^2/g), high mechanical properties, and gas impermeability. Many researchers have been used graphene as primarily or secondarily filler to fabricate BPs.

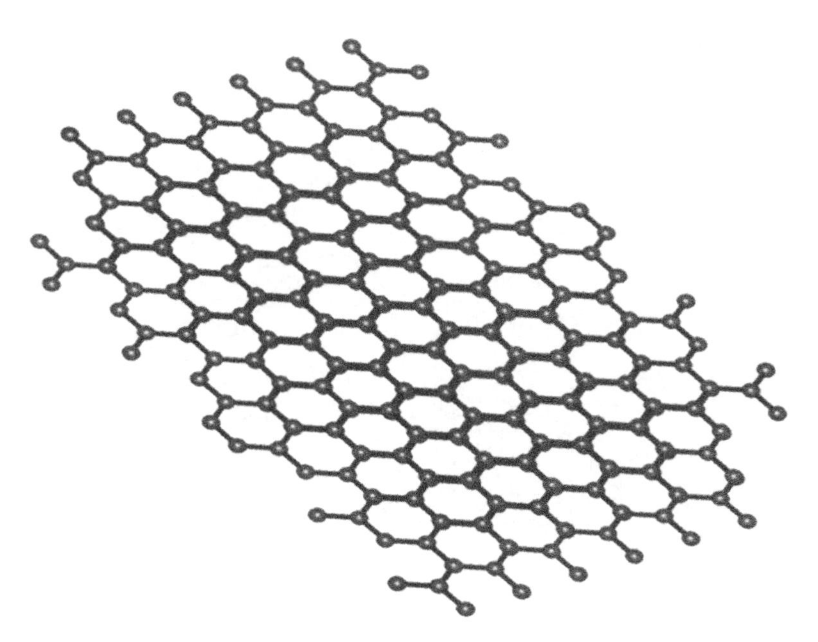

Figure 7.8 Structure of a single graphene sheet.

7.3.2.3 Expanded Graphite

EG is a special kind of graphite, which contains many structural layers with interlayer space. Figure 7.9 shows the difference between the layers structure of graphite and EG. EG possesses layered structures similar to natural flake graphite, but it expanded the distance between the layers of graphite. EG can be expanded to more than 150 times graphite, in which appropriate macromolecules and monomers have the ability to enter into porosities, voids, and interlayer spaces of EG to fabricate EG/polymer composites. The change in the distance between the layers of the graphite due to the high expansion of EG results in a reduction of the density between 10^{-3} and 10^{-2} g cm^{-3}, while the area increases noticeably, in addition, the aspect ratio increases to 40 and 15 m^{2} g^{-1}, respectively. EG provides very high electrical conductivity (see Table 7.5).

7.3.2.4 Carbon Black

CB belongs to a family of small particle sizes, essentially amorphous, or the aggregates of different particles sizes and shapes which are formed by growing the paracrystalline carbon particles together. CB has an aggregate structure [152], and its particles have a form near to spherical shape of colloidal sizes. Figure 7.10 shows the structure of CB. CB is a material produced synthetically by the partial combustion of heavy petroleum

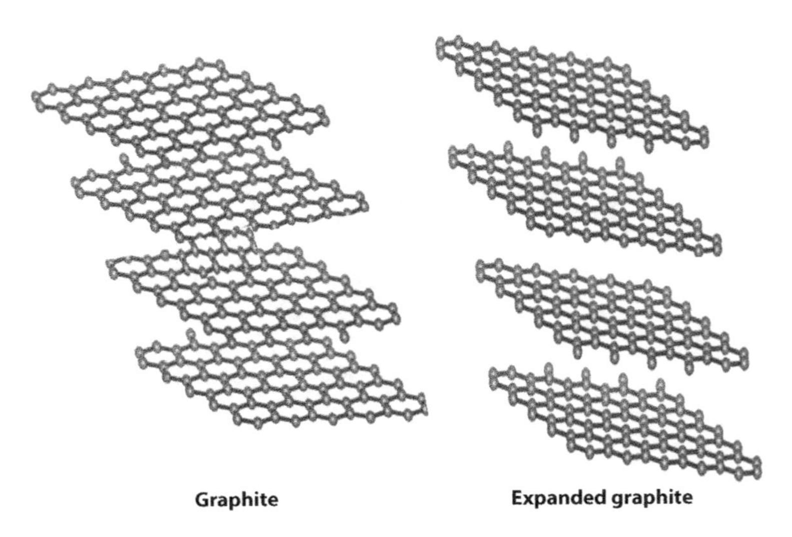

Figure 7.9 The structure of natural flake and expanded graphite.

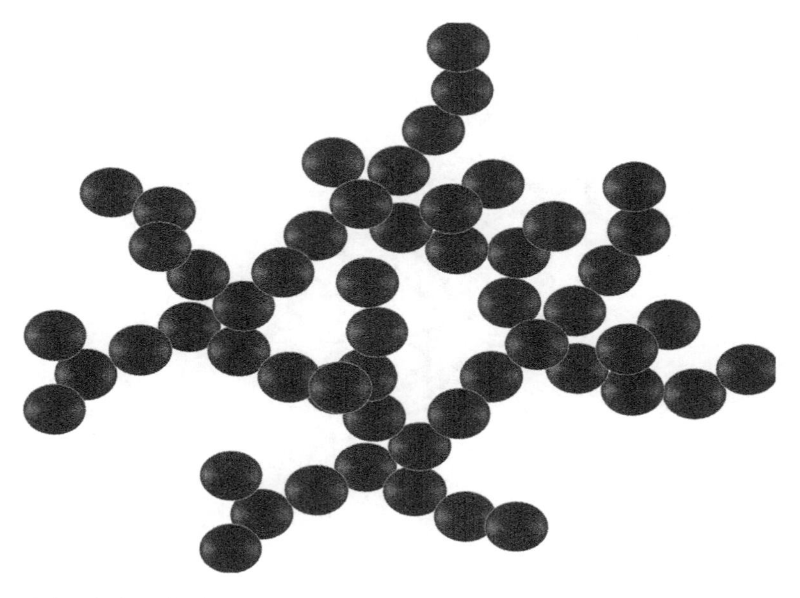

Figure 7.10 Carbon black structure.

products (coal tar, fluid catalytic cracking, and ethylene cracking tar). On the other hand, in the gas phase, the CB is produced by the process of thermal decomposition of hydrocarbons in the presence or absence of oxygen in substoichiometric amounts. CB is produced by five processes impingement, lampblack, furnace, acetylene, and thermal processes. Although CB has low electrical and mechanical properties, it could be used in the composition of composite BPs, and it has limited applications in the commercial BPs field.

7.3.2.5 Carbon Nanotube

CNTs are nanosized cylindrical molecules of carbon atoms with micrometer-sized length and nanometer-sized diameter (where the length to diameter ratio exceeds 1,000), which is consisting of a hexagonal arrangement, the same arrangement as in graphite. Figure 7.11 shows the structure of the CNT. In general, CNTs divide into multiwalled CNTs (MWNTs) and single-walled CNTs (SWNTs). MWNTs and SWNTs are manufactured mostly by laser-ablation, arc-discharge, and catalytic growth techniques [153–155]. CNTs have high electrical conductivity, mechanical strength, high surface area, large aspect ratio, and chemical stability, (see Table 7.5).

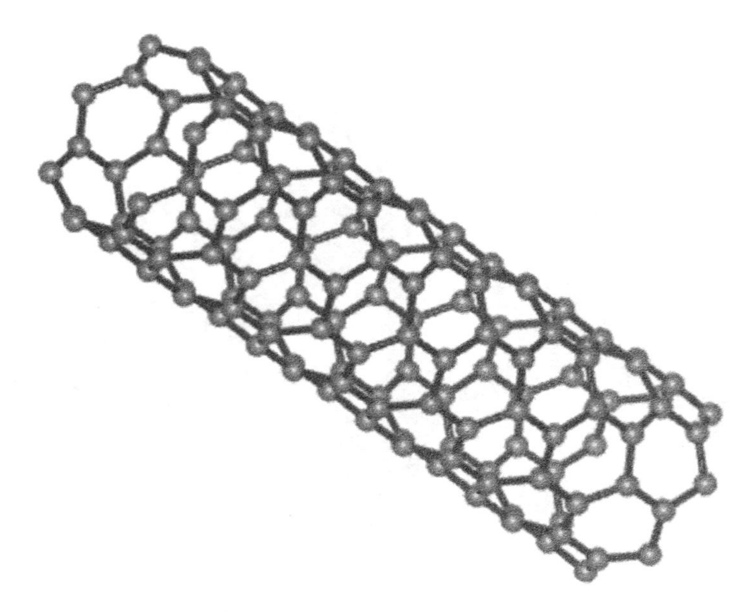

Figure 7.11 Carbon nanotube structure.

These characteristics cause CNTs to be a suitable conductive filler for BP applications.

7.3.2.6 Carbon Fiber

Carbon fibers indicate fibers that contain at least 92 wt.% carbon in composition. CF structure can be amorphous, crystalline, or partially crystalline. Figure 7.12 shows the structure of carbon fiber. CF consists of about 5–10 μm in diameters. It is made of interlocked graphene sheets, which in role contain carbon atoms. Carbon fibers are thousands of times larger than nanotubes. Around 90% of carbon fibers manufactured can be produced from polyacrylonitrile (PAN) and 10% from petroleum pitch and rayon [156]. Carbon fibers are considered the main source of enhancing mechanical strength, such as flexural strength, hardness, and fracture toughness for composite BPs. On the other hand, the hydrogen permeability of the composite BPs increases with the increasing concentration of carbon fibers [157, 158]. Thus, hydrogen permeation affects the performance and safety operation of a FC. Consequently, the concentration of CF in composite BPs must be appropriate.

SEM micrographs images for different carbon fillers are shown in Figure 7.13.

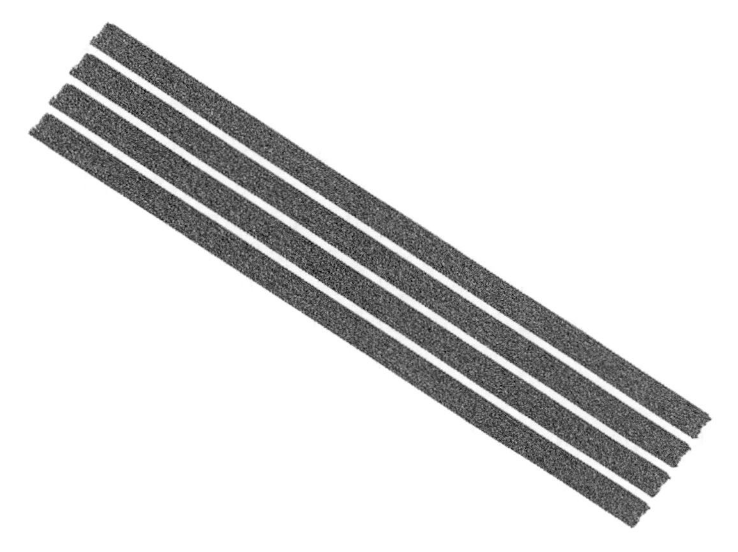

Figure 7.12 Carbon fiber structure.

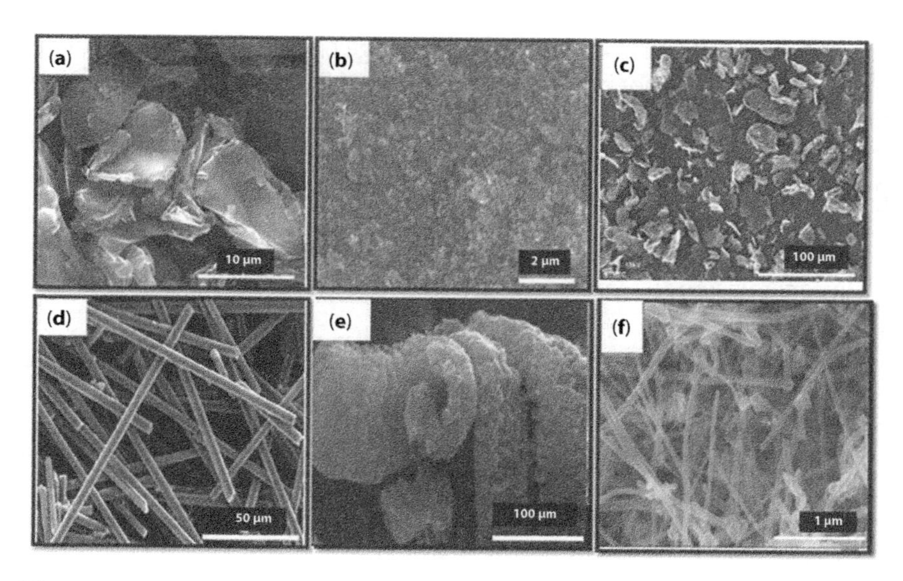

Figure 7.13 SEM micrographs images for (a) graphite (reproduced with permission from Ref. [195], copyright of Elsevier), (b) carbon black (reproduced with permission from Ref. [182], copyright of Elsevier), (c) graphene (reproduced with permission from Ref. [175], copyright of Elsevier), (d) carbon fiber (reproduced with permission from Ref. [196], copyright of Elsevier), (e) expended graphite [197], and (f) carbon nano tube (reproduced with permission from Ref. [198], copyright of Elsevier).

7.4 The Manufacturing Process of Thermoset-Based Composite BPs

In order to fabricate the thermoset-based composite BPs, different methods are used in the literature. Compression molding [78, 93, 94, 120, 159–161] is the common method for producing thermoset-based composite BPs. Recently, the selective laser sintering (SLS) process [162–165], resin vacuum impregnation, and hot press method [166, 167], wet method, and dry method [168] are used for the fabrication of thermoset-based composite BPs.

7.4.1 Compression Molding

Compression molding is a very successful processing method for manufacturing thermoset-based composite BPs. In general, compression molding can produce plates with high mechanical, electrical, and thermal properties and dimensional stability. In addition, the compression molding creates the desired shape for BPs. Because of short cycle time, structural integrity, and volatile emission, compression molding is more suitable for thermoset composites. The most important advantage of compression molding is that the viscosity is not required to be very low in the final formulation of this process. In the compression molding method, it is possible at operating conditions to maintain the die under high temperatures with constant pressure. Furthermore, in compression molding, a very high quantity of filler can be added to the thermoset resin, leading to improve the electrical conductivity of thermoset-based composite BPs.

In general, compression molding requires for a curved and complex side two parts [169]:

(a) Female (lower), which consists of a cavity.
(b) Male (upper), which is possessing a projection and totally fit into a cavity in a lower part.

For the preparation of the BP by using the compression molding, follow the steps below [170, 171]:

1. Prepare a mixture of polymer resin (binder), a conductive filler, and additional materials.

2. Feed the mixture into the mold, whereas the surface of the mold includes the flow field geometry.
3. Compress the mixture between the molds under pressure and temperature using a hydraulic press with a heating facility.
4. Remove the fabricated BP from the mold after sufficient curing.

A lot of researchers used compression molding for the fabrication of thermoset-based composites BP as we mentioned previously. The compression molding process for thermoset-based composite BP is illustrated in Figure 7.14.

7.4.2 The Selective Laser Sintering Process

SLS is a speedy fabrication technique; computer-aided design (CAD) model can be built a three-dimensional (3D) object via an additive process. In this process, the laser is used for scanning the mixture of matrix and fillers in a powder bed. The matrix is melted by a laser, which has a low melting point and binds the particles of filler for building 3D parts of composite BPs layer by layer, in which after scanning the first layer, one layer of thickness lowered the powder bed, and a new layer of the mixture powder was replaced on the top. This process repeats unto completed the fabrication

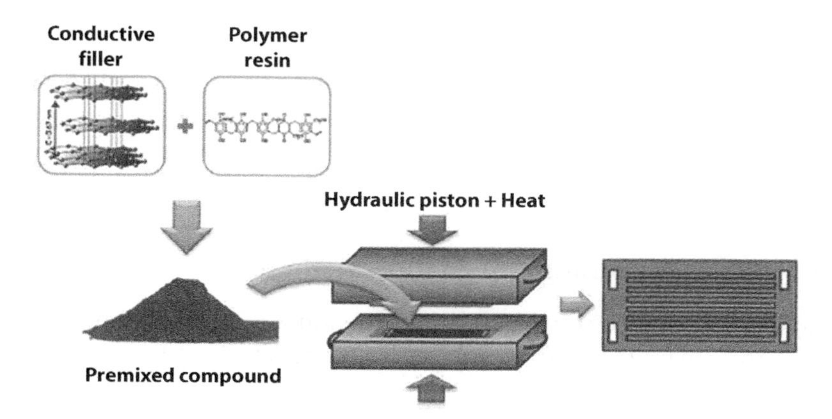

Figure 7.14 Compression molding process for the manufacturing of thermoset-based composite bipolar plates. Reprinted with permission from Ref. [199], copyright of Elsevier.

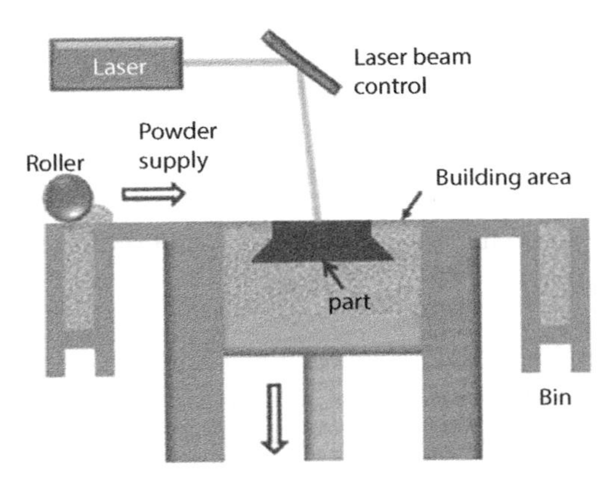

Figure 7.15 Manufacturing process of Selective Laser Sintering (SLS) process. Reprinted from Ref. [162] with permission, copyright of Elsevier.

of bipolar blate. Figure 7.15 depicts the SLS process. With a comparison between SLS and other manufacturing processes, the SLS process does not consume longer time for the design and research stage as well as reduce the cost of composite BPs, where the SLS has many advantages such as its ability to reduce fabrication time, build complex flow channels, and reduce cost resource in fabricating composite BPs from various designs.

7.4.3 Wet and Dry Method

By using dry and wet methods, Chaiwan and Pumchusak [168] fabricated thermoset-based composite BPs contain MWCNTs/reinforced graphite/PR. The authors reported that via the dry method the higher mechanical and electrical properties are observed. Figures 7.16a and b present a schematic of the manufacturing of the composite BPs, which were prepared by the wet and the dry methods.

7.4.4 Resin Vacuum Impregnation Method

Recently, by using resin vacuum impregnation and hot press method, Li *et al.* [166] manufactured thermoset-based composite BP consisting of EG and PR. As compared with compression molding, the preparation of composites via resin vacuum impregnation is dependent on the raw compressed EG (CEG) sheets. Firstly, on the CEG sheet, the flow field was embossed, and then, the resin was used for impregnating the CEG sheet

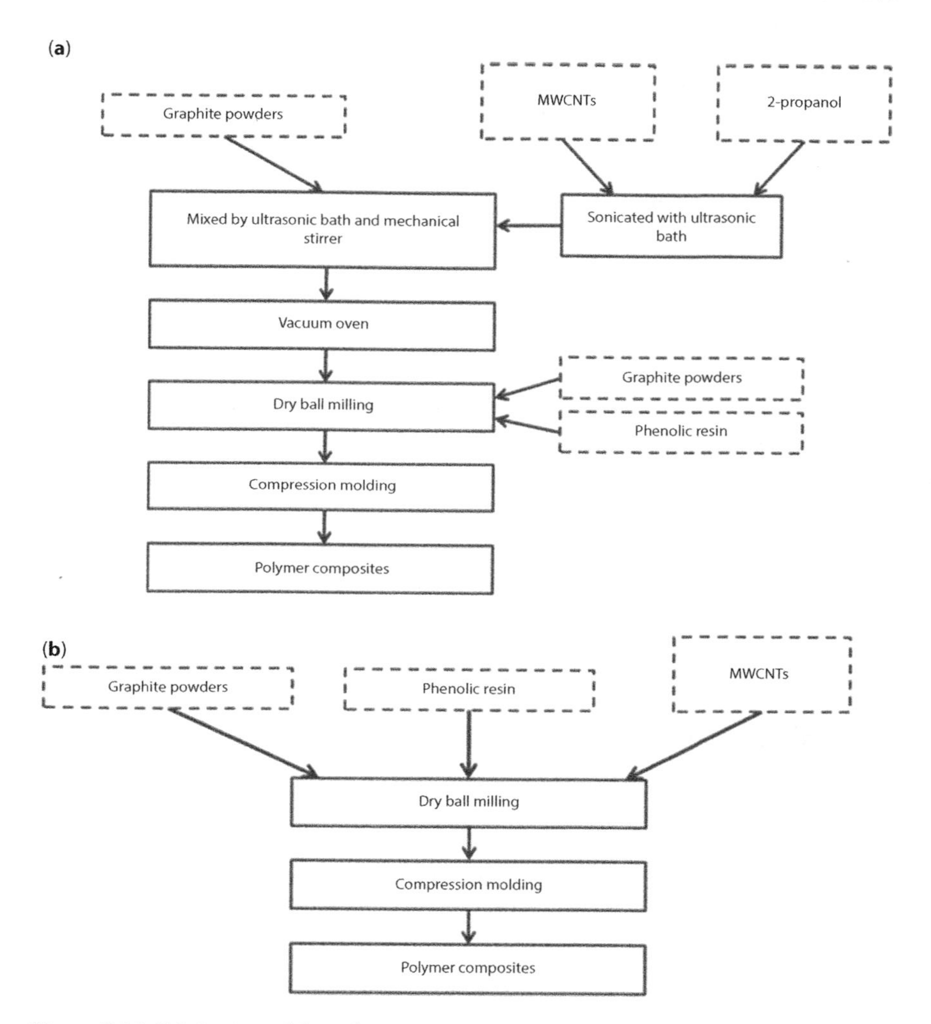

Figure 7.16 Fabrication of the polymer composite BPs by (a) wet nanoparticle dispersion method and (b) dry nanoparticle dispersion method. Reprinted from Ref. [168] with permission, copyright of Elsevier.

and cured. The resin vacuum impregnation process for the preparation of the composite BPs is shown in Figure 7.17.

Hot press method also is used for manufacturing composite BPs. Lee *et al.* [77] manufactured thermoset-based composite BP consisting of graphite particles and epoxy resin by using the hot press method.

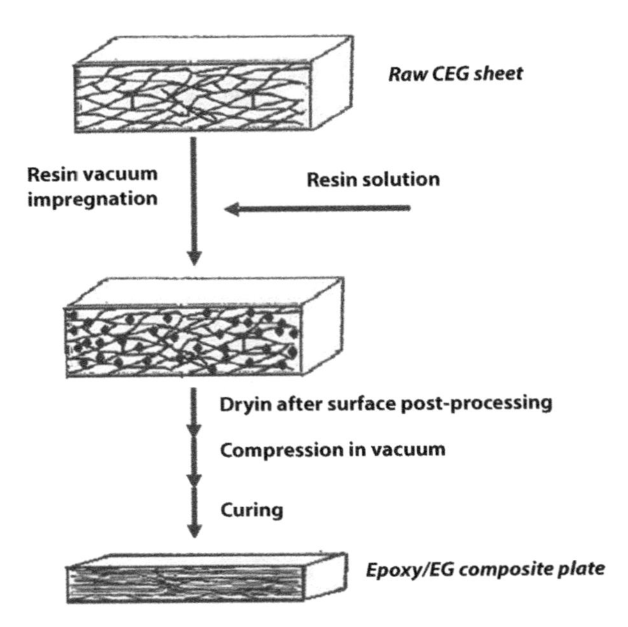

Figure 7.17 Resin vacuum impregnation process for preparing the composite BPs. Reprinted from Ref. [167] with permission, copyright of Elsevier.

7.5 Effect of Processing Parameters on the Properties Thermoset-Based Composite BPs

Many parameters play a major role in the properties of the BP, such as the type of polymer matrix and conductive fillers, as well as production methods.

In this section, the effect of different processing parameters on electrical, mechanical, and thermal properties of thermoset-based composite BPs are introduced.

7.5.1 Compression Molding Parameters

Compression molding parameters such as pressure, temperature, and time play a significant role in through-plane and in-plane conductivities, flexural and tensile strength, water absorption, porosity, and density of composite BPs.

7.5.1.1 Pressure

If the compression molding pressure increases, then the electrical and mechanical properties of the composite BPs would increase; this occurs for many reasons as follows. The increasing of molding pressure causes a reduction of the gap between particles, which is leading to increasing the composite electrical conductivity, as shown in Figure 7.18. When the pressure of compression molding increases, the voids of the composite decrease. Consequently, the porosity decreases. As a consequence, the flexural strength of the polymer composite BPs increases with the increasing pressure of compression molding due to the following Knibbs formula.

$$\sigma = kD^{-1/2} e^{-bp} \tag{7.6}$$

where σ is a flexural strength, D and P are the maximum diameter of grains and porosity, respectively, while k and b are empirical constants.

Gautam *et al.* [25] fabricated the composite BPs via compression molding technique with various mold pressure. The composites of 67 wt.% PR, 30 wt.% EG, 5.0 wt.% CB, and 3.0 wt.% GP were prepared by using 5, 10, 15, 20, and 25 MPa of mold pressure, respectively. Their results showed that the increasing value of mold pressure from 5 to 25 MPa causing the increase in the in-plane electrical conductivity values from 200.4 to 377.8 S cm^{-1}. While the values of flexural strength and compressive strength were increased from 38.7 to 55.6 MPa and from 43.4 to 64.8 MPa, respectively, with increasing the value of mold pressure from 5 to 15 MPa. After 15 MPa of the mold pressure, the mechanical properties values reached the saturation value. The applied high mold pressure is resulting in the destruction of the structure of composite BPs internally, which obstructs

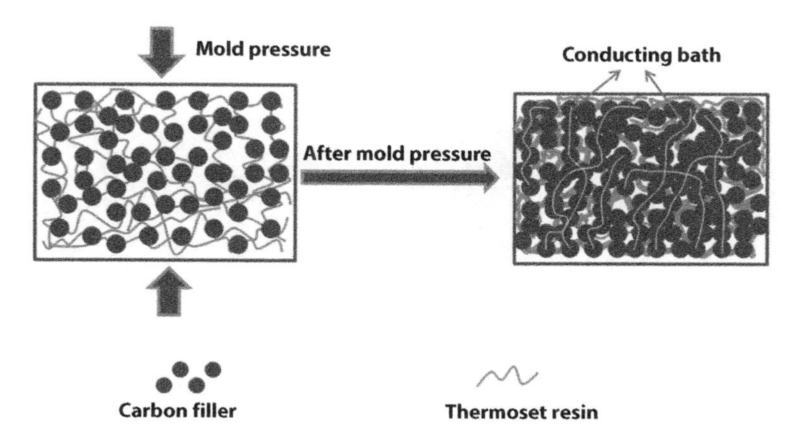

Figure 7.18 Schematic illustration of the mold pressure of thermoset-based composite BP.

the cross-linking between PR and carbon fillers, resulting in generating weak mechanical properties.

7.5.1.2 *Temperature*

The in-plane and through-plane electrical conductivities of the composite BPs increase with the temperature of the compression molding process. The increasing of the molding temperature improves the generation of the conductive path within the polymer matrix system between the fillers. As a result, the composite BPs electrical conductivity is enhanced. However, the high temperature of the compression molding leads to creating the surface deformation of the composite BPs; this defect on the surface results in the slow removal of the gas and fast surface curing process. Consequently, the compression molding temperature must be sufficient.

Sherman *et al.* [172] studied the influence of compression molding parameters (pressure, temperature, and time) on the composite BPs electrical properties. For the preparation of composites BPs, they used 20/5/75 (wt.%) EP, MWCNTs, and SG, respectively. For enhancing the electrical conductivity of composite BPs, they used different compression molding parameters: molding pressure, 1,200, 1,500, and 1,800 psi; molding temperature, 110°C, 130°C, and 150°C; and molding time, 60, 75, and 90 min. From their results, they found that, with increasing the molding temperature from 110°C to 130°C, the epoxy resin viscosity is decreased, that leads to form a network of electrical conductivity easily between the flake shaped synthetic graphite (SG) as primary conductive filler and the nanosized and tubular elongated shaped MWCNTs as a second conductive filler. As a consequence, the electrical conductivity is improved. They reached that

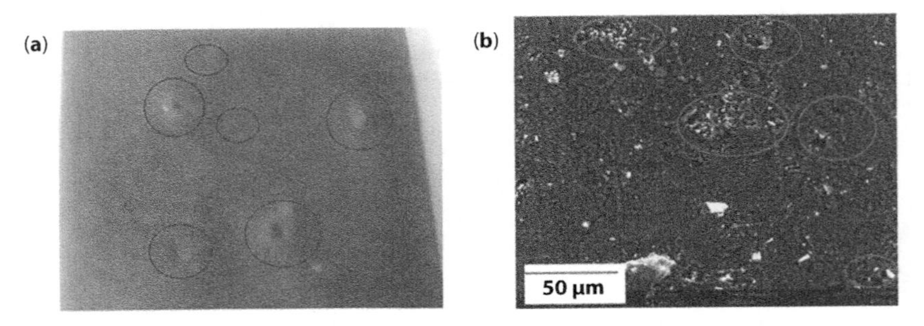

Figure 7.19 (a) Photo of plate surface and (b) SEM image of the plate with 1–5 bubbles 3–10 mm diameter. Reproduced from Ref. [173] with permission, copyright of Elsevier.

optimum combination of MWCNT/SG/EP nanocomposite was at molding pressure of 1,800 psi, molding temperature of 130°C, and molding time of 60 min. Hence, the optimum value of electrical conductivity was 163 S/cm.

Boyaci San *et al.* [173] observed that the use of unsuitable temperature and low pressure leads to create gas bubbles at the surface of the plate, as shown in Figures 7.19a and b.

7.5.1.3 *Time*

The molding time can be affected by the electrical and mechanical properties of composite BPs. Sherman *et al.* [172] stated that the molding time affects the electrical conductivity; 60 min is better than 75 and 90 min to reach the maximum conductivity (163 S/cm). Boyaci San *et al.* [173] observed that, at low temperatures with increasing time, the electrical conductivity was increased. On the other hand, at high temperatures, the conductivity decreased with increasing time; the maximum electrical conductivity of 107.4 S/cm is obtained at a time of 5 min, temperature of 187°C, and pressure of 119 bar. Kim *et al.* [98] investigated the influence of compression molding time and temperature on the electrical and mechanical properties of graphite/PB-a composite BPs, as shown in Figure 7.20. They found that the good conditions of compression molding parameters meet the US DOE requirement when the temperature and time of 40 min and 220°C.

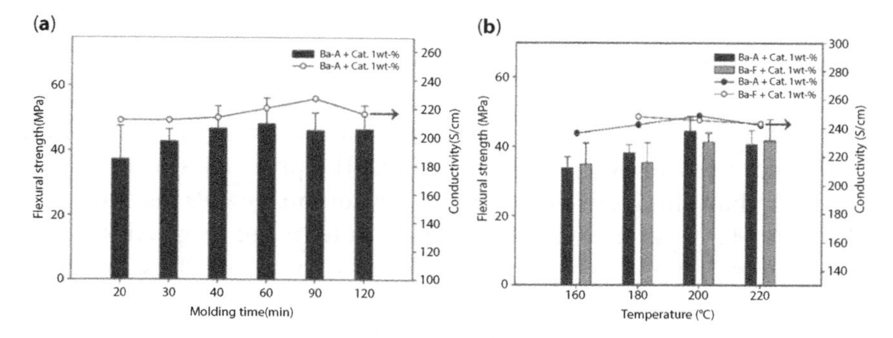

Figure 7.20 The variation of the electrical conductivity and flexural strength of (a) G/benzoxazine resin (Ba-F) as a function of compression molding time, where the temperature and pressure were kept at 220°C and 30 MPa (b) G/(Ba-A and Ba-F) as a function of molding temperature, where the time and pressure were kept at 60 min and 30 MPa. Reprinted from Ref. [98] with permission, copyright of Springer Nature.

7.5.2 The Mixing Time Effect on the Properties of Composite Bipolar Plates

The mixing time can influence the mechanical and electrical properties of polymer composite BPs, in which the increasing mixing time results in an increase in the values of electrical and mechanical properties of composite BPs. For the appropriate distribution of the thermoset resin between filler particles, the required mixing time must be sufficient.

Kim *et al.* [174] discussed the influence of mixing time on electrical and mechanical properties for composite BPs. They prepared the composite of graphite 85 and 15 wt.% of the matrics system epoxy resin, Cresol Novolac phenol resin, triphenylphosphine (TPP), and acetone with the various mixing times 5, 30, 60, and 120 min, respectively. The mechanical mixer (SFM-555SP, Shinil electric mixer) was used for stirring the materials of composite. The values of electrical conductivity and tensile strength were increased from 14.61 to 36.07 S/cm and 3.51 to 6.13 MPa with increasing the mixing time from 5 to 60 min. At the same time, the electrical conductivity and tensile strength values were similar after 60 min of mixing time. The results of this experiment reached that the sufficient mixing time was at 60 min.

7.6 Effect of Polymer Type, Filler Type, and Composition on Properties of Thermoset Composite BPs

The important functions of the BP in PEMFC are the electrical performance (collecting and transport electrons into the PEMFCs) as well as providing structural support (mechanical strength) for the thin membrane. For achieving these functions, the properties of BPs must be higher than the DOE-2025 targets for composite BPs, as shown in Table 7.1. Many parameters play a major role in the properties of the BP, such as the type of polymer resins and conductive fillers, and production methods.

In this section, the effect of content and type of polymer resins and conductive fillers, as well as particle size of fillers and production method on electrical, mechanical, and thermal properties of thermoset-based composite BPs are introduced.

7.6.1 Electrical Properties

Basically, polymer composite BPs electrical conductivity depends on the contact space between the filler particles, conducting channels, and the electrical conductivity of filler itself [25, 141, 158]. Multiple carbon fillers, in comparison to a single filler, are producing a strong conducting network as shown in Figure 7.21.

Witpathomwong *et al.* [175] found that the best in-plane electrical conductivity of polymer composite BP was 364 S/cm, and it was obtained in 2 wt.% CNT, 7.5 wt.% graphene, 74.5 wt.%. graphite, and 16 wt.%. PB-a and compression molding under the pressure of 150 MPa and temperature of 200°C for 3.0 h. Taherian *et al.* [176] studied the effect of single-filler and double-filler on electrical properties of polymer composite BP. They reported that the maximum electrical conductivity of EG/PR was 110 S/cm. This result is more than the conductivity of CF/PR (90 S/cm). Dhakate *et al.* [177] have been investigated the effect of the addition of MWNTs in graphite composite plates. They reported that on the addition of 0.5 vol.% MWCNTs, the in-plane electrical conductivity of the polymer composite increased from 80 to 165 S/cm, and with the increment of quantity of MWCNTs up to 1.0 vol.%, the conductivity increased up to 178 S/cm.

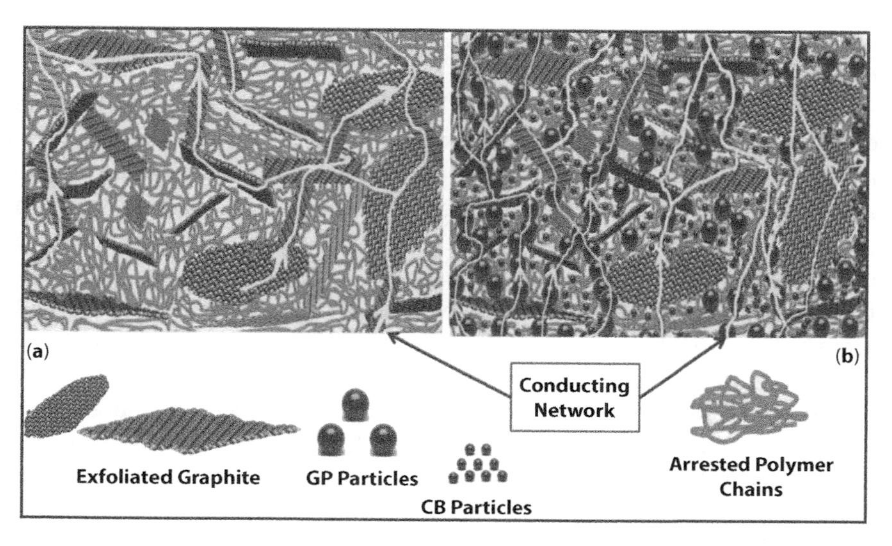

Figure 7.21 Schematic drawing of electrical conductivity path in composite BPs consisting of carbon fillers: (a) single and (b) multiple. Reprinted with permission from Ref. [25], copyright of John Wiley and Sons.

However, the conductivity decreased up to 145 S/cm on further increasing the MWCNTs content up to 2 vol.%.

Kim *et al.* [178] used epoxy resin as a matrics system for the preparation of composite BPs. The composites were prepared with different volumes of epoxy 40, 35, 30, 25, and 20 vol.%. According to their results, a composite of 25 vol.% EP and 75 vol.% graphite powder (GP) was acceptable in order to achieve the highest through-plane and in-plane electrical conductivity of the composite BPs.

Hui *et al.* [179] analyzed the effect of novolac epoxy resin content on the electrical conductivity of composite BPs. They used different rates of epoxy resin content (10, 15, 20, 25, 30, and 35 wt.%) for the preparation of novolac epoxy/graphite composites BPs. The results of electrical conductivity according to the rate of epoxy content which was found by the experiment were equal to 82, 80, 57, 44.5, 26, and 22.5 S/cm, respectively. Accordingly, the electrical conductivity decreased rapidly when the resin content was over 15 wt.%.

Masand *et al.* [180] investigated the effect of filler content on the properties of EG-based composite BPs. They reported the composite, which was prepared using EG/NG/CB/NPR (20/16/4/60 wt.%), respectively, and they found that the electrical conductivity increased from 162.23 to 401.60 S/cm with increasing content of CB from 4 to 6 wt.%. In the composite BPs, CB generates the additional conductive pathway due to its particle size. Whereas the particle size of CB is smaller than the particle size of natural graphite and EG. Consequently, the electrical conductivity of composite is improved.

Radzuan *et al.* [181] prepared CF/EP composite for the BP by using the compression molding technique. They analyzed the electrical conductivity for the single filler CF/EP composite and reported that the in-plane and through-plane electrical conductivity values were unavailable when the filler content was less than 80 wt.% due to the conductive path, which unformed in the structure of composite below the percolation threshold. However, the through-plane and in-plane electrical conductivity were 6.34 S/cm and 4.26 S/cm, respectively, at 80 wt.% of CF content. Hence, CB and CNT were used to enhance the composite electrical conductivity.

Alo *et al.* [182] mixed thermoplastic polymer PP and thermoset resin EP with graphite and CB by melt mixing followed by compression molding for fabrication composite BPs. It was determined that 11 wt.% PP/9 wt.% EP/73 wt.% G/7 wt.% CB composite had an electrical conductivity of 88.5 and 8.52 S/cm in the in-plane and through-plane, respectively. However, these electrical conductivity values are still below the DOE-US targets. Dhakate *et al.* [149] studied the influence of EG particle size on the electrical properties of 50 wt.% EG/50 wt.% NPR composite BPs. When the particle size of EG was 300 μm, the composite electrical conductivity was 150 S/cm. While the composite electrical conductivity was 40 S/cm when the particle size of EG was 50 μm. Ghosh *et al.* [183] investigated the effect of CF length on the electrical conductivity of the thermoset composite BPs as shown in Figure 7.22.

7.6.2 Mechanical Properties

Guo *et al.* [162] studied the influence of different types of fillers on the mechanical properties of thermoset composite BPs. They observed that the CF increased the flexural strength significantly, and SG had a slightly negative influence on flexural strength. Liao *et al.* [83] and Chaiwan *et al.* [168] reported that the addition of the MWCNTs could enhance the tensile and flexural strength of the thermoset-based composite BPs. Ryszkowska *et al.* [184] found that increasing the number of nanotubes in epoxy composite decreased the elastic modulus, elongation endurance, and hardness. Gautam *et al.* [185] reported that the compressive and flexural strengths are decreased with increasing NFG wt.% in PR.

Hui *et al.* [179] discussed the influence of novolac epoxy resin content on the flexural strength of composite BPs. The various rates of epoxy resin content (10, 15, 20, 25, 30, and 35 wt.%) were used for the preparation of novolac epoxy/graphite composite BPs. The flexural strength values were 26, 39, 45.6, 48, 48.5, and 50.8 MPa, respectively. These results cleared that the flexural strength increased gradually when the resin content was over 20 wt.%. Witpathomwong *et al.* [175] used PB-a as the matrix system and two types of fillers (aniline graphite and graphene) to study the effect of combinations of different fillers on mechanical properties. They also used different concentrations of CNT to improve the composite BPs mechanical properties. The best flexural strength and flexural modulus of the composite were reported as 41.5 and 49.7 GPa, respectively. This result was obtained in 2 wt.% CNT, 7.5 wt.% graphenes, 74.5 wt.% graphite, and 16 wt.% PB-a and compression molding under the pressure of 150 MPa and temperature of 200°C for 3.0 h. Ghosh *et al.* [183] studied the effect of CF length on the mechanical properties of thermoset composite BPs as shown in Figure 7.22.

Gautam *et al.* [25] found that the compressive and flexural strength, as well as compressive and flexural modulus of polymer composite were increased with EG quantity from 10 to 35 wt.% as shown in Figure 7.23.

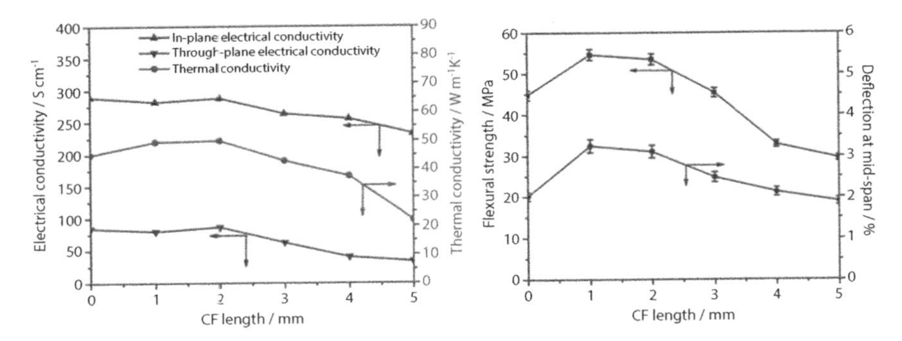

Figure 7.22 Effect of CF length on electrical conductivity, thermal conductivity, and flexural strength of (CF/CB/NG/NPFR 5/5/60/30%) composite BP. Reprinted with permission from Ref. [183], copyright of Springer Nature.

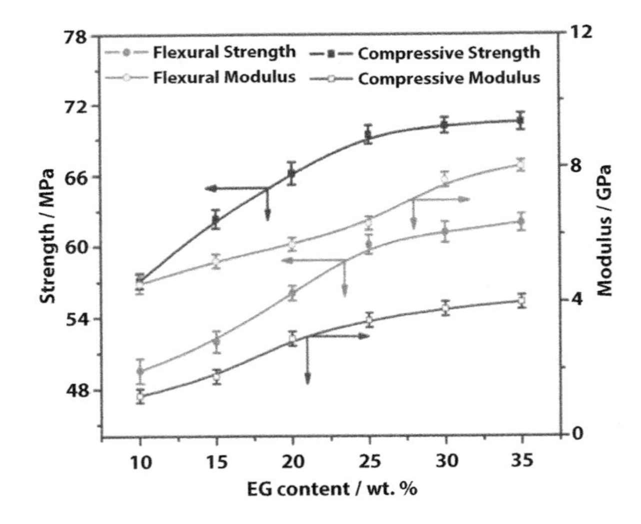

Figure 7.23 Compressive and flexural properties of composite BP (EG/CB/GP/PR) with different EG content. Reprinted with permission from Ref. [25], copyright of John Wiley and Sons.

7.6.3 Thermal Properties

In general, thermal conductivity is considered as one of the important properties of BPs in PEMFCs. Another substantial function of the BP is regulating the temperature by removing the heat out of the activated area. Thus, the essential parameter correlating to this function is thermal conductivity.

Conductive filler content must be high for generation continuously conductive network, resulting in significantly increase of the thermal conductivity of the polymer matrix as shown in Figure 7.24. Three phenomena: scattering, defect scattering, and boundary scattering, cause the generation of conductive networks and minimization of heat resistance [186–190].

Thermal conductivity (κ) can be calculated using the following:

$$k = \alpha \times Cp \times \rho \tag{7.7}$$

where k is thermal conductivity, α is thermal diffusivity, Cp is a specific heat capacity, and ρ the measured density of the sample.

Many researchers have reported the thermal properties of thermoset composite BPs. Plengudomkit *et al.* [191] fabricated graphene/PBA composite BP, and the thermal conductivity was 8.03 W/m K. Dhakate *et al.* [177] prepared the graphite/CNTs/phenolic composite, and the value of thermal conductivity was 13 W/m K. Phuangngamphan *et al.* [28] prepared the graphite/graphene/PBA, and the thermal conductivity was 14.5 W/m K. Hsiao *et al.* [192] investigated the improvement of the characteristics of 70 NG/30 VR (wt.%) composite BPs by incorporating the different amount of graphene. The results carried out that with adding of 0.2 phr of graphene in composite, the thermal conductivity of composite increased from 18.4 to 27.2 W/m K. Witpathomwong *et al.* [175] found that the graphite/graphene/PBA composite shown the value of through-plane thermal conductivity was smaller than the thermal conductivity value of composite with 2 wt.% of CNTs as shown in Figure 7.25. Yao *et al.*

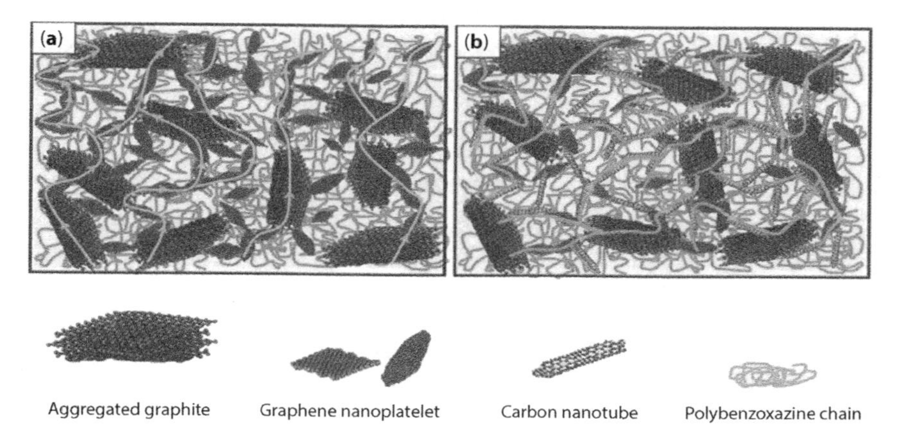

Figure 7.24 Schematic of thermal conductivity path in composites (a) G/GP/PBA. Reprinted from Ref. [28] with permission, copyright of John Wiley & Sons. (b) G/GP/CNT/PBA. Reprinted from Ref. [175] with permission, copyright of Elsevier.

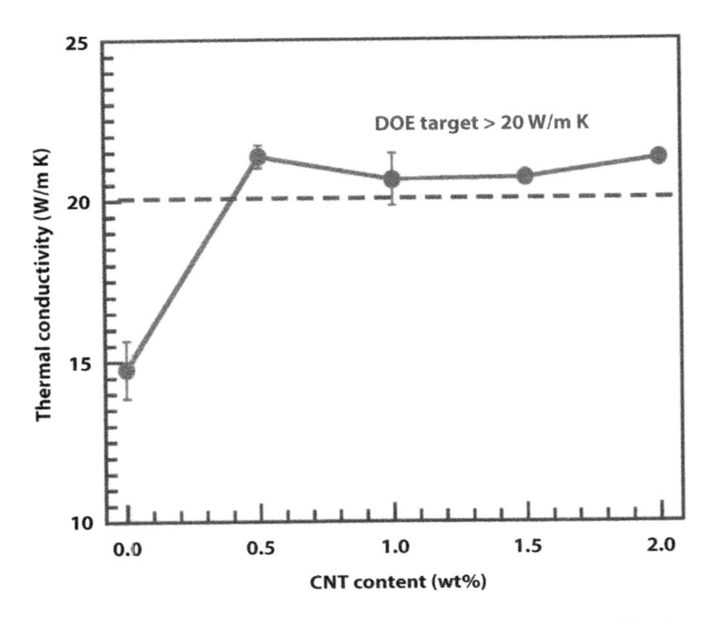

Figure 7.25 Thermal conductivities of G/GP/CNT/PBA composites at 25°C. Reprinted from Ref. [175] with permission, copyright of Elsevier.

[193] discussed the influence of MWNTs content on the thermal conductivity at RT of 80 EG/20 NPR (wt.%) composite BPs. The various rates of MWNTs content (0, 0.5, 1, 1.5, and 2 wt.%) were used for preparation of 80 EG/20 NPR (wt.%) composite BPs. The results of the thermal conductivity according to the rate of MWNTs content which was found by the experiment were equal to 12.41, 14.84, 13.23, 12.11, and 13.99 W/m/K, respectively. The materials, properties and manufacturing processes of different thermoset composite BPs for the previous studies were Summarized in Table 7.6.

7.7 Testing and Characterization of Polymer Composite-Based BPs

7.7.1 Electrical Analysis

7.7.1.1 In-Plane Electrical Conductivity

Usually, the most common technique used for measuring the electrical conductivity of BPs is the four-probe method. A four-point probe is a simple device, which is used to measure the conductivity or resistivity of semiconductor specimens. The conductivity or resistivity is measured when the

Table 7.6 Summarized description of materials, properties, and manufacturing processes of different thermoset composite BPs for the previous studies.

Composition	Electrical conductivity (S cm^{-1})		Flexural strength (MPa)	Tensile strength (MPa)	Thermal conductivity W/mK	Process	References
	In-plane	Through-plane					
MWCNTs/SG/NPR 1.0phr/80/20 wt.%	196.7	-na-	57.5	30	-na-	Dry method	[168]
NG/CB/RPFR 70/5/25vol.%	424.96	115.71	45.97	-na-	-na-	Compression molding	[230]
NG/CB/CF/RPFR 65/5/5/25vol.%	415.05	99.70	54.23	-na-	-na-	Compression molding	[230]
NG/CB/CF/GP/RPFR 64/5/5/1/25vol.%	435.31	130.17	57.28	-na-	-na-	Compression molding	[230]
EG/NPR 80/20 wt.%	182	29	109	-na-	-na-	Compression molding	[231]
MWCNTs/EG/NPR 1.0 phr/80/20 wt.%	178	33	100	-na-	-na-	Compression molding	[231]

(Continued)

Table 7.6 Summarized description of materials, properties, and manufacturing processes of different thermoset composite BPs for the previous studies. (*Continued*)

Composition	Electrical conductivity ($S\ cm^{-1}$)		Flexural strength (MPa)	Tensile strength (MPa)	Thermal conductivity W/mK	Process	References
	In-plane	Through-plane					
G/PB-a 80/20 wt.%	245	-na-	52	-na-	10.2	Compression-molded by hot pressing	[232]
G/EG/NPR 30/40/30 wt.%	-na-	15.93	14	-na-	-na-	Compression molding	[233]
GP/G/CNTs/PB-a 7.5/74.5/2/16 wt.%	364	-na-	41.5	-na-	21.3 (through plane)	Compression molding	[175]
MWCNT/NG/EP 70/20/10 wt.%	129	16	27	-na-	-na-	Compression molding	[234]
GP/PB-a 60/40 wt.%	357	-na-	42	-na-	8.0	Compression molding	[235]

(*Continued*)

Table 7.6 Summarized description of materials, properties, and manufacturing processes of different thermoset composite BPs for the previous studies. (*Continued*)

Composition	Electrical conductivity ($S\ cm^{-1}$)		Flexural strength (MPa)	Tensile strength (MPa)	Thermal conductivity W/mK	Process	References
	In-plane	Through-plane					
NG/CNTs/PR 65/1.0 phr/35vol.%	178	30	56	-na-	50 (in plane) 13 (through plane)	Compression molding	[177]
MWCNT/SG/EP 5/75/20 wt.%	163	-na-	-na-	-na-	-na-	Compression molding	[172]
G/EP 80/20 wt.%	28	-na-	-na-	18.5 N/mm^2 (65 wt.% of graphite)	-na-	Hot processing	[236]
G/PPS/VE 55/5/40 wt.%	55	-na-	~34.5	73 (Compressive strength)	-na-	Compression molding	[37]
CF/SG/NG/PR 10/10/45/35vol.%	120	-na-	40	-na-	-na-	Selective laser sintering process	[162]

(*Continued*)

Table 7.6 Summarized description of materials, properties, and manufacturing processes of different thermoset composite BPs for the previous studies. (*Continued*)

Composition	Electrical conductivity (S cm^{-1})		Flexural strength (MPa)	Tensile strength (MPa)	Thermal conductivity W/mK	Process	References
	In-plane	Through-plane					
G/CF/EG/PR + CFC in middle of composite 45/10/5/40 wt.%	-na-	101	74	-na-	9.6	Hot pressing	[237]
GP/CF/CB/NG/RPFR 1.5/5/5/63.5/25 wt.%	409.23	98	56.42	-na-	-na-	Compression molding	[238]
NFG/NCB/RPFR 45/4.5/50.5 wt.%	358	-na-	29.3	-na-	-na-	Compression molding	[185]
G/CNTs/EP/PR + 1 layer CFC 75/0.5phr/12.5/12.5 wt.%	48.41	-na-	41.13	-na-	-na-	Melt blending and compression–curing process	[239]
EG/MWNTs/NPR 80/1.0phr/20 wt.%	180	33.38	100	-na-	13.23	Compression molding	[193]

(*Continued*)

Table 7.6 Summarized description of materials, properties, and manufacturing processes of different thermoset composite BPs for the previous studies. (*Continued*)

Composition	Electrical conductivity (S cm^{-1})		Flexural strength (MPa)	Tensile strength (MPa)	Thermal conductivity W/mK	Process	References
	In-plane	Through-plane					
NG(40-μm)/PR 85/15 wt.%	146	-na-	44	-na-	-na-	Hot press	[199]
SG/MCF/EP 80/2.0phr/20 wt.%	69.79	50.34	36.28	-na-	-na-	Compression molding	[23]
SG/EP/PP 80/9/11 wt.%	68.03	3.211	40.16	-na-	-na-	By melt mixing followed by compression molding	[240]
CB/SG/EP 10/65/25 vol%	150	55	38.8	-na-	-na-	Compression molding	[241]
G/CF/EG/PR + 1 layer CFC 45/10/5/40 wt.%	74	-na-	74	-na-	9.6	Compression molding	[176]

(*Continued*)

Table 7.6 Summarized description of materials, properties, and manufacturing processes of different thermoset composite BPs for the previous studies. (*Continued*)

Composition	Electrical conductivity ($S\ cm^{-1}$)		Flexural strength (MPa)	Tensile strength (MPa)	Thermal conductivity W/mK	Process	References
	In-plane	Through-plane					
CF/EP 80/20 wt.%	4.26	6.34	64.37	-na-	-na-	Compression molding	[181]
CF/CNT/EP 74/6/20 wt.%	~ 10	40.31	80.5	-na-	-na-	Compression molding	[181]
6-mm flake-like natural graphite particles/EP (90/10 wt.%)	172	38 S	-na-	-na-	-na-	Compression moulding	[242]
EG/NG/NPR 30/30/40 wt.%	416.66	-na-	42	-na-	-na-	Compression molding	[180]

(*Continued*)

Table 7.6 Summarized description of materials, properties, and manufacturing processes of different thermoset composite BPs for the previous studies. (*Continued*)

Composition	Electrical conductivity ($S\ cm^{-1}$)		Flexural strength (MPa)	Tensile strength (MPa)	Thermal conductivity W/mK	Process	References
	In-plane	Through-plane					
EG/NG/NPR 25/25/50 wt.%	~ 280	-na-	49	-na-	-na-	Compression molding	[180]
G/EG/CB/NPR 52.5/7.5/20/20wt.%	269	-na-	48	-na-	-na-	Compression molding	[243]
EG/CB/G/RPFR 35/5/3/57 wt.%	374.4	97	61.82	-na-	-na-	Compression molding	[25]
SG/CB/EP/PP 73/7/15/5 wt.%	88.5	8.52	28.53	-na-	-na-	By melt mixing and compression molding	[182]
G/EG/CF/NPR 40/5/5/50 wt.%	1518	76	84	-na-	-na-	Hot pressing	[244]

(*Continued*)

Table 7.6 Summarized description of materials, properties, and manufacturing processes of different thermoset composite BPs for the previous studies. (*Continued*)

Composition	Electrical conductivity ($S\ cm^{-1}$)		Flexural strength (MPa)	Tensile strength (MPa)	Thermal conductivity W/mK	Process	References
	In-plane	Through-plane					
G/(2EtIm)/PB-a 85/1/14 wt.%	242	-na-	42.1	-na-	-na-	Compression molding	[98]
NGP/Ag/EP 30/35/35 wt.%	297	-na-	-na-	-na-	-na-	Sheet molding compound	[245]
Ag/VE 50/50 wt.%	294	-na-	-na-	-na-	-na-	sheet molding compound	[245]
G/POE 66/34 wt.%	4.52	-na-	5.96	-na-	-na-	Compression molding	[246]

(*Continued*)

Table 7.6 Summarized description of materials, properties, and manufacturing processes of different thermoset composite BPs for the previous studies. (*Continued*)

Composition	Electrical conductivity (S cm^{-1})		Flexural strength (MPa)	Tensile strength (MPa)	Thermal conductivity W/mK	Process	References
	In-plane	Through-plane					
NG/PR-EP 90/10 wt.%	134	-na-	26	-na-	-na-	By solution intercalation mixing, compression molding, and curing	[247]
NG/PR-EP 70/30 wt.%	54	-na-	48	-na-	-na-	By solution intercalation mixing, compression molding, and curing	[247]
NG/NEP 85/15 wt.%	123.6	-na-	39.4	-na-	-na-	Compression molding	[248]
NG/CB/CF/NPFR 60/5/5/30 vol.%	285.54	91.79	-na-	-na-	-na-	Compression molding	[62]

(*Continued*)

Table 7.6 Summarized description of materials, properties, and manufacturing processes of different thermoset composite BPs for the previous studies. (*Continued*)

Composition	Electrical conductivity ($S\ cm^{-1}$)		Flexural strength (MPa)	Tensile strength (MPa)	Thermal conductivity W/mK	Process	References
	In-plane	Through-plane					
NG/NTRR 80/20 wt.%	248.14	-na-	28	-na-	-na-	Hot press	[249]
NG/GP/VR 70/0.2phr/30 wt.%	286.4	-na-	49.2	-na-	27.2	Compression molding	[192]
SG/MWNTs/EP 73/2/25 vol.%	250.8	35	45	-na-	-na-	Compression molding	[250]
EG/MTMS/EP 45/5/50 wt.%	131	-na-	20	-na-	-na-	Solution impregnation, followed by compression molding and curing	[251]

(*Continued*)

Table 7.6 Summarized description of materials, properties, and manufacturing processes of different thermoset composite BPs for the previous studies. (*Continued*)

Composition	Electrical conductivity (S cm^{-1})		Flexural strength (MPa)	Tensile strength (MPa)	Thermal conductivity W/mK	Process	References
	In-plane	Through-plane					
G/GP/PB-a 75.5/7.5/17 wt.%	323	-na-	55	-na-	14.5	Compression molding	[28]
G/EP 90/10 wt.%	107.7813	-na-	9.08	5.96	-na-	Compression molding	[252]
G/EP 70/30 wt.%	79.95	-na-	54.70	15.64	-na-	Compression molding	[252]

current passes through two outer probes, and through the inner probes, the voltage is measured.

The four-probe device consists of four equally spaced made of tungsten metal tips with finite radius. For reducing the sample damage in the probing duration, each tip is supplied by springs on the other end. The four metal tips are the portion of an auto-mechanical stage, which are moving up and down during measurements. In this device, to supply current through the outer two probes, a high impedance current source is used, whereas the voltage across the inner two probes is measured by a voltmeter to determine the resistivity of the sample, as depicted in Figure 7.26.

The in-plane electrical conductivity of composite BPs is defined as

$$\sigma = \frac{i \times d}{V \times x \times y} \ \text{Scm}^{-1} \tag{7.8}$$

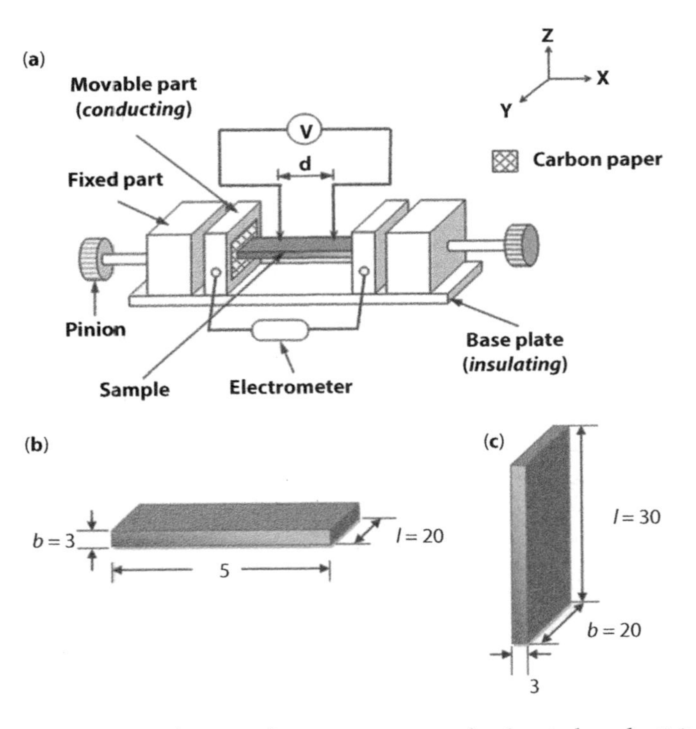

Figure 7.26 (a) Schematic diagram of measurement setup for electrical conductivity. Reprinted from Ref. [79] with permission, copyright of American Chemical Society. Size and orientation of the specimen for (b) in-plane and (c) through-plane electrical conductivity measurements. Reprinted from Ref. [95] with permission, copyright of Elsevier.

where σ is the in-plane electrical conductivity of composite BPs, x is the width of the sample, y is the thickness of the sample, i is the constant current supplied through the sample, and V is the voltage drop between two points separated by a distance d (cm).

7.7.1.2 Through-Plane Electrical Conductivity

The procedure used for measurement the through-plane conductivity is as follows: the specimen of composite BPs is placed between two (nickel-copper-gold) plates. Through those plates, a current source is applied. The voltage across the two plates is measured by a multimeter to determine the resistivity of the sample. In addition, the specimen of composite BPs must be set under pressure, as seen in Figure 7.27.

The through-plane electrical conductivity of composite BPs is defined as

$$\sigma = \frac{1}{\rho[\Omega \text{cm}]} = \frac{\text{L}}{\text{AR}} \text{ [S/cm]} \tag{7.9}$$

where σ is the through-plane electrical conductivity of composite BPs, ρ is the resistivity of the sample, L is the thickness of the sample, A is the surface area of the sample, and R is the electrical resistance of the sample.

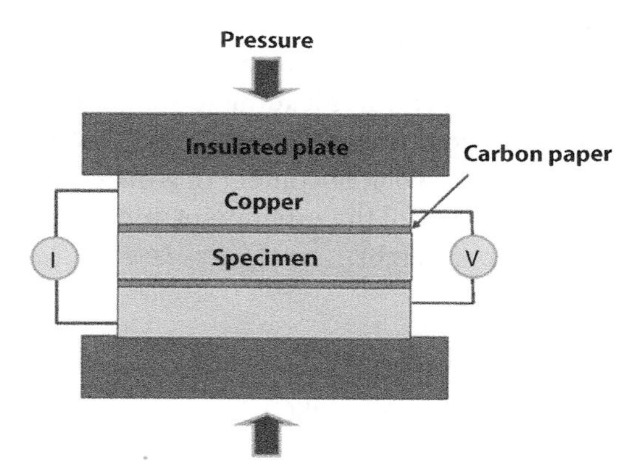

Figure 7.27 Schematic of the measurement setup for the through-plane electrical conductivity. Reprinted from Ref. [178] with permission, copyright of Elsevier.

7.7.2 Thermal Analysis

7.7.2.1 Thermal Gravimetric Analysis

Thermal gravimetric analysis (TGA) is a method of thermal analysis, which is used for measuring the thermal stability of the specimen by determining weight changes in the specimen as a function of temperature (or time). TGA utilizes to validate the real filler contents in each composite BPs due to occurring the inconsistencies with the material rate during the extrusion and batch mixing processes. The precision of filler loading values refers to the efficacy of filler dispersion in the matrix system. TGA is also utilized for measuring the thermal stability of composite BPs, where the FC performance can be affected by thermal property. A TGA instrument usually consists of a precision balance with a pan made up of ceramic or platinum on which a sample is placed. The amount of sample used is small in the range of milligrams; the sample size requirement is depending on the weight change percentage. The pan holds the sample and is placed in an electrically heated furnace with a programmable control temperature. For purging the electrically heated furnace, oxidizing gas such as pure oxygen or an inert gas such as nitrogen or compressed air is used. TGA analysis is carried out in the temperature range up to 1,000°C, and during it, the sample mass is continuously observed and recorded. Because of a thermal lag phenomenon, as known, the accuracy is reduced by high ramp rates, so it is important to choose the temperature ramp rate carefully during analysis. This is a delay between changes in the temperature of the furnace and the sample response, which possess a low relative thermal conductivity, meaning that, under high ramp rate conditions, there might be a totally difference between the sample temperature and recorded temperature. Such an error at higher temperatures leading to weight loss than actually is occurring. The recorded alterations in mass could give information in the thermal degradation rates or volatile content of a material.

A computer is used to control the instrument and to process the output curves (temperature versus weight).

7.7.2.2 Differential Scanning Calorimetry

Differential scanning calorimetry (DSC) is a technique for determining the difference in the amount of heat required to raise the sample temperature and reference as a function of temperature. It provides information about the reversing and non-reversing properties of thermal events. DSC is a thermal analysis method for obtaining the degree of cure, melting point

(Tm), and/or glass transition temperature (Tg) of the composite material as well as it is commonly used in the improvement and characterization of the different composite materials. This method is depending on a sample that is heated at constant speed together with a reference sample in different ovens for a sample and a reference sample separately. The energy supply required for the different temperatures between the samples to be zero is registered. Application areas of the DSC are the determination of the degree of cure, the determination of glass transition temperature (Tg), and melting point (Tm), and the determination of crystallinity.

From the DSC thermograms, thermal transitions analysis of conductive BPs composites is related to other properties, which are relevant to the application of BPs. The crystallinity of samples can be calculated using the following equation:

$$\text{crystallinity},\% = \frac{(\Delta\text{Ha} - \Delta\text{H})}{(\Delta\text{Ha} - \Delta\text{Hc})} \times 100 \tag{7.10}$$

$$\text{crystallinity},\% = \frac{\Delta\text{H}}{\Delta\text{Hc}} \times 100 \tag{7.11}$$

where ΔH is the enthalpy of fusion for the sample, ΔHa is the enthalpy of fusion for a 100% amorphous standard, and ΔHc is the enthalpy of fusion for crystallization.

7.7.2.3 Thermal Conductivity

The thermal conductivity is an essential characteristic of the BP. The BP for PEMFCs must be thermally conductive for conducting the heat to the cooling channel in the FC, which is generated in the active part. Thermal conductivity (k) of the composite BP can be determined by measuring the density (ρ), thermal diffusivity (α), and specific heat capacity (Cp) of the composite BP specimen. This thermal conductivity calculation is expressed by Equation (7.7). Thermal diffusivity is a specific property of a material that describes the characteristic of the material to be used as a heat conductor or an insulator. It is a measurement of the rate of material able to conduct thermal energy relative to the stored thermal energy ability in it. The "laser flash" technique is an accurate and fast method for measuring the thermal diffusivity of polymer composite BP specimens. In this method, the thermal diffusivity of a polymer composite BP is performed by speedily heating one of the sample sides and, on the other opposite side, is measured the rising in the temperature curve. The time takes for transferring heat through the specimen and causes the rising temperature is used for calculation of the through-plane thermal conductivity by determined the measuring through-plane diffusivity.

7.7.3 Mechanical Analysis

7.7.3.1 Flexural Strength

The flexural strength of the composite BPs is considered as a significant property, where during the clamping in the FC hardware, the BPs might undergo high bending force. It is one of the mechanical properties that is used to find the ability of the specimen polymer composite BP to resist the deformation under the load. The rigidity of the polymer composites can be measured by flexural strength. A common method for obtaining flexural strength is the three-point flexural test technique. The testing specimens of the flexural strength must be prepare as per the ASTM D790 standards. In the process of testing the flexural strength, the test specimen is placed in the universal testing machine, and force is applied to the specimen until it gets fractures and breaks. The flexural strength is determined using the following equation:

$$\sigma_F = \frac{3FL}{2bh^2} \tag{7.12}$$

where σ_F is the flexural strength, F is the maximum load at the point of fracture, L is the span length of the sample, b is the width of the sample, and h is the height of the sample.

7.7.3.2 Tensile Strength

The tensile strength is an important mechanical property that is used to evaluate the behavior of strength and deformation as well as the mechanical performance of the material. Tensile strength plays a significant role in the evaluation of fundamental properties of engineering materials. Furthermore, it can be used to control the quality of materials for the purpose of designing and construction as well as in developing new materials. If a material is to be used under stress, then it must be to know and examine the strength and rigidity of the materials, if it is enough to withstand

and resist the loads applied on it. The tensile strength is the maximum mechanical tensile stress which is defined as a force (load) at the fracture point for the specimen relative to the cross-section. The measurement of tensile strength is especially important for the newly prepared polymer composite BP, which can be used in FC applications. The testing specimens of the tensile strength must be prepare as per the ASTM D638 standards. The tensile strength is determined using the following equation:

$$\delta_T = \frac{F}{bd} \tag{7.13}$$

where δ_T is the tensile strength, F is the force at the point of fracture, b is the width of the sample, and d is the thickness of the sample.

7.7.3.3 Compressive Strength

Compressive strength or compression strength is defined as a maximum load that a specimen/material can withstand before fracture or deformed, divided by its cross-sectional area. BPs require good mechanical strength to withstand heavily loaded and vibrations during applications. In the FC application, a BP must undergo compressive force, compressive strength, compressive yield, and compressive modulus must be determined. The compressive strength and compressive modulus of composite BP are measured as per ASTM C695 [25].

The compressive strength can be calculated using the following formula:

$$\mathrm{F} = \frac{\mathrm{P}}{\mathrm{A}} \tag{7.14}$$

where F is The compressive strength, P is the maximum load (failure load) applied to the specimen, and A is the initial cross-sectional area of the specimen resisting the load.

7.8 Conclusions

In this work, the current materials, fabricating methods, the effect of different conductive fillers, and parameters of manufacturing methods on the properties of thermoset-based composites BPs are systematically reviewed.

Thermoset-based composites BPs have good processing, high resistance to corrosion, low cost, low weight, and diverse functional application; it is promising for PEM FC commercialization. The drawback of such type of BP is the difficulties for meeting together good mechanical, thermal, and electrical properties of it, which can be overcome by selecting the suitable composition of thermoset-based composites BPs such as appropriate content of fillers and thermoset resins. The following significant points were gathered from the previous research studies:

- If thermoset resin has large curing shrinkage, then that leads to causing sink marks on the BP surface, so it might not be suitable for composite BPs.
- G, CF, CB, EG, CNT, and GP as conductive fillers have novel characteristics that can improve the performance of the BP by enhancing the electrical, thermal, and mechanical properties.
- Graphene is considered as an appropriate conductive filler for the fabrication of composite BPs, in which at the low quantity of this filler can be enhanced the properties of BPs.
- Compression molding is a very successful processing method for manufacturing thermoset-based composite BPs. Compression molding parameters such as pressure, temperature, and time play a significant role in the properties of composite BPs.
- To reduce the cost and make it suitable for mass production, thermoset-based composite BPs must achieve the DOE targets with less amount conductive fillers.

Abbreviations

Novolac phenolic resin	NPR
Multiwalled carbon nanotubes	(MWCNTs)
Synthetic graphite	(SG)
natural graphite	NG
Carbon black	CB
Carbonfiber	CF
Resole-type phenol-formaldehyde resin	RPFR
Grapheme	GP
Expanded graphite	EG

Polybenzoxazine	PB-a
Graphite powder	G
Novolac phenolic resin powder	NPR
Epoxy	EP
Phenolic resin	PR
Carbon nanotubes	CNTs
Carbon fiber cloth	CFC
Resole phenol-formaldehyde resin	RPFR
Natural flake graphite	NFG
Nanosized carbon black	NCB
Milled carbon fiber	MCF
Polypropylene	PP
2-ethylimidazole (catalyst)	(2EtIm)
Silver nanoparticles	Ag
Nanographene plate	NGP
Vinyl ester resin	VE
Polyesterresin	POE
The mixture of phenolic resin and epoxy resin	PR-EP
Novolacepoxy	NEP
Novolactypephenol formaldehyde resin	NPFR
Novolac thermo-rigid resin	NTRR
Methyltrimethoxysilane	MTMS
Not available	-na-

References

1. Iqbal, M.Z., Rehman, A., Siddique, S., Prospects and challenges of graphene based fuel cells. *J. Energy Chem.*, 39, 217, 2019.
2. Vakulchuk, R., Overland, I., Scholten, D., Renewable energy and geopolitics: A review. *Renew. Sust. Energ. Rev.*, 122, 109547, 2020.
3. Deng, X. and Lv, T., Power system planning with increasing variable renewable energy: A review of optimization models. *J. Clean. Prod.*, 246, 118962, 2020.
4. Daud, W.R.W., Rosli, R.E., Majlan, E.H., Hamid, S.A.A. *et al.*, PEM fuel cell system control: A review. *Renew. Energy*, 113, 620, 2017.
5. Balat, M. and Balat, M., Political, economic and environmental impacts of biomass-based hydrogen. *Int. J. Hydrogen Energy*, 34, 3589, 2009.
6. Cattani-cavalieri, I. and Valença, S., Nanodomains in cardiopulmonary disorders and the impact of air pollution. *Biochem. Soc. Trans.*, 48, 799, 2020.

7. Raji, H., Riahi, A., Borsi, S.H., Masoumi, K. *et al.*, Acute effects of air pollution on hospital admissions for asthma, COPD, and bronchiectasis in Ahvaz, Iran. *Int. J. Chron. Obstruct. Pulmon. Dis.*, 15, 501, 2020.
8. Cox Jr., L.A., Communicating more clearly about deaths caused by air pollution. *Glob. Epidemiol.*, 1, 100003, 2019.
9. Radzuan, N.A.M., Sulong, A.B., Irwan, M., Firdaus, M. *et al.*, Fabrication of multi-filler MCF/MWCNT/SG-based bipolar plates. *Ceram. Int.*, 45, 7413, 2019.
10. Rosli, R.E., Sulong, A.B., Daud, W.R.W., Zulkifley, M.A., The design and development of an HT-PEMFC test cell and test station. *Int. J. Hydrogen Energy*, 44, 30763, 2019.
11. Sharma, S. and Krishna, S., Hydrogen the future transportation fuel: From production to applications. *Renew. Sust. Energ. Rev.*, 43, 1151, 2015.
12. Majlan, E.H., Rohendi, D., Daud, W.R.W., Husaini, T. *et al.*, Electrode for proton exchange membrane fuel cells: A review. *Renew. Sust. Energ. Rev.*, 89, 117, 2018.
13. Al-Mufti, S.M.S., Ali, M.M., Rizvi, S.J.A., Synthesis and structural properties of sulfonated poly ether ether ketone (SPEEK) and Poly ether ether ketone (PEEK). *3Rd Int. Conf. Condens. Matter Appl. Phys*, vol. 2220, p. 020009, 2020.
14. Turoń, K., Hydrogen-powered vehicles in urban transport systems - current state and development. *Transp. Res. Proc.*, 45, 835, 2020.
15. Tanç, B., Arat, H.T., Baltacıoğlu, E., Aydın, K., Overview of the next quarter century vision of hydrogen fuel cell electric vehicles. *Int. J. Hydrogen Energy*, 44, 10120, 2019.
16. Staffell, I., Scamman, D., Velazquez Abad, A., Balcombe, P. *et al.*, The role of hydrogen and fuel cells in the global energy system. *Energy Environ. Sci.*, 12, 463, 2019.
17. Kargar-Pishbijari, H., Hosseinipour, S.J., Aval, H.J., A novel method for manufacturing microchannels of metallic bipolar plate fuel cell by the hot metal gas forming process. *J. Manuf. Process.*, 55, 268, 2020.
18. Singh, R.S., Gautam, A., Rai, V., Graphene-based bipolar plates for polymer electrolyte membrane fuel cells. *Front. Mater. Sci.*, 13, 217, 2019.
19. Jin, C.K. and Kang, C.G., Fabrication by vacuum die casting and simulation of aluminum bipolar plates with micro-channels on both sides for proton exchange membrane (PEM) fuel cells. *Int. J. Hydrogen Energy*, 37, 1661, 2012.
20. Radzuan, N.A.M., Sulong, A.B., Somalu, M.R., Abdullah, A.T. *et al.*, Fibre orientation effect on polypropylene/milled carbon fiber composites in the presence of carbon nanotubes or graphene as a secondary filler: Application on PEM fuel cell bipolar plate. *Int. J. Hydrogen Energy*, 44, 30618, 2019.
21. Napporn, T.W., Karpenko-Jereb, L. *et al.*, Polymer electrolyte fuel cells, in: *Fuel Cells and Hydrogen*, V. Hacker, and S. Mitsushima, (Eds.), pp. 63–89, Elsevier, Amsterdam, 2018.

22. Wilberforce, T., Ijaodola, O. *et al.*, Effect of bipolar plate materials on performance of fuel cells, in: *Reference Module in Materials Science and Materials Engineering*, M.S.J. Hashmi, (Ed.), pp. 1–15, Elsevier, Oxford, 2018.
23. Zakaria, M.Y., Sulong, A.B., Sahari, J., Suherman, H., Effect of the addition of milled carbon fiber as a secondary filler on the electrical conductivity of graphite/epoxy composites for electrical conductive material. *Compos. B Eng.*, 83, 75, 2015.
24. Hermann, A., Chaudhuri, T., Spagnol, P., Bipolar plates for PEM fuel cells: A review. *Int. J. Hydrogen Energy*, 30, 1297, 2005.
25. Gautam, R.K. and Kar, K.K., Synergistic effects of carbon fillers of phenolic resin based composite bipolar Plates on the performance of PEM fuel cell. *Fuel Cells*, 16, 179, 2016.
26. Wu, M. and Shaw, L.L., A novel concept of carbon-filled polymer blends for applications in PEM fuel cell bipolar plates. *Int. J. Hydrogen Energy*, 30, 373, 2005.
27. Cho, E.A., Jeon, U.S., Ha, H.Y., Hong, S.A. *et al.*, Characteristics of composite bipolar plates for polymer electrolyte membrane fuel cells. *J. Power Sources*, 125, 178, 2004.
28. Phuangngamphan, M., Okhawilai, M., Hiziroglu, S., Rimdusit, S., Development of highly conductive graphite-/graphene-filled polybenzoxazine composites for bipolar plates in fuel cells. *J. Appl. Polym. Sci.*, 136, 47183, 2019.
29. Feng, K., Shen, Y., Sun, H., Liu, D. *et al.*, Conductive amorphous carbon-coated 316L stainless steel as bipolar plates in polymer electrolyte membrane fuel cells. *Int. J. Hydrogen Energy*, 34, 6771, 2009.
30. Barzegari, M.M. and Khatir, F.A., Study of thickness distribution and dimensional accuracy of stamped metallic bipolar plates. *Int. J. Hydrogen Energy*, 44, 31360, 2019.
31. Antunes, R.A., Oliveira, M.C.L., Ett, G., Ett, V., Corrosion of metal bipolar plates for PEM fuel cells: A review. *Int. J. Hydrogen Energy*, 35, 3632, 2010.
32. André, J., Antoni, L., Petit, J.P., Corrosion resistance of stainless steel bipolar plates in a PEFC environment: A comprehensive study. *Int. J. Hydrogen Energy*, 35, 3684, 2010.
33. Ehteshami, S.M.M., Taheri, A., Chan, S.H., A review on ions induced contamination of polymer electrolyte membrane fuel cells, poisoning mechanisms and mitigation approaches. *J. Ind. Eng. Chem.*, 34, 1, 2016.
34. Jayasayee, K., Van Veen, J.A.R., Hensen, E.J.M., De Bruijn, F.A., Influence of chloride ions on the stability of PtNi alloys for PEMFC cathode. *Electrochim. Acta*, 56, 7235, 2011.
35. Yang, Z., Peng, H., Wang, W., Liu, T., Crystallization behavior of poly(ε-caprolactone)/layered double hydroxide nanocomposites. *J. Appl. Polym. Sci.*, 116, 2658, 2010.
36. Yu, H.N., Hwang, I.U., Kim, S.S., Lee, D.G., Integrated carbon composite bipolar plate for polymer-electrolyte membrane fuel cells. *J. Power Sources*, 189, 929, 2009.

37. Rafi-Ud-Din, Arshad, M., Saleem, A., Shahzad, M. *et al.*, Fabrication and characterization of bipolar plates of vinyl ester resin/graphite-based composite for polymer electrolyte membrane fuel cells. *J. Thermoplast. Compos. Mater.*, 29, 1315, 2016.
38. Al-Saleh, M.H. and Sundararaj, U., Review of the mechanical properties of carbon nanofiber/polymer composites. *Compos. Part A Appl. Sci. Manuf.*, 42, 2126, 2011.
39. Antunes, R.A., De Oliveira, M.C.L., Ett, G., Ett, V., Carbon materials in composite bipolar plates for polymer electrolyte membrane fuel cells: A review of the main challenges to improve electrical performance. *J. Power Sources*, 196, 2945, 2011.
40. Planes, E., Flandin, L., Alberola, N., Polymer composites bipolar plates for PEMFCs. *Energy Procedia*, 20, 311, 2012.
41. Karimi, S., Fraser, N., Roberts, B., Foulkes, F.R., A review of metallic bipolar plates for proton exchange membrane fuel cells: Materials and fabrication methods. *Adv. Mater. Sci. Eng.*, 2012, 1, 2012.
42. San, F.G.B. and Tekin, G., A review of thermoplastic composites for bipolar plate applications. *Int. J. Energ. Res.*, 37, 283, 2013.
43. Taherian, R., A review of composite and metallic bipolar plates in proton exchange membrane fuel cell: Materials, fabrication, and material selection. *J. Power Sources*, 265, 370, 2014.
44. Kausar, A., Rafique, I., Muhammad, B., Review of applications of polymer/carbon nanotubes and epoxy/CNT composites. *Polym. Plast. Technol. Eng.*, 55, 1167, 2016.
45. Wilberforce, T., El Hassan, Z., Ogungbemi, E., Ijaodola, O. *et al.*, A comprehensive study of the effect of bipolar plate (BP) geometry design on the performance of proton exchange membrane (PEM) fuel cells. *Renew. Sust. Energ. Rev.*, 111, 236, 2019.
46. Yi, P., Zhang, D., Qiu, D., Peng, L. *et al.*, Carbon-based coatings for metallic bipolar plates used in proton exchange membrane fuel cells. *Int. J. Hydrogen Energy*, 44, 6813, 2019.
47. Song, Y., Zhang, C., Ling, C.Y., Han, M. *et al.*, Review on current research of materials, fabrication and application for bipolar plate in proton exchange membrane fuel cell. *Int. J. Hydrogen Energy*, 45, 29832, 2020.
48. Lux, F., Models proposed to explain the electrical conductivity of mixtures made of conductive and insulating materials. *J. Mater. Sci.*, 28, 285, 1993.
49. Ardanuy, M., Rodríguez-Perez, M.A., Algaba, I., Electrical conductivity and mechanical properties of vapor-grown carbon nanofibers/trifunctional epoxy composites prepared by direct mixing. *Compos. B Eng.*, 42, 675, 2011.
50. Lu, C. and Mai, Y.W., Anomalous electrical conductivity and percolation in carbon nanotube composites. *J. Mater. Sci.*, 43, 6012, 2008.
51. Stauffer, D. and Aharony, A., *Introduction to Percolation Theory*, pp. 89–94, Taylor & Francis, London, 1994.

52. Yan, K.Y., Xue, Q.Z., Zheng, Q.B., Hao, L.Z., The interface effect of the effective electrical conductivity of carbon nanotube composites. *Nanotechnology*, 18, 255705, 2007.
53. Alig, I., Lellinger, D., Engel, M., Skipa, T. *et al.*, Destruction and formation of a conductive carbon nanotube network in polymer melts: In-line experiments. *Polymer*, 49, 1902, 2008.
54. Skipa, T., Lellinger, D., Böhm, W., Saphiannikova, M. *et al.*, Influence of shear deformation on carbon nanotube networks in polycarbonate melts: Interplay between build-up and destruction of agglomerates. *Polymer*, 51, 201, 2010.
55. Castillo, F.Y., Socher, R., Krause, B., Headrick, R. *et al.*, Electrical, mechanical, and glass transition behavior of polycarbonate-based nanocomposites with different multi-walled carbon nanotubes. *Polymer*, 52, 3835, 2011.
56. Alig, I., Pötschke, P., Lellinger, D., Skipa, T. *et al.*, Establishment, morphology and properties of carbon nanotube networks in polymer melts. *Polymer*, 53, 4, 2012.
57. Bauhofer, W. and Kovacs, J.Z., A review and analysis of electrical percolation in carbon nanotube polymer composites. *Compos. Sci. Technol.*, 69, 1486, 2009.
58. McLachlan, D.S., An equation for the conductivity of binary mixtures with anisotropic grain structures. *J. Phys. C Solid State Phys.*, 20, 865, 1987.
59. McLachlan, D.S., Blaszkiewicz, M., Newnham, R.E., Electrical Resistivity of Composites. *J. Am. Ceram. Soc.*, 73, 2187, 1990.
60. Clingerman, M.L., King, J.A., Schulz, K.H., Meyers, J.D., Evaluation of electrical conductivity models for conductive polymer composites. *J. Appl. Polym. Sci.*, 83, 1341, 2002.
61. Radzuan, N.A.M., Sulong, A.B., Rao Somalu, M., Electrical properties of extruded milled carbon fibre and polypropylene. *J. Compos. Mater.*, 51, 3187, 2017.
62. Kakati, B.K., Sathiyamoorthy, D., Verma, A., Semi-empirical modeling of electrical conductivity for composite bipolar plate with multiple reinforcements. *Int. J. Hydrogen Energy*, 36, 14851, 2011.
63. Kirkpatrick, S., Percolation and Conduction. *Rev. Mod. Phys.*, 45, 574, 1973.
64. Zallen, R., *The Physics of Amorphous Solids*, pp. 172–179, John Wiley & Sons, New York, 1983.
65. Mamunya, E.P., Davidenko, V.V., Lebedev, E.V., Effect of polymer-filler interface interactions on percolation conductivity of thermoplastics filled with carbon black. *Compos. Interfaces*, 4, 169, 1997.
66. Mamunya, Y.P., Davydenko, V.V., Pissis, P., Lebedev, E.V., Electrical and thermal conductivity of polymers filled.pdf. *Eur. Polym. J.*, 38, 1887, 2002.
67. Merzouki, A. and Haddaoui, N., Electrical Conductivity Modeling of Polypropylene Composites Filled with Carbon Black and Acetylene Black. *ISRN Polym. Sci.*, 2012, 1, 2012.

68. Taherian, R., Hadianfard, M.J., Golikand, A.N., A new equation for predicting electrical conductivity of carbon-filled polymer composites used for bipolar plates of fuel cells. *J. Appl. Polym. Sci.*, 128, 1497, 2013.
69. Zhang, J., Zou, Y.W., He, J., Influence of graphite particle size and its shape on performance of carbon composite bipolar plate. *J. Zhejiang Univ. Sci. A*, 6, 1080, 2005.
70. Richie, N.J., *Development of hybrid composite bipolar plates for proton exchange membrane fuel cells*, p. 7137, Masters Theses, Missouri, 2011.
71. Chanda, M., *Introduction to polymer science and chemistry: A problem-solving approach*, pp. 18–19, Taylor & Francis Group, Boca Raton, 2013.
72. Bank, L.C., *Composites for construction: Structural design with FRP materials*, pp. 46–47, John Wiley & Sons, New Jersey, 2007.
73. Du, C., Ming, P., Hou, M., Fu, J. *et al.*, The preparation technique optimization of epoxy/compressed expanded graphite composite bipolar plates for proton exchange membrane fuel cells. *J. Power Sources*, 195, 5312, 2010.
74. Aleksendrić, D. and Carlone, P., *Soft computing in the design and manufacturing of composite materials: Applications to brake friction and thermoset matrix composites*, pp. 16–17, Woodhead Publishing, Oxford, 2015.
75. Biron, M., *Thermosets and composites, technical information for plastic users*, pp. 17–18, Elsevier Science & Technology Books, London, 2004.
76. Biron, M., *Thermosets and composites: Material selection, applications, manufacturing, and cost analysis*, William Andrew, Cambridge, 2014.
77. Lee, H.S., Kim, H.J., Kim, S.G., Ahn, S.H., Evaluation of graphite composite bipolar plate for PEM (proton exchange membrane) fuel cell: Electrical, mechanical, and molding properties. *J. Mater. Process. Technol.*, 187, 425, 2007.
78. Lee, J.H., Jang, Y.K., Hong, C.E., Kim, N.H. *et al.*, Effect of carbon fillers on properties of polymer composite bipolar plates of fuel cells. *J. Power Sources*, 193, 523, 2009.
79. Kakati, B.K. and Deka, D., Effect of resin matrix precursor on the properties of graphite composite bipolar plate for PEM fuel cell. *Energ. Fuel.*, 21, 1681, 2007.
80. Du, L. and Jana, S.C., Hygrothermal effects on properties of highly conductive epoxy/graphite composites for applications as bipolar plates. *J. Power Sources*, 182, 223, 2008.
81. Busick, D.N. and Wilson, M.S., Low-cost composite materials for PEFC bipolar plates. *Fuel Cells Bull.*, 2, 6, 1999.
82. Liao, S.H., Hsiao, M.C., Yen, C.Y., Ma, C.C.M. *et al.*, Novel functionalized carbon nanotubes as cross-links reinforced vinyl ester/nanocomposite bipolar plates for polymer electrolyte membrane fuel cells. *J. Power Sources*, 195, 7808, 2010.
83. Liao, S.H., Hung, C.H., Ma, C.C.M., Yen, C.Y. *et al.*, Preparation and properties of carbon nanotube-reinforced vinyl ester/nanocomposite bipolar plates for polymer electrolyte membrane fuel cells. *J. Power Sources*, 176, 175, 2008.

84. Kuan, H.C., Ma, C.C.M., Chen, K.H., Chen, S.M., Preparation, electrical, mechanical and thermal properties of composite bipolar plate for a fuel cell. *J. Power Sources*, 134, 7, 2004.
85. Avasarala, B. and Haldar, P., Effect of surface roughness of composite bipolar plates on the contact resistance of a proton exchange membrane fuel cell. *J. Power Sources*, 188, 225, 2009.
86. Hsiao, M.C., Liao, S.H., Yen, M.Y., Su, A. *et al.*, Effect of graphite sizes and carbon nanotubes content on flowability of bulk-molding compound and formability of the composite bipolar plate for fuel cell. *J. Power Sources*, 195, 5645, 2010.
87. C.C. Ma, K.H. Chen, H.C. Kuan, S.M. Chen, M.H. Tsai, Preparation of fuel cell composite bipolar plate. US Patent 7090793B2, assigned to Industrial Technology Research Institute, 2006.
88. Yen, C.Y., Liao, S.H., Lin, Y.F., Hung, C.H. *et al.*, Preparation and properties of high performance nanocomposite bipolar plate for fuel cell. *J. Power Sources*, 162, 309, 2006.
89. Kang, S.J., Kim, D.O., Lee, J.H., Lee, P.C. *et al.*, Solvent-assisted graphite loading for highly conductive phenolic resin bipolar plates for proton exchange membrane fuel cells. *J. Power Sources*, 195, 3794, 2010.
90. Maheshwari, P.H., Mathur, R.B., Dhami, T.L., Fabrication of high strength and a low weight composite bipolar plate for fuel cell applications. *J. Power Sources*, 173, 394, 2007.
91. Hui, C., Liu, H.B., Li, J.X., Li, Y., *et al.*, Characteristics and preparation of polymer/graphite composite bipolar plate for PEM fuel cells. *J. Compos. Mater.*, 43, 755, 2009.
92. Heo, S.I., Oh, K.S., Yun, J.C., Jung, S.H. *et al.*, Development of preform moulding technique using expanded graphite for proton exchange membrane fuel cell bipolar plates. *J. Power Sources*, 171, 396, 2007.
93. Dhakate, S.R., Mathur, R.B., Kakati, B.K., Dhami, T.L., Properties of graphite-composite bipolar plate prepared by compression molding technique for PEM fuel cell. *Int. J. Hydrogen Energy*, 32, 4537, 2007.
94. Mathur, R.B., Dhakate, S.R., Gupta, D.K., Dhami, T.L. *et al.*, Effect of different carbon fillers on the properties of graphite composite bipolar plate. *J. Mater. Process. Technol.*, 203, 184, 2008.
95. Kakati, B.K., Sathiyamoorthy, D., Verma, A., Electrochemical and mechanical behavior of carbon composite bipolar plate for fuel cell. *Int. J. Hydrogen Energy*, 35, 4185, 2010.
96. Pengdam, A. and Rimdusit, S., Preparation and Characterization of Highly Filled Graphite-Based Polybenzoxazine Composites. *J. Met. Mater. Miner.*, 22, 83, 2012.
97. H.J. Kim, Y.C. Eun, S.Y. Cho, H.J. Kweon *et al.*, Composite material for bipolar plate. US Patent 7510678B2, assigned to Samsung SDI Co., Ltd, 2009.

98. Kim, S.G., Kim, J.H., Yim, J.H., A study on the physicochemical properties of a graphite/polybenzoxazine composite for bipolar plate of polymer electrolyte membrane fuel cells. *Macromol. Res.*, 21, 1226, 2013.
99. Dornbusch, M., Christ, U., Rasing, R., *Epoxy resins: Fundamentals and applications*, pp. 21–23, Vincentz Network, Hannover, 2016.
100. Sprenger, S., *The effects of silica nanoparticles in toughened epoxy resins and fiber-reinforced composites*, pp. 22–26, Hanser Publishers, Munich, 2015.
101. Schiraldi, A. and Baldini, P., Epoxy polymers. *J. Therm. Anal. Calorim.*, 28, 295, 1983.
102. Ellis, B. (Ed.), *Chemistry and technology of Epoxy Resins*, Springer, Netherlands, Dordrecht, 1993.
103. Dusek, K. (Ed.), *Epoxy resins and composites III*, Springer-Verlag, Berlin, 1986.
104. Ji, J., Zhu, H., Jiang, X., Qi, C., Zhang, X.M., Mechanical strengths of epoxy resin composites reinforced by calcined pearl shell powders. *J. Appl. Polym. Sci.*, 114, 3168, 2009.
105. Yorkgitis, E.M., Adhesive compounds, in: *ofand*, p. 256, 2002.
106. May, C.A. (Ed.), *Epoxy Resins: Chemistry and Technology*, 2nd edition, Marcel Dekker, New York, 1988.
107. Azeez, A.A., Rhee, K.Y., Park, S.J., Hui, D., Epoxy clay nanocomposites - Processing, properties and applications: A review. *Compos. B Eng.*, 45, 308, 2013.
108. Murias, P., MacIejewski, H., Galina, H., Epoxy resins modified with reactive low molecular weight siloxanes. *Eur. Polym. J.*, 48, 769, 2012.
109. Lem, K.W. and Han, C.D., Chemorheology of thermosetting resins. II. Effect of particulates on the chemorheology and curing kinetics of unsaturated polyester resin. *J. Appl. Polym. Sci.*, 28, 3185, 1983.
110. Lu, W., Lin, H., Wu, D., Chen, G., Unsaturated polyester resin/graphite nanosheet conducting composites with a low percolation threshold. *Polymer*, 47, 4440, 2006.
111. Thomas, S., Chirayil, C.J., Hosur, M. (Eds.), *Unsaturated polyester resins: Fundamentals, Design, Fabrication, and Applications*, Elsevier, Amsterdam, 2019.
112. Dholakiya, B., Unsaturated polyester resin for speciality applications, in: *Polyester*, H.E.M. Saleh, (Ed.), pp. 167–202, InTech, Rijeka, 2012.
113. Davallo, M., Pasdar, H., Mohseni, M., Mechanical properties of unsaturated polyester resin. *Int. J. ChemTech Res.*, 2, 2113, 2010.
114. Cook, W.D., Simon, G.P., Burchill, P.J., Lau, M. *et al.*, Curing kinetics and thermal properties of vinyl ester resins. *J. Appl. Polym. Sci.*, 64, 769, 1997.
115. Kandelbauer, A., Tondi, G. *et al.*, Unsaturated Polyesters and Vinyl Esters, in: *Handbook of thermoset plastics*, H. Dodiuk, and S.H. Goodman, (Eds.), pp. 111–172, William Andrew Publishing, San Diego, 2014.

116. Karger-Kocsis, J. and Gryshchuk, O., Morphology and fracture properties of modified bisphenol a and novolac type vinyl ester resins. *J. Appl. Polym. Sci.*, 100, 4012, 2006.
117. Robinette, E.J., Ziaee, S., Palmese, G.R., Toughening of vinyl ester resin using butadiene-acrylonitrile rubber modifiers. *Polymer*, 45, 6143, 2004.
118. Thostenson, E.T., Ziaee, S., Chou, T.W., Processing and electrical properties of carbon nanotube/vinyl ester nanocomposites. *Compos. Sci. Technol.*, 69, 801, 2009.
119. Bishop, G.R. and Sheard, P.A., Fire-resistant composites for structural sections. *Compos. Struct.*, 21, 85, 1992.
120. Kakati, B.K. and Deka, D., Differences in physico-mechanical behaviors of resol(e) and novolac type phenolic resin based composite bipolar plate for proton exchange membrane (PEM) fuel cell. *Electrochim. Acta*, 52, 7330, 2007.
121. Asim, M., Saba, N., Jawaid, M., Nasir, M. *et al.*, A review on Phenolic resin and its Composites. *Curr. Anal. Chem.*, 13, 185, 2017.
122. Trick, K.A. and Saliba, T.E., Mechanisms of the pyrolysis of phenolic resin in a carbon/phenolic composite. *Carbon*, 33, 1509, 1995.
123. Zaks, Y., Lo, J., Raucher, D., Pearce, E.M., Some structure-property relationships in polymer flammability: Studies of phenolic-derived polymers. *J. Appl. Polym. Sci.*, 27, 913, 1982.
124. Agag, T. and Takeichi, T., Novel benzoxazine monomers containing p-phenyl propargyl ether: Polymerization of monomers and properties of polybenzoxazines. *Macromolecules*, 34, 7257, 2001.
125. Ghosh, N.N., Kiskan, B., Yagci, Y., Polybenzoxazines-New high performance thermosetting resins: Synthesis and properties. *Prog. Polym. Sci.*, 32, 1344, 2007.
126. Ishida, H., Overview and historical background of polybenzoxazine research, in: *Handbook of Benzoxazine Resins*, I. Hatsuo, and A. Tarek, (Eds.), pp. 3–81, Elsevier, Amsterdam, 2011.
127. Takeichi, T., Kawauchi, T., Agag, T., High performance polybenzoxazines as a novel type of phenolic resin. *Polym. J.*, 40, 1121, 2008.
128. Zhang, K. and Ishida, H., Thermally stable polybenzoxazines via ortho-norbornene functional benzoxazine monomers: Unique advantages in monomer synthesis, processing and polymer properties. *Polymer*, 66, 240, 2015.
129. Dhakate, S.R., Sharma, S., Borah, M., Mathur, R.B. *et al.*, Expanded graphite-based electrically conductive composites as bipolar plate for PEM fuel cell. *Int. J. Hydrogen Energy*, 33, 7146, 2008.
130. Dweiri, R. and Sahari, J., Electrical properties of carbon-based polypropylene composites for bipolar plates in polymer electrolyte membrane fuel cell (PEMFC). *J. Power Sources*, 171, 424, 2007.
131. Cavaille, J.Y., Anomalous percolation transition in carbon-black–epoxy composite materials. *Phys. Rev. B Condens. Matter Mater. Phys.*, 59, 14349, 1999.

132. De Oliveira, M.C.L., Sayeg, I.J., Ett, G., Antunes, R.A., Corrosion behavior of polyphenylene sulfide-carbon black-graphite composites for bipolar plates of polymer electrolyte membrane fuel cells. *Int. J. Hydrogen Energy*, 39, 16405, 2014.
133. Zou, J.F., Yu, Z.Z., Pan, Y.X., Fang, X.P. *et al.*, Volume-exclusion effects in polyethylene blends filled with carbon black, graphite, or carbon fiber. *J. Polym. Sci. B Polym. Phys.*, 40, 1013, 2002.
134. Choi, B.C., Lee, J.J., Lee, J.Y., Lee, H.K., Cure condition of epoxy/graphite/CNT system for the preparation of bipolar plate by press molding. *10th IEEE Int. Conf. Nanotechnol*, vol. 5, p. 507, 2010.
135. Lee, H.K., Rim, H.R., Lee, J.Y., Lee, J. *et al.*, Improvements of electrical properties containing carbon nanotube in epoxy/graphite bipolar plate for polymer electrolyte membrane fuel cells. *J. Nanosci. Nanotechnol.*, 8, 5464, 2008.
136. Liao, S.H., Weng, C.C., Yen, C.Y., Hsiao, M.C. *et al.*, Preparation and properties of functionalized multiwalled carbon nanotubes/polypropylene nanocomposite bipolar plates for polymer electrolyte membrane fuel cells. *J. Power Sources*, 195, 263, 2010.
137. Pan, Y., Li, L., Chan, S.H., Zhao, J., Correlation between dispersion state and electrical conductivity of MWCNTs/PP composites prepared by melt blending. *Compos. Part A Appl. Sci. Manuf.*, 41, 419, 2010.
138. Royan, N.R.R., Sulong, A.B., Sahari, J., Suherman, H., Effect of acid- and ultraviolet/ozonolysis-treated MWCNTs on the electrical and mechanical properties of epoxy nanocomposites as bipolar plate applications. *J. Nanomater.*, 2013, 1, 2013.
139. Zhang, Q., Rastogi, S., Chen, D., Lippits, D. *et al.*, Low percolation threshold in single-walled carbon nanotube/high density polyethylene composites prepared by melt processing technique. *Carbon*, 44, 778, 2006.
140. Zeng, Y., Liu, P., Du, J., Zhao, L. *et al.*, Increasing the electrical conductivity of carbon nanotube/polymer composites by using weak nanotube-polymer interactions. *Carbon*, 48, 3551, 2010.
141. Liao, S.H., Yen, C.Y., Weng, C.C., Lin, Y.F. *et al.*, Preparation and properties of carbon nanotube/polypropylene nanocomposite bipolar plates for polymer electrolyte membrane fuel cells. *J. Power Sources*, 185, 1225, 2008.
142. Hwang, I.U., Yu, H.N., Kim, S.S., Lee, D.G. *et al.*, Bipolar plate made of carbon fiber epoxy composite for polymer electrolyte membrane fuel cells. *J. Power Sources*, 184, 90, 2008.
143. Lim, J.W., Kim, M., Lee, D.G., Conductive particles embedded carbon composite bipolar plates for proton exchange membrane fuel cells. *Compos. Struct.*, 108, 757, 2014.
144. Mighri, F., Huneault, M.A., Champagne, M.F., Electrically conductive thermoplastic blends for injection and compression molding of bipolar plates in the fuel cell application. *Polym. Eng. Sci.*, 44, 1755, 2004.

145. Simaafrookhteh, S., Shakeri, M., Baniassadi, M., Sahraei, A.A., Microstructure reconstruction and characterization of the porous GDLs for PEMFC based on fibers orientation distribution. *Fuel Cells*, 18, 160, 2018.
146. Yeetsorn, R., Fowler, M., Tzoganakis, C., Wang, Y. *et al.*, Polypropylene composites for polymer electrolyte membrane fuel cell bipolar plates. *Macromol. Symp.*, 264, 34, 2008.
147. Carter, R.L.B., *Development and modeling of electrically conductive resins for fuel cell bipolar*, Ph.D. thesis, Michigan University, 2008.
148. Balandin, A.A., Ghosh, S., Bao, W., Calizo, I. *et al.*, Superior thermal conductivity of single-layer graphene. *Nano Lett.*, 8, 902, 2008.
149. Huang, X., Qi, X., Boey, F., Zhang, H., Graphene-based composites. *Chem. Soc. Rev.*, 41, 666, 2012.
150. Du, X., Skachko, I., Barker, A., Andrei, E.Y., Approaching ballistic transport in suspended graphene. *Nat. Nanotechnol.*, 3, 491, 2008.
151. Kim, H., Abdala, A.A., MacOsko, C.W., Graphene/polymer nanocomposites. *Macromolecules*, 43, 6515, 2010.
152. Cabot, G.L., Carbon, C., Carbon, U., Petroleum, P., Carbon black. *Ind. Eng. Chem.*, 52, 25A, 1960.
153. Hornbostel, B., Haluska, M., Cech, J., Dettlaff, U. *et al.*, Arc discharge and laser ablation synthesis of singlewalled carbon nanotubes, in: *Carbon Nanotubes*, vol. 1, p. 1, 2006.
154. Meyyappan, M. (Ed.), *Carbon nanotubes: Science and applications*, CRC Press, Boca Raton, 2005.
155. Popov, V.N., Carbon nanotubes: Properties and application. *Mater. Sci. Eng. R Rep.*, 43, 61, 2004.
156. Bhatt, P. and Goe, A., Carbon fibres: Production, properties and potential use. *Mater. Sci. Res. India*, 14, 52, 2017.
157. Blunk, R., Zhong, F., Owens, J., Automotive composite fuel cell bipolar plates: Hydrogen permeation concerns. *J. Power Sources*, 159, 533, 2006.
158. Wolf, H. and Willert-Porada, M., Electrically conductive LCP-carbon composite with low carbon content for bipolar plate application in polymer electrolyte membrane fuel cell. *J. Power Sources*, 153, 41, 2006.
159. Dhakate, S.R., Mathur, R.B., Sharma, S., Borah, M. *et al.*, Influence of expanded graphite particle size on the properties of composite bipolar plates for fuel cell application. *Energ. Fuel.*, 23, 934, 2009.
160. Du, L. and Jana, S.C., Highly conductive epoxy/graphite composites for bipolar plates in proton exchange membrane fuel cells. *J. Power Sources*, 172, 734, 2007.
161. Heo, S.I., Yun, J.C., Oh, K.S., Han, K.S., Influence of particle size and shape on electrical and mechanical properties of graphite reinforced conductive polymer composites for the bipolar plate of PEM fuel cells. *Adv. Compos. Mater. Off. J. Japan Soc. Compos. Mater.*, 15, 115, 2006.

162. Guo, N. and Leu, M.C., Effect of different graphite materials on the electrical conductivity and flexural strength of bipolar plates fabricated using selective laser sintering. *Int. J. Hydrogen Energy*, 37, 3558, 2012.
163. Chen, S., Murphy, J., Herlehy, J., Bourell, D.L. *et al.*, Development of SLS fuel cell current collectors. *Rapid Prototyp. J.*, 12, 275, 2006.
164. Chen, S., Bourell, D.L., Wood, K.L., Improvement of electrical conductivity of SLS PEM fuel cell bipolar plates. *16th Solid Free. Fabr. Symp. SFF 2005*, vol. 78712, p. 458, 2005.
165. Chen, S., Bourell, D.L., Wood, K.L., Fabrication of PEM fuel cell bipolar plates by indirect SLS. *Proc. Solid Free. Fabr. Symp*, vol. 1, p. 244, 2004.
166. Li, W., Jing, S., Wang, S., Wang, C. *et al.*, Experimental investigation of expanded graphite/phenolic resin composite bipolar plate. *Int. J. Hydrogen Energy*, 41, 16240, 2016.
167. Du, C., Ming, P., Hou, M., Fu, J. *et al.*, Preparation and properties of thin epoxy/compressed expanded graphite composite bipolar plates for proton exchange membrane fuel cells. *J. Power Sources*, 195, 794, 2010.
168. Chaiwan, P. and Pumchusak, J., Wet vs. dry dispersion methods for multiwall carbon nanotubes in the high graphite content phenolic resin composites for use as bipolar plate application. *Electrochim. Acta*, 158, 1, 2015.
169. Gibson, G., Epoxy Resins, in: *Brydson's Plastics Materials*, Eighth Edition, M. Gilbert, (Ed.), pp. 773–797, Elsevier, Amsterdam, 2017.
170. Dixit, D., Pal, R., Kapoor, G., Stabenau, M., Lightweight composite materials processing, in: *Lightweight Ballistic Composites*, A. Bhatnagar, (Ed.), pp. 157–216, Woodhead Publishing, Amsterdam, 2016.
171. Gautam, R.K K.K., Bipolar plate materials for proton exchange membrane fuel cell application. *Recent Pat. Mater. Sci.*, 8, 15, 2015.
172. Suherman, H., Irmayani, Optimization of compression moulding parameters of multiwall carbon nanotube/synthetic graphite/epoxy nanocomposites with respect to electrical conductivity. *AIMS Mater. Sci.*, 6, 621, 2019.
173. Boyacı San, F.G. and Okur, O., The effect of compression molding parameters on the electrical and physical properties of polymer composite bipolar plates. *Int. J. Hydrogen Energy*, 42, 23054, 2017.
174. Kim, D.H., Kim, J.H., Lyu, Y.Y., Yim, J.H., Effects of hybrid hardener on properties of a composite bipolar plate for polymer electrolyte membrane fuel cells. *Macromol. Res.*, 20, 1124, 2012.
175. Witpathomwong, S., Okhawilai, M., Jubsilp, C., Karagiannidis, P. *et al.*, Highly filled graphite/graphene/carbon nanotube in polybenzoxazine composites for bipolar plate in PEMFC. *Int. J. Hydrogen Energy*, 45, 30898, 2020.
176. Taherian, R., Hadianfard, M.J., Golikand, A.N., Manufacture of a polymer-based carbon nanocomposite as bipolar plate of proton exchange membrane fuel cells. *Mater. Des.*, 49, 242, 2013.
177. Dhakate, S.R., Sharma, S., Chauhan, N., Seth, R.K. *et al.*, CNTs nanostructuring effect on the properties of graphite composite bipolar plate. *Int. J. Hydrogen Energy*, 35, 4195, 2010.

178. Kim, J.W., Kim, N.H., Kuilla, T., Kim, T.J. *et al.*, Synergy effects of hybrid carbon system on properties of composite bipolar plates for fuel cells. *J. Power Sources*, 195, 5474, 2010.
179. Hui, C., Hong-bo, L., Li, Y., Jian-xin, L. *et al.*, Study on the preparation and properties of novolac epoxy/graphite composite bipolar plate for PEMFC. *Int. J. Hydrogen Energy*, 35, 3105, 2010.
180. Masand, A., Borah, M., Pathak, A.K., Dhakate, S.R., Effect of filler content on the properties of expanded- graphitebased composite bipolar plates for application in polymer electrolyte membrane fuel cells. *Mater. Res. Express*, 4, 095604, 2017.
181. Mohd Radzuan, N.A., Yusuf Zakaria, M., Sulong, A.B., Sahari, J., The effect of milled carbon fibre filler on electrical conductivity in highly conductive polymer composites. *Compos. B Eng.*, 110, 153, 2017.
182. Alo, O.A., Otunniyi, I.O., Pienaar, H., Sadiku, E.R., Electrical and mechanical properties of polypropylene/epoxy blend-graphite/carbon black composite for proton exchange membrane fuel cell bipolar plate. *Mater. Today Proc.*, 38, 658, 2020.
183. Ghosh, A., Goswami, P., Mahanta, P., Verma, A., Effect of carbon fiber length and graphene on carbon-polymer composite bipolar plate for PEMFC. *J. Solid State Electrochem.*, 18, 3427, 2014.
184. Ryszkowska, J., Jurczyk-Kowalska, M., Szymborski, T., Kurzydłowski, K.J., Dispersion of carbon nanotubes in polyurethane matrix. *Physica E Low Dimens. Syst. Nanostruct.*, 39, 124, 2007.
185. Gautam, R.K. and Kar, K.K., Synthesis and properties of highly conducting natural flake graphite/phenolic resin composite bipolar plates for PEM fuel cells. *Adv. Compos. Lett.*, 25, 87, 2016.
186. Rimdusit, S., Jubsilp, C., Tiptipakorn, S., *Alloys and Composites of Polybenzoxazines:Properties and Applications*, pp. 70–71, Springer, Switzerland, 2013.
187. Ishida, H. and Rimdusit, S., Very high thermal conductivity obtained by boron nitride-filled polybenzoxazine. *Thermochim. Acta*, 320, 177, 1998.
188. Sadeghi, E., Bahrami, M., Djilali, N., Analytic determination of the effective thermal conductivity of PEM fuel cell gas diffusion layers. *J. Power Sources*, 179, 200, 2008.
189. Sadeghifar, H., Djilali, N., Bahrami, M., Thermal conductivity of a graphite bipolar plate (BPP) and its thermal contact resistance with fuel cell gas diffusion layers: Effect of compression, PTFE, micro porous layer (MPL), BPP out-of-flatness and cyclic load. *J. Power Sources*, 273, 96, 2015.
190. Tritt, T.M. (Ed.), *Thermal Conductivity Theory, Properties, and Applications*, Plenum Publishers, New York, 2004.
191. Plengudomkit, R., Okhawilai, M., Rimdusit, S., Highly filled graphene-benzoxazine composites as bipolar plates in fuel cell applications. *Polym. Compos.*, 37, 1715, 2016.

192. Hsiao, M.C., Liao, S.H., Yen, M.Y., Teng, C.C. *et al.*, Preparation and properties of a graphene reinforced nanocomposite conducting plate. *J. Mater. Chem.*, 20, 8496, 2010.
193. Yao, K., Adams, D., Hao, A., Zheng, J.P. *et al.*, Highly conductive and strong graphite-phenolic resin composite for bipolar plate applications. *Energ. Fuel.*, 31, 14320, 2017.
194. Vargas-Bernal, R., Herrera-Pérez, G., Calixto-Olalde, M.E., Tecpoyotl-Torres, M., Analysis of DC electrical conductivity models of carbon nanotube-polymer composites with potential application to nanometric electronic devices. *J. Electr. Comput. Eng.*, 2013, 1, 2013.
195. Li, W., Liu, J., Yan, C., Graphite-graphite oxide composite electrode for vanadium redox flow battery. *Electrochim. Acta*, 56, 5290, 2011.
196. Taherian, R., Experimental and analytical model for the electrical conductivity of polymer-based nanocomposites. *Compos. Sci. Technol.*, 123, 17, 2016.
197. Tjong, S.C., Li, Y.C., Li, R.K.Y., Frequency and temperature dependences of dielectric dispersion and electrical properties of polyvinylidene fluoride/expanded graphite composites. *J. Nanomater.*, 2010, 1, 2010.
198. Esawi, A.M.K., Morsi, K., Sayed, A., Taher, M. *et al.*, The influence of carbon nanotube (CNT) morphology and diameter on the processing and properties of CNT-reinforced aluminium composites. *Compos. Part A Appl. Sci. Manuf.*, 42, 234, 2011.
199. Kang, K., Park, S., Ju, H., Effects of type of graphite conductive filler on the performance of a composite bipolar plate for fuel cells. *Solid State Ionics*, 262, 332, 2014.
200. Hao, W., Ma, H., Lu, Z., Sun, G. *et al.*, Design of magnesium phosphate cement based composite for high performance bipolar plate of fuel cells. *RSC Adv.*, 6, 56711, 2016.
201. U.S. Department of Energy Fuel Cell Technologies Office (FCTO, in: *Bipolar Plate Workshop Summary Report*, 2017.
202. Taherian, R., Application of polymer-based composites: Bipolar plate of PEM fuel cells, in: *Electrical Conductivity in Polymer-Based Composites Experiments, Modelling, and Application*, R. Taherian, and A. Kausar, (Eds.), pp. 183–237, William Andrew, Oxford, 2018.
203. Pal, R., *Electromagnetic, Mechanical, and Transport Properties of Composite Materials*, pp. 13–23, CRC Press, Florida, 2014.
204. Maxwell, J.C., *A Treatise on Electricity & Magnetism*, pp. 435–436, Clarendon Press, Oxford, 1881.
205. Pal, R., On the electrical conductivity of particulate composites. *J. Compos. Mater.*, 41, 2499, 2007.
206. Clingerman, M.L., Weber, E.H., King, J.A., Schulz, K.H., Development of an additive equation for predicting the electrical conductivity of carbon-filled composites. *J. Appl. Polym. Sci.*, 88, 2280, 2003.

207. Keith, J.M., King, J.A., Barton, R.L., Electrical conductivity modeling of carbon-filled liquid-crystalline polymer composites. *J. Appl. Polym. Sci.*, 102, 3293, 2006.
208. McCullough, R.L., Generalized combining rules for predicting transport properties of composite materials. *Compos. Sci. Technol.*, 22, 3, 1985.
209. Taherian, R., Development of an equation to model electrical conductivity of polymer-based carbon nanocomposites. *ECS J. Solid State Sci. Technol.*, 3, M26, 2014.
210. Weber, M. and Kamal, M.R., Estimation of the volume resistivity of electrically conductive composites. *Polym. Compos.*, 18, 711, 1997.
211. Nielsen, L.E., The thermal and electrical conductivity of two-phase systems. *Ind. Eng. Chem. Fundam.*, 13, 17, 1974.
212. Voet, A., Whitten, W.N., Cook, F.R., Electron Tunneling in Carbon Blacks. *Colloid Polym. Sci.*, 1, 39, 1964.
213. Pramanik, P.K., Khastgir, D., Saha, T.N., Conductive nitrile rubber composite containing carbon fillers: Studies on mechanical properties and electrical conductivity. *Composites*, 23, 183, 1992.
214. Tiusanen, J., Vlasveld, D., Vuorinen, J., Review on the effects of injection moulding parameters on the electrical resistivity of carbon nanotube filled polymer parts. *Compos. Sci. Technol.*, 72, 1741, 2012.
215. Zare, Y., Rhee, K.Y., Park, S.J., A developed equation for electrical conductivity of polymer carbon nanotubes (CNT) nanocomposites based on Halpin-Tsai model. *Results Phys.*, 14, 102406, 2019.
216. Nair, C.P.R., Advances in addition-cure phenolic resins. *Prog. Polym. Sci.*, 29, 401, 2004.
217. Wang, R.M., Zheng, S.R., Zheng, Y.P., *Polymer Matrix Composites and Technology*, pp. 113–114, Woodhead Publishing, Cambridge, 2011.
218. Dodiuk, H. and Goodman, S.H. (Eds.), *Handbook of Thermoset Plastics*, William Andrew, San Diego, 2014.
219. Li, H., *Synthesis, characterization and properties of vinyl ester matrix resins*, PhD diss, Virginia Tech, 1998.
220. Chaudhary, V. and Ahmad, F., A review on plant fiber reinforced thermoset polymers for structural and frictional composites. *Polym. Test.*, 91, 106792, 2020.
221. Zaske, O.C. and Goodman, S.H., Unsaturated polyester and vinyl ester resins, in: *Handbook of Thermoset Plastics (Second Edition)*, pp. 97–168, Noyes Publications, New Jersey, 1998.
222. Alemour, B., Yaacob, M.H., Lim, H.N., Hassan, M.R., Review of electrical properties of graphene conductive composites. *Int. J. Nanoelectron. Mater.*, 11, 371, 2018.
223. Ali, A.H., Benmokrane, B., Mohamed, H.M., Manalo, A. *et al.*, Statistical analysis and theoretical predictions of the tensile-strength retention of glass fiber-reinforced polymer bars based on resin type. *J. Compos. Mater.*, 52, 2929, 2018.

224. Rodriguez, E.L., Microdelamination due to resin shrinkage in filament-wound fibreglass composites. *J. Mater. Sci. Lett.*, 8, 116, 1989.
225. Probst, N. and Grivei, E., Structure and electrical properties of carbon black. *Carbon*, 40, 201, 2002.
226. Saito, R., Fujita, M., Dresselhaus, G., Dresselhaus, M.S., Electronic structure of chiral graphene tubules. *Appl. Phys. Lett.*, 60, 2204, 1992.
227. Kaneko, S., Mele, P., Endo, T., Tsuchiya, T. *et al.*, *Carbon-related materials in recognition of nobel lectures by Prof. Akira Suzuki in ICCE*, Springer, New York, 2017.
228. Kuilla, T., Bhadra, S., Yao, D., Kim, N.H. *et al.*, Recent advances in graphene based polymer composites. *Prog. Polym. Sci.*, 35, 1350, 2010.
229. Spitalsky, Z., Tasis, D., Papagelis, K., Galiotis, C., Carbon nanotube-polymer composites: Chemistry, processing, mechanical and electrical properties. *Prog. Polym. Sci.*, 35, 357, 2010.
230. Kakati, B.K., Ghosh, A., Verma, A., Efficient composite bipolar plate reinforced with carbon fiber and graphene for proton exchange membrane fuel cell. *Int. J. Hydrogen Energy*, 38, 9362, 2013.
231. Yao, K., Adams, D.L., Hao, A., Zheng, J.P. *et al.*, Highly conductive, strong, thin and lightweight graphite-phenolic resin composite for bipolar plates in proton exchange membrane fuel cells. *ECS Trans.*, 77, 1303, 2017.
232. Dueramae, I., Pengdam, A., Rimdusit, S., Highly filled graphite polybenzoxazine composites for an application as bipolar plates in fuel cells. *J. Appl. Polym. Sci.*, 130, 3909, 2013.
233. Simaafrookhteh, S., Khorshidian, M., Momenifar, M., Fabrication of multi-filler thermoset-based composite bipolar plates for PEMFCs applications: Molding defects and properties characterizations. *Int. J. Hydrogen Energy*, 45, 14119, 2020.
234. Akhtar, M.N., Sulong, A.B., Umer, A., Yousaf, A. *et al.*, Multi-component MWCNT/NG/EP-based bipolar plates with enhanced mechanical and electrical characteristics fabricated by compression moulding. *Ceram. Int.*, 44, 14457, 2018.
235. Plengudomkit, R., Okhawilai, M., Rimdusit, S., Highly filled graphene-benzoxazine composites as bipolar plates in fuel cell applications. *Polym. Polym. Compos.*, 37, 1715, 2016.
236. Suherman, H., Mahyoedin, Y., Septe, E., Rizade, R., Properties of graphite/epoxy composites: The in-plane conductivity, tensile strength and Shore hardness. *AIMS Mater. Sci.*, 6, 165, 2019.
237. Taherian, R., Golikand, A.N., Hadianfard, M.J., Preparation and properties of a phenolic/graphite nanocomposite bipolar plate for proton exchange membrane fuel cell. *ECS J. Solid State Sci. Technol.*, 1, M39, 2012.
238. Ghosh, A. and Verma, A., Carbon-polymer composite bipolar plate for HT-PEMFC. *Fuel Cells*, 14, 259, 2014.
239. Chen, J., Zhang, Q.Z., Hou, Z.S., Jin, F.L. *et al.*, Preparation and characterization of graphite/thermosetting composites. *Bull. Mater. Sci.*, 42, 1, 2019.

240. Alo, O.A. and Otunniyi, I.O., Pienaar, Hc.Z., Development of graphite-filled polymer blends for application in bipolar plates. *Polym. Compos.*, 41, 3364, 2020.
241. Suherman, H., Sahari, J., Sulong, A.B., Effect of small-sized conductive filler on the properties of an epoxy composite for a bipolar plate in a PEMFC. *Ceram. Int.*, 39, 7159, 2013.
242. Kang, K., Park, S., Jo, A., Lee, K. *et al.*, Development of ultralight and thin bipolar plates using epoxy-carbon fiber prepregs and graphite composites. *Int. J. Hydrogen Energy*, 42, 1691, 2017.
243. Li, K.C., Zhang, K., Wu, G., Fabrication of electrically conductive polymer composites for bipolar plate by two-step compression molding technique. *J. Appl. Polym. Sci.*, 130, 2296, 2013.
244. Taherian, R., Golikand, A.N., Hadianfard, M.J., The effect of mold pressing pressure and composition on properties of nanocomposite bipolar plate for proton exchange membrane fuel cell. *Mater. Des.*, 32, 3883, 2011.
245. B.Z. Jang, Sheet molding compound flow field plate, bipolar plate and fuel cell. US Patent 8518603B2, assigned to Nanotek Instruments, Inc, 2013.
246. Bhlapibul, S. and Pruksathorn, K., Preparation of graphite composite bipolar plate for PEMFC. *Korean J. Chem. Eng.*, 25, 1226, 2008.
247. Chen, H., Xia, X.H., Yang, L., He, Y. *et al.*, Preparation and characterization of graphite/resin composite bipolar plates for polymer electrolyte membrane fuel cells. *Sci. Eng. Compos. Mater.*, 23, 21, 2016.
248. Hui, C., Hong-Bo, L., Li, Y., Yue-De, H., Effects of resin type on properties of graphite/polymer composite bipolar plate for proton exchange membrane fuel cell. *J. Mater. Res.*, 26, 2974, 2011.
249. Negrea, A., Bacinschi, Z., Bucurica, I.A., Teodorescu, S. *et al.*, A new material for bipolar plates used in fuel cells. *Rom. J. Phys.*, 61, 527, 2016.
250. Lee, J.H., Lee, J.S., Kuila, T., Kim, N.H. *et al.*, Effects of hybrid carbon fillers of polymer composite bipolar plates on the performance of direct methanol fuel cells. *Compos. B Eng.*, 51, 98, 2013.
251. Dursun, B., Yaren, F., Unveroglu, B., Yazici, S. *et al.*, Expanded graphite-epoxy-flexible silica composite bipolar plates for PEM fuel cells. *Fuel Cells*, 14, 862, 2014.
252. Mohammed, A.M., Nafaty, E.--., A., U., Bugaje, I.M., Mukthar, B., Effect of Filler Concentration on Properties of Graphite Composite Bipolar Plate. *Int. J. Sci. Eng. Technol.*, 3, 2395, 2015.

8

Metal-Organic Framework Membranes for Proton Exchange Membrane Fuel Cells

Yashmeen, Gitanjali Jindal and Navneet Kaur*

Department of Chemistry, Panjab University, Chandigarh, India

Abstract

Proton exchange membrane fuel cells (PEMFCs) have caught the attention due to many of the extraordinary features that they possessed such as fast start-up, low operational temperature, large energy conversion rate, and environmental friendliness over other types of fuel cells. For PEMFCs, usually, a hydrated proton exchange membrane acts as an electrolyte and protons (H^+) are used as mobile ions. For better efficiency of the proton transport, proton conductivity value of the electrolyte must lie in the range of 0.02 S cm^{-1} or higher. Metal-organic frameworks (MOFs) are the porous hybrids of organic-inorganic materials and have emerged as the promising proton conducting materials because of many advantages like large surface area, controllable cavity structure, tenability, modifiable functional groups, and good stability. This books chapter will throw light on different types of MOFs employed to function as PEMFCs.

Keywords: Metal-organic frameworks, Grotthuss mechanism, proton conduction, PEMFCs

8.1 Introduction

The world energy consumption is increasing day by day due to number of reasons such as rapid urbanization, population growth, and large numbers of industries. It has now become the necessity to shift from non-renewable resources (as fossil fuel) to alternative renewable energy resources (as fuel cells). Fuel cells work as of electrochemical devices that produce electrical

**Corresponding author*: neet_chem@yahoo.co.in; neet_chem@pu.ac.in

Inamuddin, Omid Moradi and Mohd Imran Ahamed (eds.) Proton Exchange Membrane Fuel Cells: Electrochemical Methods and Computational Fluid Dynamics, (213–244) © 2023 Scrivener Publishing LLC

energy by conversion of chemical energy via controlled oxidation-reduction (REDOX) processes. Electrodes (anode and cathode), electrolyte, and fuel are the basic components of fuel cells. Depending upon the type of electrolyte and fuel used, there are several kinds of fuel cells, namely, proton exchange membrane fuel cells (PEMFCs), microbial fuel cell (MFC), direct methanol fuel cells (DMFCs), alkaline fuel cells (AFCs), phosphoric acid fuel cells (PAFC), solid oxide fuel cells (SOFCs), and molten carbonate fuel cells (MCFCs) [1].

Irrespective of the availability of various types of fuel cells mentioned above, recent attention has been made to the PEMFCs [2]. In 1960s, General Electric of the United States developed PEMFCs for the first time for execution of the Gemini space program [3]. In comparison to other available fuel cells, PEMFCs possess number of promising features including fast start-up, large energy conversion rate, low operational temperature, and environmental friendliness [4]. For PEMFCs, usually, a hydrated proton exchange membrane (PEM) acts as an electrolyte and protons (H^+) are used as mobile ions. When platinum is used as electro-catalyst with carbon-based electrodes, operating temperature of these fuel cells usually lies between 30°C and 100°C.

At anode, oxidation of hydrogen occurs that liberate electrons with half reaction as $H_2 \rightarrow 2H^+ + 2e^-$, whereas gain of electrons (reduction) occurs at cathode with half reaction of $½O_2 + 2e^- \rightarrow O_2^-$. Thus, the overall reaction is

$$H_2 + ½O_2 \rightarrow H_2O + Q$$

where Q is the heat produced during the reaction.

In PEMFCs, PEM acts as electrolyte inside the cell, through which protons move from anode to cathode and via the external circuit, the electrons generated at anode travel toward cathode, generating electricity. A number of literature reports highlight the use of oxides, intercalation compounds, ceramics, and many polymers for proton conduction. However, Nafion (sulfonated fluoropolymer) is the one of the successfully commercialized polymers as PEM because of the proton conductivity lying in the order of 0.1–0.01 S cm^{-1} at 85°C and activation energies in the range of 0.1–0.5 eV under humid conditions [5]. However, the major limiting factor of using Nafion is its ability to achieve high proton conducting efficiency under hydrated conditions or enough humidification only. In addition, for Nafion-based operating system, extra cost is required to power the humidifiers [6]. Moreover, if operating temperatures are taken into consideration, then, at higher temperatures (≤100°C), CO poisoning of Pt catalysts has been observed, which is responsible for sluggish reaction kinetics in Fuel cell. On the other hand, proton conduction of some of the oxo-acid salts lie within the range of 10^{-2} to 10^{-5} S cm^{-1} around moderate temperatures (120°C–300°C), but their water solubility and low conduction value at room temperature limit their usage also [7].

Hence, an ideal proton conducting material for fuel cells should have similar or higher efficiency of proton conduction as compared to Nafion under dry or anhydrous conditions and at higher temperature range (100°C–300°C). Additional desirable requirements include economic viability and good mechanical strength along with chemical and thermal stability and compatibility of the material with other components of fuel cell such as electrolyte and electrode materialsc.

Recently, metal-organic frameworks (MOFs), the porous hybrids of organic-inorganic materials, have emerged as potential materials to act as PEMs [8]. This is because of their efficiency of proton conduction due to large surface area, controllable cavity structure, tenability, modifiable functional groups, and good stability. Moreover, MOFs could accommodate more bound water because of their large specific surface area, resulting in enhanced migration of protons by which the proton hopping conductivity improved. In addition, fuel and oxidant diffusion were hindered due to the small pore size of MOFs.

Although a number of benefits are being offered by MOFs, yet, the limitation of signal transduction for MOF based sensor is one of the major stumbling features in the development of MOF-based sensors. In this chapter, we commence by examining different design strategies of MOFs for the PEMs fuel cells and their performance in various matrices. Taking into consideration the importance of MOFs, the present exertion has been carried out for their use as PEMFCs and the discussion has been categorized into various sections:

Aluminium containing MOFs for PEMFCs
Chromium containing MOFs for PEMFCs
Copper containing MOFs for PEMFCs
Cobalt containing MOFs for PEMFCs
Iron containing MOFs for PEMFCs
Nickel containing MOFs for PEMFCs
Platinum containing MOFs for PEMFCs
Zinc containing MOFs for PEMFCs
Zirconium containing MOFs for PEMFCs

8.2 Aluminium Containing MOFs for PEMFCs

Wang *et al.* developed a PEM (PEM) based on MIL-53(Al)-NH_2framework (**1**) with electro-blown spinning Al_2O_3 nanofibers that coordinated with ligand via hydrothermal reaction [9]. The flower-like structure of **1** along with having large specific area could also be modified chemically with amine groups that synergistically deliver **1** with excellent proton conduction and strength enhancement for creating high-performance hybrid PEMs. The frameworks of **1** were integrated in sulfonated poly(ether sulfone) (SPES) matrix to obtain hybrid **1**@SPES membranes. The **1** worked as proton conductor with enough hydrophilic amino groups, giving high water capacity and more "water-vehicles" to efficiently increase proton transfer under highly humid conditions. Further, the interconnection between the -SO_3H groups of SPES and -NH_2 groups of **1** could act as additional and successive proton-hopping channels to promote proton migration (Figure 8.1). Thus, the structural composition of tailored **1** assisted the protons to transfer through the membranes via both vehicle and Grotthuss mechanisms. Moreover, the methanol permeabilities of sample membranes studied for the evaluation of practical applicability pointed to the fact that, the higher the acid-base interactions between the interfaces of **1** and SPES, the lesser would be the methanol permeability. The **1**@SPES hybrid membranes hold high selectivity upto a value of 67.176×10^4 S cm^{-3}.

Alam *et al.* synthesized proton exchange nanocomposite, 2, by using sulfonic acid–functionalized silica nanomaterials and imidazoleen-capsulated

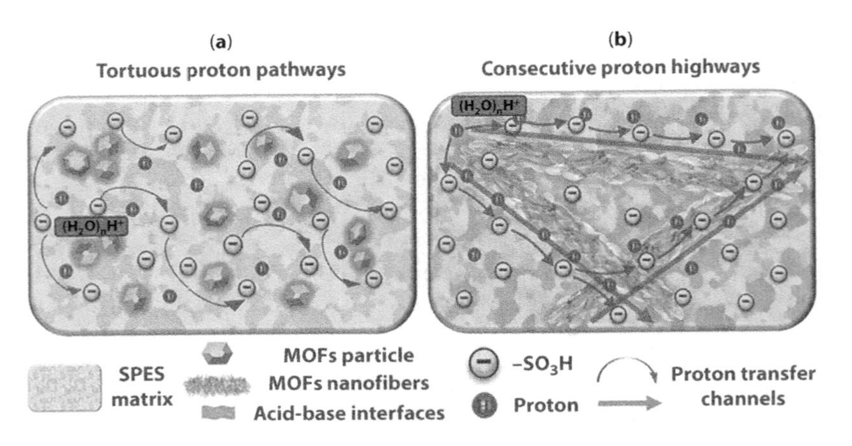

Figure 8.1 Schematic illustration of (a) tortuous and (b) consecutive proton conduction routes in hybrid SPES-matrix membrane (reproduced from ACS Appl. Mater. Interfaces, 11 (2019) 39979–39990, [9]).

MOF [NH_2-MIL-53(Al)] nanoparticles [10]. The structural characterization of 2 was done using UV-vis spectroscopy, FTIR, XRD, TGA, ^{1}HNMR, SEM, and TEM analysis. The proton conduction studies were done, which inferred that the conductivity increased by increasing 2/Si-SO_3H nanoparticles content and temperature, and these results were concurrent with the water uptake and ion exchange capacity readings. The fuel cell performance was also observed to be linearly related with the content of nanoparticles and temperature. Then, a nanocomposite, 2, showed a power density and peak current density of 40.80 mW cm^{-2} and 100.30 mA cm^{-2}, respectively.

Tsai and co-workers synthesized two Nafion-based composite PEMs by utilizing 1D channel microporous MOFs, [CPO-27(Mg), **3** and MIL-53(Al), **4**] as fillers [11]. The TGA curves unveiled the higher thermal stability of **4** framework than that of **3**, and it also suggested that the **3** had more water than **4**, signifying the presence of more number of coordinated water molecules within its composition. It was also found that both the PEMs were quite stable for working at high temperatures for PEMFC applications. The water uptakes values of both **3** and **4** were higher than normal Nafion membrane. As the water retention factor directly deal with the proton conductivities, accordingly, the membrane **3** possessed highest conductivities and water uptake properties followed by **4** and the normal Nafion membrane. It was further studied that the presence of the MOF fillers restrained the growth of **3** and **4** and would be advantageous for applying in PEMFCs.

8.3 Chromium Containing MOFs for PEMFCs

Liu *et al.* developed two ultrastable isostructural MOFs: Cr-MIL-88B-pyridinesulfonic acid (**5a**) and Cr-MIL-88B-pyridineethanesulfonic acid (**5b**), to be used as proton conductors by using modified synthesis method [12]. The thermal analysis corroborated higher structural stability upto a temperature of 400°C. The water uptake properties showed that water uptake of both **5a** and **5b** was enhanced with increasing relative humidity (RH). The water adsorption capacity of the **5a** and **5b** was found to be stronger than that of only Cr-MIL-88B, which might be credited to the presence of the hydrophilic groups of the sulfonic acid in former that have opened the pores largely and improved the water adsorption capacity with increased RH (Table 1). Very high values of proton conductivities of these MOFs, **5a** and **5b**, were evaluated as 1.58×10^{-1} and 4.50×10^{-2} S cm^{-1}, respectively, at RH of 85% and temperature of 100°C.

Abdolmaleki *et al.* reported amine functionalized PEM using MOF (Cr-MIL-101-NH2, **6**) and SPES synthesized via post grafting method [13]. The synthesized membrane **6**-SPES was characterized using FT-IR, CHNS, XRD, SEM, and TGA experiments. The TGA analysis showed that increased cross-linking in the membrane (**6**-SPES) enhanced the thermal stability at elevated temperatures (Table 8.1). In addition, the water uptake and swelling ratio were found to improve due to cross-linked structure between the polymer chains and **6**. The high value of proton conductivity of the **6**-SPES (0.041 S cm^{-1}) membrane at 0% RH and 160°C might be due to number of factors such as the voids of **6** having heteroatoms, presence of phosphoric acid molecules having ability of formation of strong hydrogen

Table 8.1 Chromium-MOF.

MOF	Conductivity	Humidity and Temperature	Ea	Mechanism	Ref.
Cr-MIL-88B-pyridinesulfonic acid (**5a**)	1.58×10^{-1} S cm^{-1}	RH = 85% and T = 100°C	< 0.4 eV	Grotthus	[12]
Cr-MIL-88B-pyridineethane-sulfonic acid (**5b**)	4.50×10^{-2} S cm^{-1}	--	<0.4 eV	Grotthus	[12]
6-SPES	0.041 S cm^{-1}	160°C and RH = 0%,	20.57 KJ mol^{-1}	combination of vehicular and Grotthuss mechanisms, particularly by the Grotthuss mechanism at higher temperatures (100°C–160°C)	[13]
BUT-8(Cr)A, **7**	1.27×10^{-1}	100% RH and 80°C	0.11 eV	Grotthus	[14]
SA, **8a**	1.89×10^{-3}	Anhydrous, at 150°C	0.262 eV	Grotthus	[15]
MSA, **8b**	1.02×10^{-4}		0.304 eV	Grotthus	[15]
PTSA, **8c**	2.78×10^{-4}		0.426eV	Vehicle	[15]
H_2SO_4@ MIL-101-SO_3H,**9**	1.82	70°C, 90% RH	<0.4 eV	Grotthus	[16]

bonds and the enhanced acid retention capacity of the **6**-SPES membrane due to ammonium phosphate generation (Figure 8.2). The chemical as well as mechanical properties of **6**-SPES were also enhanced because of the presence of **6** against the polymeric template.

Chen and co-workers synthesized a chemically stable chromium based MOF, **7**, using 4,8-disulfonaphthalene-2,6-dicarboxylateand further used it as PEM [14]. From the PXRD and water sorption isotherms, it was observed that adsorption of water molecules in the pores of **7** under different humidity levels was responsible for the structural transformation of **7**. In addition, the high polarity and plentiful $-SO_3H$ sites that enclose strong interactions with water molecules via hydrogen bonding, facilitated the MOF, **7**, to easily adsorb water molecules, hence expanding pore spaces. Moreover, the proton conductivities were investigated using alternating current impedance, and it was found that **7** preserved high proton conductivities in wide range of humidity and temperatures (Table 8.1). This could be ascribed to the structural flexibility induced by water molecules and the anticipated self-adaptation of the frameworks to sustain the networks of hydrogen-bonding interactions in their pores at a lower RH (Table 8.1).

Chen and group developed a series of imidazolium possessing IL (ionic liquids)–based MOFs (EIMS@MIL-101) using sulfate acid (SA, MOF **8a**), methanesulfonate acid (MSA, MOF **8b**), and *p*-toluenesulfonate acid (PTSA, MOF **8c**) [15]. The TGA analysis of **8a-8c** showed the resistance of high temperature upto 240°C even heating in air, pointing to their enough stability for the application as PEM at moderate temperature. The proton

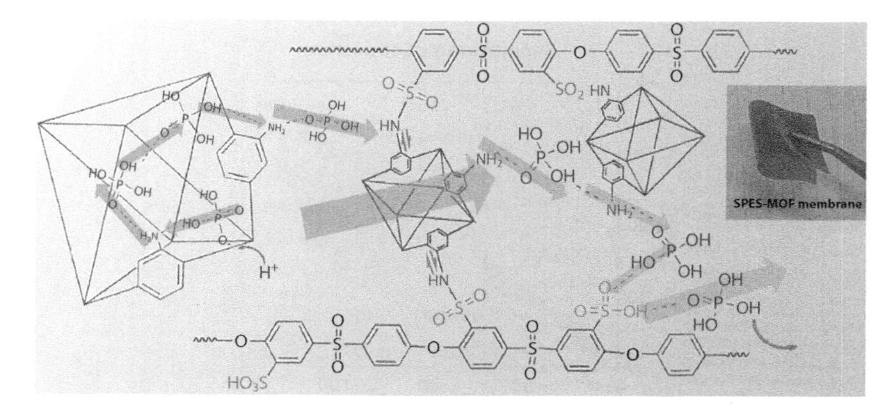

Figure 8.2 Schematic representation of proton transport occurring via Grotthuss mechanism in hydrogen bonding networks of 6-SPES membrane (reproduced from J. Membrane Sci., 365 (2018) 281–292, [13]).

conductivities were also evaluated through alternating current impedance spectroscopy at different temperatures, which suggested that MOF **8a** (1.89×10^{-3} S cm^{-1}) possessed highest conductivity followed by **8b** (1.02×10^{-4} S cm^{-1}) and **8c** (2.78×10^{-4} S cm^{-1}) at 150°C. This was attributed to the different van der Waals volume of acid counter ions of the three MOFs, according to which, the higher the volume, the lesser was the conduction value (Table 8.1). Further, at elevated temperatures, the MOF **8a** was found to be highly durable for upto 15 days.

Li and group synthesized a water-stable MOF composite (H_2SO_4@MIL-101-SO_3H,**9)** as super proton conduction membrane [16]. The RH and proton conduction studies were carried out, and it was observed that the conductivity reached its maximum value at 90% humidity. The presence of sulfonic acid groups enhanced the hydrophilic nature of the membrane and hence increased the proton transfer. The super proton conducting nature of **9** was ascribed to the water molecules based number of proton units, high acidic nature of H_2SO_4 acid (guest medium) and high-density Brønsted acidic $-SO_3H$ groups (Figure 8.3). It was seen that, at sub-zero temperature (−40°C), the material demonstrated the commendable proton conductivity of 0.92×10^{-2} S cm^{-1} (Table 8.1). Moreover, the MOF, **9**, could sustain sub-zero temperature for at least 20 h with high proton conductivity, which clearly implied its exceptional durability and utilization in harsh fuel cell conditions.

Zhang *et al.* synthesized MOF (HPW@MIL-9, **10)** by encapsulating phosphortungstic acid into the cavity of MIL-101(Cr) using precursors Na_2WO_4•$2H_2O$ and Na_2HPO_4 [17]. The prepared composite was doped

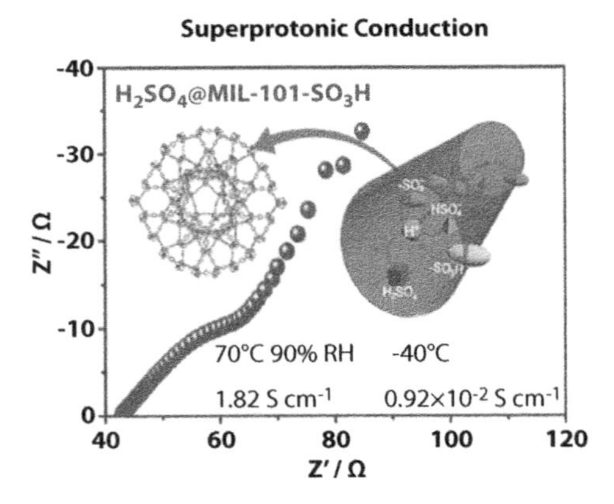

Figure 8.3 Structure of MOF **9** and graph displaying high proton conductivity (reproduced from ACS Energy Lett., 2 (2017) 2313–2318, [16]).

into sulfonated poly(ether ether ketone) (SPEEK) to form the proton conducting membrane (SPEEK/**10**). The nanocomposite was characterized by SEM, XRD, XPS, and TGA analyses. From the TGA curves, it was demonstrated that the membrane **10** possessed high thermal stability at higher temperatures due to the presence of strained polymeric chains. In addition, it was observed that the swelling degree and water uptake was reduced because of rigid structure of **10**. Further, the proton conductivity of **10** was found to increase at lower humid levels. The membrane with 9 wt.% of **10** exhibited proton conductivity of 6.51 mS cm^{-1} at 60°C (40% RH) and 272 mS cm^{-1} at 65°C (100% RH), which were 7.25 and 45.5 times higher than those of only SPEEK membrane (0.898 and 187 mS cm^{-1}). The H_2/O_2 single fuel cell performance was investigated with membrane (SPEEK/**10**), and it was noted that because of increased proton conductivity and decreased electrolyte resistance, the membrane displayed excellent fuel cell application.

Dong and co-workers synthesized three hybrid membranes: pure MIL-101 (**11a**, $[Cr_3O(H_2O)_3(bdc)_3]$, where bdc is terephthalic acid); the ligand-modified MIL-101 (**11b**, S-MIL-101($[Cr_3O(H_2O)_3(STA)_3].nH_2O$, where STA is 2-sulfoterephthalic acid); and the non-volatile acid-loaded MIL-101 [**11c**, acids@MIL-101 (acids are H_2SO_4, H_3PO_4 or CF_3SO_3H)], based on MOFs and chitosan [18]. The TGA results indicated high thermal stability of **11b** and **11c** upto 300°C. In addition, the water uptake was observed to decrease upon incorporation of more functional groups in the composite. Hence, **11a** acquired high water uptake than **11b** and **11c**, which was in accordance with the swelling extent of the membranes. The proton conductivity studies were carried out by alternating current impedance, and it was corroborated that the presence of sulfonic groups in the ligand efficiently slowed down the proton transfer process, which was responsible for low value of conductivity for **11b**. Further, the acid addition in cavities of MOFs found to enhance the proton concentration, and thus, higher value of conduction for **11c** was observed.

Li and group infused phytic acid onto MIL101 (**12**) using vacuum-assisted impregnation method to synthesize phytic@**12** that was further fabricated to generate hybrid PEM [19]. The structural characterization of **12** was done using SEM, TEM, FTIR, XPS, EDS, TGA, and XRD experiments. The leakage ratio was studied to evaluate phytic acid retention property of **12** in water at 80°C and was found to increase with time along with decrease in leakage rate of phytic acid. The TGA analysis suggested high thermal stability upto 300°C. The rigidity of the membrane caused the swelling degree and water uptake to decrease, owing to restricted motion of polymer chains. Further, it was examined that the incorporation of phytic acid to **12** (phytic@**12**) lowered the activation energy and demonstrated higher proton conductivity attributed to strong acidity of phosphate groups, which are responsible for enhanced transfer of protons. Moreover, using the filler content less than 12%, homogeneous dispersion ofthe phytic@**12** was observed in the Nafion matrix. The proton conductivity of hybrid membrane, **Nafion/** phytic@**12**was evaluated up to 0.0608 and 7.63×10^{-4} S cm^{-1} at 57.4% and 10.5% RH, respectively, which was almost 2.8 and 11.0 times higher, when only Nafion membrane was used.

Wu *et al.* reported an imidazole based nanocomposite, which acted as proton carrier molecule and was encapsulated in Fe-MIL-101-NH_2 containing MOF (**13**) [20]. This MOF was further modified to hybrid proton conductive membrane by reacting with 2,6-dimethyl-1,4-phenyleneoxide. From the TGA measurements, the MOF **13** was found to be thermally stable between 200°C and 300°C and enclosed high mechanical strength, which was inversely related to temperature. The protons were transferred from sulphonamide to the surface of MOFs because of hydrogen bonding interactions and further transferred to the next crystal of the MOFs by imidazole moiety to generate the complete conductive pathway (Figure 8.4). The proton transfer was enhanced due to the presence of N-aminopropylimidazole in the membrane, which also participated in the proton conductivity via intermolecular proton transfer at high temperatures. The best proton conductivity value of 2.5×10^{-4} S/cm was obtained by choosing a membrane with a loading of 6% encapsulated **13** at temperature of 25°C and RH value of 25%.

Dybtsev and co-workers developed MOF based structures, **14a-14b**, by incorporating triflic (**14a**) and toluenesulfonic acids (**14b**) into frameworks of chromium and terephthalic acid possessing MOF, MIL-101, and further utilized them to produce hybrid proton-conducting membranes [21]. The TGA analysis revealed the large stability of both **14a** and **14b**

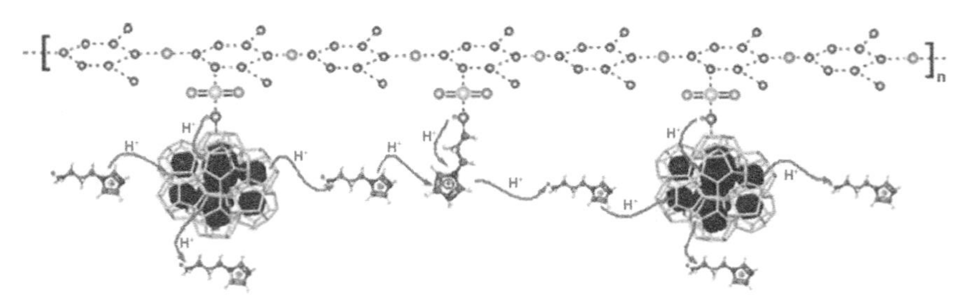

Figure 8.4 A possible pathway for proton conduction in hybrid MOF **13** (reproduced from J. Membrane Sci. 458 (2014) 86–95, [20]).

frameworks toward acidic media, signifying efficient utilization of these compounds in proton exchange materials. The alternating current impedance experiments suggested that, with the increase in temperature upto 60°C, the proton conductivity values of **14a** were enhanced; whereas in the case of **14b**, the conductivity decreased with rise in temperature and fall in humidity. This could be attributed to the decrease in acid and water content at higher temperatures, which hindered the proton transfer. Hence, it could be inferred that the dependency of the proton conduction on the temperature was more pronounced in **14b** than in **14a**. However, **14a** demonstrated the highest proton conductivity value of 0.08 S cm^{-1} at 15% RH and a temperature of 60°C. Further, values of the activation energies of **14a** and **14b** for proton transport were remarkably lower (0.23 eV) in high humidity. This was found to be critical for **14b**, where the enhanced hydration in a humid atmosphere might have fixed the broken network created due to hydrogen-bonding and thus enhanced the proton transfer between adjoiningH_3O^+··H_2O··TsO^- triads via the relay mechanism (Figure 8.5).

Ponomareva and group synthesized two mesoporous MOFs based on MIL-101, **15a** and **15b**, by instilling non-volatile acids H_2SO_4 (**15a**) and H_3PO_4 (**15b**), that resulted in formation of solid materials possessing excellent proton-conduction around moderate temperatures [22]. The TGA analysis unveiled the high stability upto a temperature of 150°C. With the increase in temperature, the proton conduction of **15a** was not affected and decreased only at temperatures above 80°C. In contrast, there was observable gradual increase in the proton conductivity of **15b** at elevated temperatures. The proton conductivities of **15a** and **15b** at temperature of 150°C and RH of 0.13% were measured as1 × 10^{-2} and 3 × 10^{-3} S cm^{-1}, respectively. The higher proton conductivity value for **15a** (higher K_a) as compared to **15b** (lower K_a) powerfully hold the fact that, the higher the

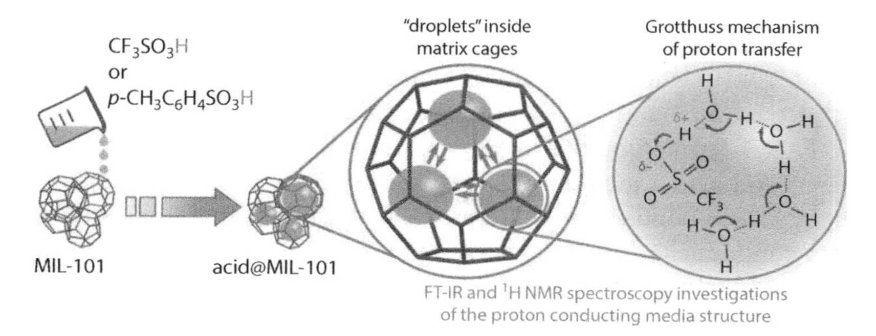

Figure 8.5 A possible pathway for proton conduction of the PEMs in **14a** and **14b** (reproduced from ACS Appl. Mater. Interfaces, 6 (2014) 5161–5167, [21]).

acidity of medium, hybrid material would have better conducting properties. Furthermore, it was observed that moisture content played a vital role in enhancing the proton concentration by dissociating phosphoric acid and by intensifying the proton-conducting routes via the network of hydrogen bonds, which ultimately increased the mobility of the protons. This was confirmed by measuring the proton conductivity for **15b** at 140°C and 5 mol %H_2O in air and was evaluated to be 1.0×10^{-2} S cm^{-1}.

Tran *et al.* synthesized MOF via encapsulating silico-tungstic acid (SiWA) into the MIL-101 (**16**) to obtain hybrid PEM using polyvinyl alcohol (PVA) [23]. It was characterized using FTIR, XRD, TEM, Mass, and TGA experiments. The TGA results indicated three mass loss steps for **16** upto 400°C, and it was observed that the SiWA occupied the porosity of the framework. The proton conductivity of membrane **16**/SiWA-1.9/PVA was measured to be 2.4×10^{-2} S cm^{-1}, which was almost double than SiWA/PVA membrane without **16**. This improvement in proton conductivity could be due to the encapsulation of SiWA into the cavities of **16** that prevented the heteropoly acid from leaching. Moreover, **16** possessed coordinatively unsaturated Metal sites (CUSs), which release numerous hydroxyl groups and protons upon hydrolysis and facilitated the formation of networks of hydrogen bonding interactions, which further enhanced the proton transfer within the pores of **16**. The membrane **16**/SiWA-1.9/PVA exhibited no remarkable change in proton conductivity even after 4 weeks, pointing toward durable proton conducting stability of the composite. Moreover, the electrochemical window of **16** was calculated as 3V, which was suitable for applications in fuel cell.

8.4 Copper Containing MOFs for PEMFCs

Moi *et al.* developed Cu(II)-based MOF composite (Cu-SAT, **17**) and used to form the proton conducting membrane $\{[Cu_2(\mu_2OH)_2$ (NDS) $(T4A)_2]$.2DMF, where NDS is 1,5-naphthalenedisulfonic acid and T4A is 1,2,4-triazol- 4-amine$\}_n$, bearing amine and sulfonate functionalities [24]. PXRD results pointed to the high stability of the membrane **17** in aqueous and acidic medium. In membrane **17**, free amino groups could form hydrogen bonds with the sulfonate oxygen, resulting in its increased proton conductivity. The impedance studies indicated the enhancement of proton conductivity from 1.57×10^{-4} to 5.26×10^{-4} S cm^{-1}, when temperature was raised from 303 to 353 K at fixed RH of 98%, attributable to thermal activation of ions. The E_a value was found to be 0.230 eV for proton conduction that corresponded to the Grotthuss mechanism of proton conduction. Further, to evaluate the practicality of the membrane **17**, simple slurry casting technique was employed to form its four mixed matrix membranes (MMMs) by combining its micro particles with a PVP-PVDF mixture. It was then noticed that the membranes loaded with around 60% of **17** was more effective than one with 100%. This was ascribed to the formation of mixed matrix that provided more facile route for proton transfer by hydrogen bonding via the MOF and water molecules acted as the carrier for conduction to the next MOF particle.

Niluroutu and co-workers reported a copper containing framework ([**18**, Cu-TMA MOF, where TMA is trimesic acid (hydrophilic organic ligand)] along with SPEEK functionality (SP/**18**) for DMFC applications [25]. The thermogravimetric analysis of **18** showed high stability between temperature of 500°C and 600°C. The proton conductivity of **18** was found to be high due to the combined interactions of sulfonic and carboxylic groups. It was observed that the dispersion of **18** in the SPEEK matrix offered easy transport of protons through the carboxylic acid groups and also confined the methanol transport through the MOF pores and therefore resulted in large electrochemical selectivity. In addition, the presence of intermolecular hydrogen bonding between the sulfonic acid group ($-SO_3H$) of the SPEEK and the carboxylic acid ($-COOH$) of **18** provided the better connectivity for enhanced proton transfer.

Wu *et al.* fabricated PEMFC by the salt-recrystallization-fixing a MOF material (HKUST-1) template from porous carbon support material possessing high-specific surface area (**19**), where NaCl crystals acted as closed nanoreactor that facilitated graphitization and encapsulation of chemicals in a confined space led to structure inheritance of carbon materials [26]. Further, using a sol method, Pt nanoparticles of diameters of 3 nm were loaded on the **19**, which exhibited high oxygen reduction reaction (ORR) activity and outstanding electrochemical performance in acidic electrolytes as revealed by the electrochemical studies. With ultralow Pt loading (cathode 0.1 mg/cm^2) and H_2/air testing conditions, the PEMFC fabricated from **19** loaded with Pt-nanoparticles displayed a maximum power density of 780 mW/cm^2.

8.5 Cobalt Containing MOFs for PEMFCs

Liang *et al.* synthesized three crystalline frameworks: **20a-20c** named PPA (4-(3-pyridinyl)-2-aminopyrimidine), **20a**; Cu(PPA)I, **20b**; and Co, PPA, and benzenedicarboxylic acid possessing MOF, **20c**, and proposed an approach to enhance proton conductivities in metal-organic coordination polymers (MOCPs) by utilizing strategy to form MOCP-cocrystal composite due to the potential property of organic ligand [27]. The composites were characterized using CHN, FTIR, PXRD, and TGA analyses. The thermogravimetric studies demonstrated that the frameworks **20a-20c** were stable thermally upto 150°C/200°C. The composites **20a-20c** were also tested for water stability by keeping them in 97% humidity at elevated temperatures and in water at room temperature for months. This experiment showed high tolerance of all the structures **20a-20c** to water and humidity. Further, the proton conductivities of **20a-20c** increased to 1.12×10^{-5}, 9.90×10^{-5}, and 1.20×10^{-4} S cm^{-1} from 1.03×10^{-10}, 4.72×10^{-10}, and 5.92×10^{-10} S cm^{-1}, respectively, as RH was increased to ~97% from ~53%, which indicated that water molecules had a fundamental role in proton transport process. Increased RH led to formation of hydrogen-bonded networks in the hydrophilic region by adsorption of large number of water molecules that provided additional protons or formed increased continuous hydrogen-bonding connectivity to enhance transfer of protons. In addition, with increasing temperatures, the proton conduction was observed to enhance regularly. This was attributed to higher mobility of protons at high temperature in **20a-20b** and, in the case of **20c**, it might be due to the strong hydrogen bonding coordination, which made proton transfer easier.

Wang and co-workers prepared zeolitic imidazolate framework (ZIF)–nanocrystal catalyst precursor by doping Co (0%–30%) into ZIF (**21**), which was later converted via one-step thermal activation into the required Co-N-C catalyst, labeled as *n*Co-NC-heating temperature (e.g., 20Co-NC-1100) and utilized in formation of PEMFCs [28]. The characterization of **21** was done by aberration-corrected electron microscopy and X-ray absorption spectroscopy, which confirmed CoN_4 interactions at an atomic level in the structure. Compared to reversible hydrogen electrode (RHE), the half-wave potential of composite was found to be 0.80 V in acidic media and was found to have attained significant activity and stability for the ORR. **21** was also tested for fuel cell applications, and it was inferred that the catalytic activity and stability could be converted to high-performance cathodes in PEMs. The extremely improved ORR performance was accredited to the well-dispersed CoN_4 active sites present in **21**-derived carbon particles. Out of all synthesized catalysts, 20Co-NC-1100 was found to have maximum ORR activity. The minimum temperature required for ORR activity to be carried out by the catalyst was 800°C. When used in H_2/air cell, the power density achieved using this catalyst was 0.28 Wcm^{-2} and, for the H_2/O_2 fuel cell, it was 0.56 W cm^{-2}.

8.6 Iron Containing MOFs for PEMFCs

Ru *et al.* reported IL-impregnated iron-based MOF to achieve (IL@NH_2-MIL-101, **22**) and introduced into polymer matrix containing sulfonated poly(aryleneether ketone) with the carboxyl pendant group (SPAEK), where SPAEK acted as nanofillers to fabricate hybrid proton exchange membrane via formation of amide linkage using amino group of **22** and pendent carboxyl group of SPAEK [29]. The TGA studies revealed high thermal, dimensional, and oxidative stability attributed to the cross-linked structure of **22**. The proton conduction experiments suggested large value for the conductivity (0.184 S cm^{-1}) than normal SPAEK membrane. This was corroborated by the fact that incorporation of IL to the membrane offered new high-speed pathways for proton conduction (Figure 8.6). Moreover, **22** showed high single-cell efficiency, where the peak power density value was measured to be 37.5 mW cm^{-2}, which was nearly 2.3-fold to only SPAEK. The ion exchange capacity results indicated low leakage ratio for **22**, which pointed toward its good performance as PEMFCs.

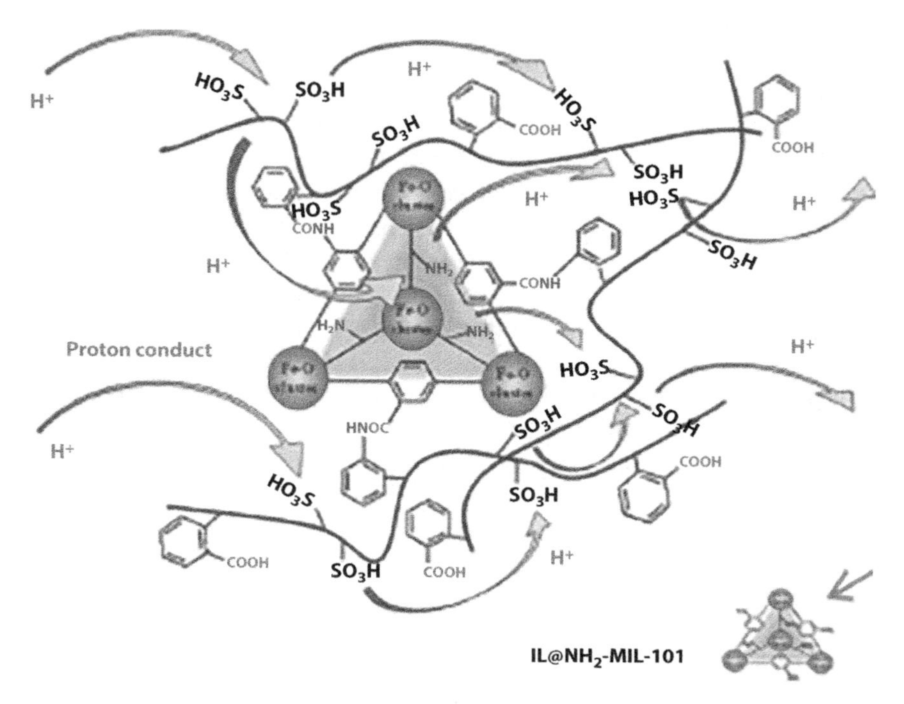

Figure 8.6 Schematic illustration of Proton conduction in **22** (reproduced from ACS Appl. Mater. Interfaces, 11 (2019) 31899–31908 [29]).

In addition, MOF **22** was found to have cross-linking between the nano-filler and matrix, so this factor improved the methanol resistance of PEM without altering proton conduction. Further, the utility of membrane **22** was evaluated in DMFC by single-cell performance investigation carried out at 80°C, which signified high selectivity and conductivity for fuel cell applications.

Wang and co-workers synthesized iron and zinc possessing MOF mixture [1MIL/40ZIF-1000, **23**] via a facile thermal activation method [30]. The MOF **23** was characterized by XRD, Raman spectroscopy, FESEM, HRTEM, XPS, and XAS experiments. It was observed that **23** possessed excellent durability in acidic medium and very low methanol permeability. In addition, the carbon graphitization in the MOF structure, **23**, facilitated fast charge transfer and increased stability against corrosion, attributable to enhanced ORR activity. Further, LSV results indicated high ORR catalytic activity and stability in Li–O_2battery attained by membrane **23**. These outcomes established MOF **23** as suitable candidate for PEMFC utilization.

Wang and group developed Fe(II)-based MOF (**24**) possessing chains of alternate bistriazolate-*p*-benzoquinone anions and iron(II) cations with four coordinated water molecules [31]. The study of magnetic properties disclosed paramagnetism nature of **24** with magnetic moment value of 5.02 μ_B. The unaffected XRPD patterns specified a high moisture durability of **24** in water over extended period of time. The proton conductivity of both the powder and the single-crystal samples of **24** were found to enhance with increasing humidity, which designated a water-mediated proton transport. In addition, at higher temperatures, the conductivity of **24** was increased due to higher proton mobility in the framework. The proton conductivity was estimated as 3.3×10^{-3} S cm^{-1} at 22°C and 94% of RH.

Zhang *et al.* synthesized different Fe containing MOFs (**25a-25c**) based on imidazoles (Im) named as Fe-MOF (**25a**), Im@Fe-MOF (**25b**) with physically adsorbed imidazole, and Im-Fe-MOF (**24c**) comprised of chemically coordinated imidazole molecules, to study the effect of their structural arrangements on proton conductivity [32]. The TGA analysis confirmed high thermal stability of all three composites. The effect of RH was evaluated on the proton conductivity, and it was noted that, at 40% RH, the conductivity of **25a** was very less, whereas with increasing RH to 98%, proton conductivity of **25a** increased. In addition, at this humidity level, the value of conductivity for **25b** was estimated as 8.41×10^{-5} S/cm, while the conductivity of **25c** was much high up to 2.06×10^{-3} S/cm. With increase in temperature, the conductivity was observed to improve for **25a** and **25b** and at 60°C, **25c** had the highest conductivity value among all the

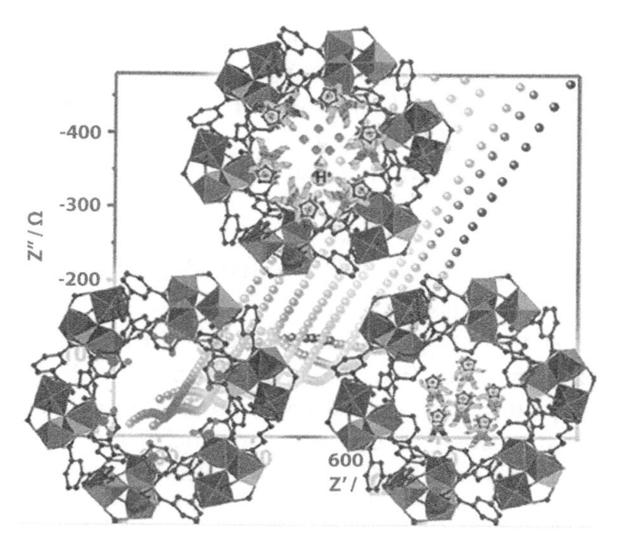

Figure 8.7 Changes in impedance spectrum of **25c** at different temperatures with RH of 98%. (Reproduced from J. Am. Chem. Soc., 139 (2017) 6183–6189, [32]).

MOFs (Figure 8.7). Thus, the coordinated imidazole not only increased the stability of membrane but also enhanced the proton concentration.

Yang and group prepared nanoporous PtFe nanoparticles supported on N-doped porous carbon sheets through an easy *in situ* thermal reaction of a Pt-modified Fe-based MOF (**26**) [33]. The linear sweep voltammetry studies of **26** corroborated lesser catalytic activities to ORR in acidic medium. The MOF, **26**, was also tested for utility in the PEMFC and the result unveiled that its eletrocatalytic activity was better than 20% Pt/C. Further, the vacant framework of Pt-modified **26** could efficiently reduce the effect of lesser binding sites, where the coordination power between oxygenated species and Pt atoms might be very strong. In addition, it could deteriorate the adsorption of non-reactive oxygenated species on the Pt surface, therefore preserving high electrocatalytic activity.

Afsahi *et al.* developed Fe-based MOF, **27**, to be utilized as the sole precursor for ORR electrocatalyst [34]. The structure was characterized by XRD, BET, and XPS experiments. From the thermal analysis, it was observed that, with the increase in temperature from 900 to 1,000°C, significant decrease of pore volume and surface area was noticed, signifying a large contraction of the microporosity. In addition, it was found that the total nitrogen content in **27** was reduced with the increase in temperature. This was attributed to the considerable provoking for the removal of nitrogen atoms by the Fe metal at elevated temperature. The membrane **27** was also examined as cathode in H_2/air in single fuel cell. Among various thermolyzed electrocatalysts, C700/950 gave best results by demonstrating an open-circuit voltage (OCV) of 0.945 V and, at 0.391 V, a maximum power density of 0.302 W cm^{-2} was reached.

8.7 Nickel Containing MOFs for PEMFCs

Kadirov and group developed an MOF (**28**) from polygalacturonates of sodium pectin with nickel (PG-NaNi) for testing in PEMFC [35]. Using inductively coupled plasma optical emission spectrometer, it was found that 25% of sodium ions were replaced by nickel ions in prepared **28**. Further, using **28**/Vulcan XC-72 as H_2/O_2 PEMFC anode catalyst, the membrane electrode assembly was built and tested. The maximum current density and the maximum power density values of the synthesized catalyst were found to be 5.2 mA cm^{-2} and 1.5 mW cm^{-2}, respectively. These values were found to be sufficiently high for PEMFC having an anode devoid of a platinum metal group catalyst and a platinum-based cathode.

Phang *et al.* reported pH-dependent Ni-based nano porous MOF [**29**, Ni_2(dobdc)] with 1D hexagonal channels of 11 Å diameter, which upon soaking in sulfuric acid solutions at different pH values resulted in formation of new proton-conducting frameworks, H^+@**29** [36]. The synthesis was carried out by microwave assisted solvothermal synthesis, reducing the preparation time from 3 days to 15 min unlike conventional solvothermal synthesis. The chemical stability of **29** was studied by PXRD, suggesting that it was stable up to pH 1.8 and the framework collapsed at pH 0.8. The proton conductivity of mentioned **29** was found to be 1.4×10^{-4} S cm^{-1} at 80°C and 95% RH, which greatly enhanced to 2.2×10^{-2} S cm^{-1} when H^+@**29** was used (at pH 1.8) under similar conditions. The *Ea* value was found to lie between 0.12 and 0.20 eV, pointing to the Grotthuss mechanism, which was lower than Nafion (0.22 eV). Hence, the protonated water clusters within the pores of H^+@**29**contributed significantly in the conduction process.

8.8 Platinum Containing MOFs for PEMFCs

Afsahi *et al.* reported the synthesis of Pt-based MOF (**30**), which acted as an electrocatalyst precursor [37]. To prepare the electrocatalyst, **30** was heated at the varied temperatures and designated as C_1 (700°C), C_2 (800°C),

C_3 (950°C), and C_4 (1,050°C) in argon atmosphere that led to the production of Pt nanoparticles and production of carbon from organic moieties in **30** (Figure 8.8). TEM images revealed the amorphous nature with porous "foam-like" texture of carbonized MOF samples and dark spherical shape of Pt nanoparticles. These electrocatalysts were further utilized to make membrane-electrode assemblies and tested them as both anode and cathode in a H_2/O_2 single cell. The anode fabricated from C_3 electrocatalyst served as the most promising electrocatalyst out of four batches that displayed an OCV of ~970 mV and power density of 0.58 W mg^{-1} Pt, which was almost comparable to the commercial electrode having OCV of ~ 960 mV and power density of 0.64 W mg^{-1} Pt at 0.6 V. However, the cathode electrode prepared from C_3 showed some deviations from the commercial electrode. At 0.6 V of cell voltage, the power density and current density values of the cathode prepared from C_3 were found to be 0.38 W mg^{-1} Pt and 317 mA cm^{-2}, respectively, which were much lesser than that of the commercial electrode (power density of 0.64W mg^{-1} Pt and current density of 537 mA cm^{-2}).

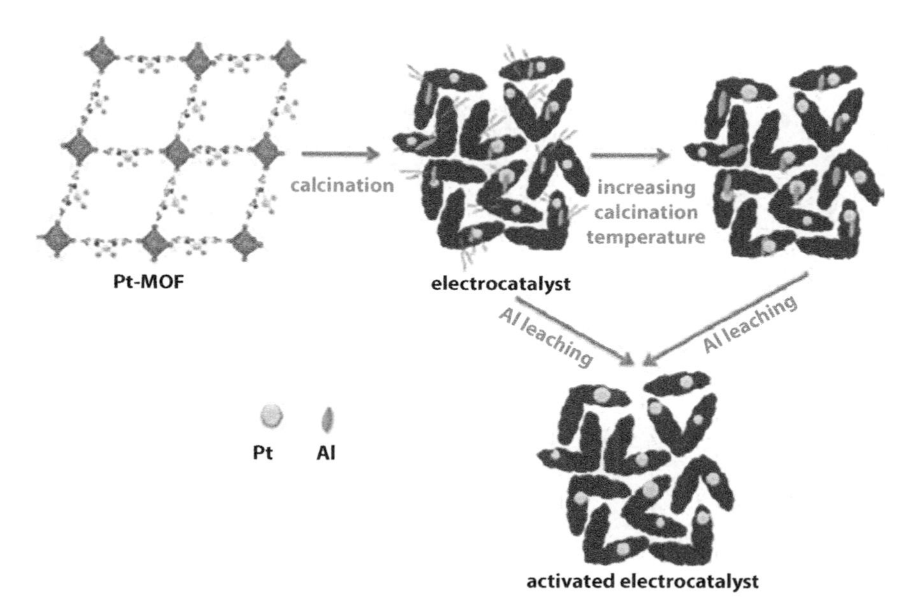

Figure 8.8 Illustration of formation of electrocatalyst from Pt-MOF, **30** (reproduced from *J. Power Sources*, 239 (2013) 415–423, [37]).

8.9 Zinc Containing MOFs for PEMFCs

Deng and co-workers reported self-humidifying PEMs for promoting water management during fuel cell operation for which perfluorosulfonic acid (PFSA) resin was suspended into the MOF (ZIF-8, **31a**, and MIL-101, **31b**)–coated stainless steel meshes, which helped in continuous proton transport (Figure 8.9) [38]. The **31**-coated stainless steel meshes were fabricated using photochemical etching method and characterized using SEM and XRD techniques. The pores of these meshes were modified with polydopamine (PDA) or polydopamine/polyethyleneimine (PDA/PEI) beforehand for promoting uniform and continuous growth of layers of **31a** and **31b**, respectively, on the mesh surface. The water uptake was found to be 8%, 9%, and 6%; whereas power densities at room temperature were measured to be 220, 190, and 140 mW cm^{-2} for PFSA/**31a**, PFSA/**31b** composite membranes, and Nafion-117, respectively, signifying better performance of PFSA/**31a** and PFSA/**31b** composite membranes as compared to Nafion-117. The excellent performance was attributed to the enhanced properties of layers of PFSA/**31a** and PFSA/**31b**, such as good water adsorption capacity and thermal stability, by which proton transfer and water management were facilitated under high temperature and dry conditions.

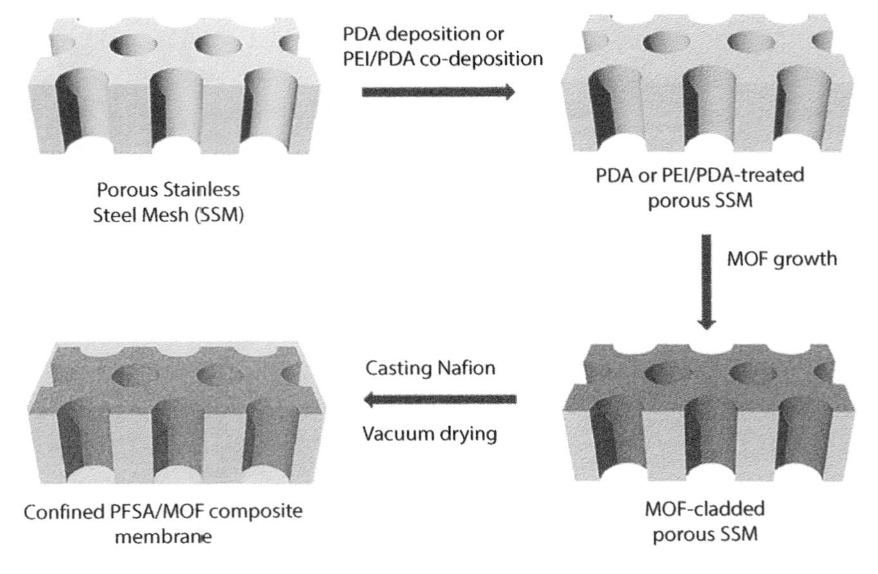

Figure 8.9 Illustration of formation of confined PFSA/**31a** or **31b** composite membrane (reproduced from Catalysis Today, 331 (2019) 12–17, [38]).

Cai and group synthesized and characterized novel MOF (**32**) by X-ray crystallography, PXRD, TGA, and SEM that showed superior stabilities in aqueous and acidic media and characterized [39]. To employ the MOF **32** in PEMFC, it was blended with PVA to produce a composite membrane [**32**@PVA-X (X = 5%, 10%, 20%, 30%, 40%, and 50%, the mass percentage of **32**)]. PVA was chosen because of its hydrophilicity and ability to act as a linker between particles of **32** that helped in proton conduction. The maximum proton conductivity of **32** was found to be 1.62×10^{-3} S/cm at 80°C. The conductivity of pure PVA was found to be 6.2×10^{-3} S/cm and, as the addition of **32** was carried out to it, the proton conduction value enhanced to 1.25×10^{-3} S/cm at 50°C up to 10% (mass percentage) addition of **32**. The proton conduction mechanism was found to be Grotthuss in the case of **32**; whereas in the case of composite membrane, both Grotthuss and vehicle mechanisms were responsible for proton conduction (Table 8.2).

Ye *et al.* obtained anhydrous proton conducting flexible MOFs (**33b-33d**) by incorporating hydroquinone (**33b**), cyclohexanol (**33c**), and butanol (**33d**) to FJU-31 (**33a**), which could operate at wide ranges of temperature and showed breathing effect [40]. The structural analysis of all the synthesized compounds was done by single-crystal X-ray diffraction studies. The proton conductivity of **33b** was found to be the maximum out of **33b-33d**; whereas it was negligible for **33a**. The proton conductivity for **33b** at −30°C was found to be 1.49×10^{-5} S/cm, which was enhanced to 2.65×10^{-4} S/cm at 125°C (Table 8.2). The activation energies (*Ea*) were found to be 0.18, 0.19, and 0.25 for **33b-33d**, respectively, indicating Grotthuss mechanism for proton conduction.

Lai *et al.* invoked the concept of host-guest chemistry to fabricate **Fe-mIm** nanoclusters (that acted as guest)@zeolite imidazole framework-8 (**ZIF-8**) (that acted as host) and its transformation to MOF-derived Fe-N/C electrocatalyst (**34**) for ORR at cathode in acidic media (Figure 8.10) [41].

Table 8.2 Zinc-MOF.

MOF	Conductivity	Humidity and Temperature	Ea	Mechanism	Ref.
JUC-200, **32**	1.62×10^{-3} S cm^{-1}	at 80°C	0.23 eV	Grotthus	[39]
33b (Hq)	3.24×10^{-6} S cm^{-1} and	−40°C (anhydrous)	0.18	Grotthuss	[40]
33c (Ch)	1.17×10^{-6}	--	0.19	Grotthus	[40]
33d (Bu)	---	--	0.25	Grotthus	[40]

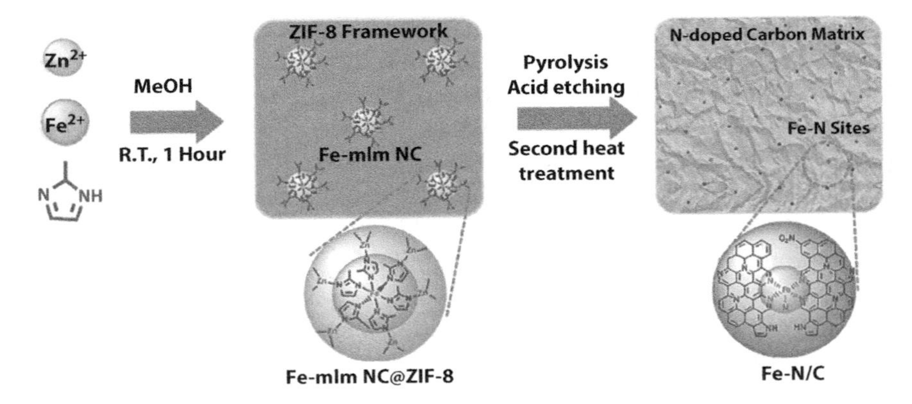

Figure 8.10 Schematic representation of fabrication of **MOF**-derived Fe-N/C catalyst (**34**) involving Host–Guest chemistry strategy (reproduced from ACS catalysis, 7 (2017) 1655–1663, [41]).

The host (**ZIF-8**) network exhibited considerable confinement effect for the guest (**Fe-mIm** NCs) at high temperature (900°C) under N_2 atmosphere, by which effective transformation into highly active Fe-Nx sites could have taken place. This led to the formation of different types of two- to five-coordinated configurations of Fe-Nx sites on the porous carbon matrix. The final Fe-N/C catalysts were obtained after acid etching and second heat treatment. It was revealed by electrochemical tests and supported by density functional theory calculations that, out of different catalysts, the five-coordinated Fe-Nx sites promoted the reaction rate to significant extent in acid media, because of the low ORR energy barrier and small adsorption energy of intermediate OH on these sites.

8.10 Zirconium Containing MOFs for PEMFCs

Wang *et al.* attached MOF nanoparticles (UiO-66-NH_2-AA, **35a-35c**) with amino acids [glutamate (**35a**), lysine (**35b**), and threonine (**35c**)], which were further incorporated into sulfonated polysulfone (SPSF) matrix (**35**/SPSF) via solution casting method that resulted in formation of the nanocomposite polymer exchange membranes (Figure 8.11) [42]. To study the morphologies of samples, SEM was employed; FT-IR and XPS techniques were used for the confirmation of production of target compounds; whereas thermal stability was studied using TGA. Results showed that, among different MOFs, glutamate-functionalized MOF (**35a**-6/SPSF; here, 6 denoted the **35a** to SPSF weight ratio) displayed promising proton conductivity of

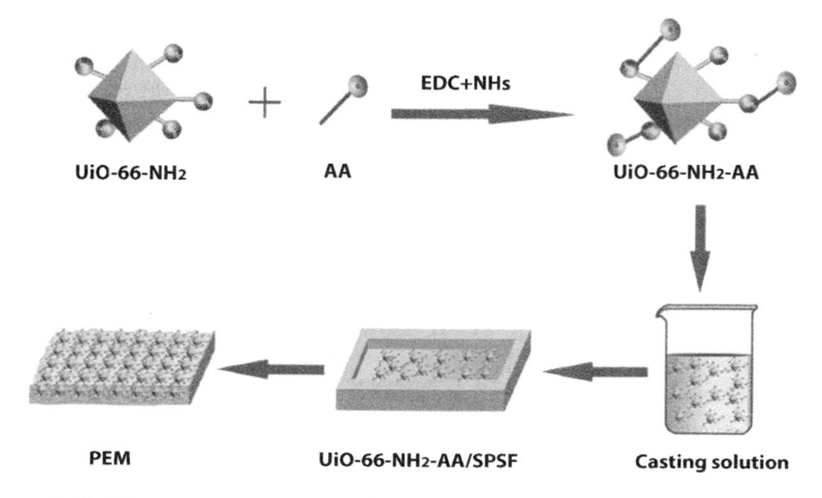

Figure 8.11 Schematic representation of formation of nanocomposite PEM using **35a-35c**/SPSF (reproduced from *Int. J. Hydrogen. Energy*, 46 (2021) 1163–1173, [42]).

0.212 S cm^{-1} at 80°C and RH of 100%. In addition, upon increasing the wt% of **35a** from 2 to 6, the methanol permeability coefficient of the nanocomposite membrane reduced from 8.5×10^{-7} to 5.8×10^{-7} $cm^2 s^{-1}$. In the methanol/O_2 fuel cell test, the **35a**-6/SPSF displayed a maximum current density of 70.45 mW cm^{-2} at 60°C.

Mukhopadhyay and co-workers synthesized nanocomposite membrane by adding post synthetically modified MOFs (**36a** and **36b**) into aryl-ether type polybenzimidazole using solution casting blending method [43]. The respective wt% of MOFs added for **36a** were 3, 5, 7, and 10 and, for **36b**, were 7 and 10. To prepare PEMs, the polybenzimidazole and **36a/36b** nanofiller loaded membranes were doped with phosphoric acid. Characterization of prepared nanocomposite membranes was done by FT-IR, PXRD, FESEM, TEM, TGA, DMA, and UTM. The proton conductivities of the composite membranes for **36a**-10% and **36b**-10% at 160°C and anhydrous environment were measured to be 0.29 and 0.308 S cm^{-1}, respectively, and were higher than that of individual constituents.

Zhang and group reported two kinds of Zr-based functionalized MOFs, Im-**37** and Im@**37**, where MOF **37** was used as a template [44]. In Im@**37**, free imidazole molecules were present in the pores of the **37**; whereas in Im-**37**, imidazole molecules were chemically coordinated. The hybrid membranes were fabricated using synthesized **MOFs** as fillers and sulfonated poly(aryleneether ketone sulfone) (C-SPAEKS) as organic matrix. The characterization of all of the three compounds, Im-**37**, Im@**37**, and

C-SPAEKS/Im@**37**, was carried out by FT-IR, PXRD, XPS, and TGA. Results from the PXRD technique showed that Im@**37** and Im-**37** exhibited excellent stabilities in both water and acid. The proton conductivity values of C-SPAEKS/Im@**37** increased from 0.050 to 0.068 S cm^{-1} under 100% RH and 90°C and that of C-SPAEKS/Im-**37** increased from 0.071 to 0.128 S cm^{-1} at the same conditions upon increasing the content of Im@**37** and Im- **37** from 2% to 4%. The hybrid membranes showed much higher proton conductivities as compared to that of pure C-SPAEKS (0.024 S cm^{-1} at 90°C) (Figure 8.12). C-SPAEKS/Im-**37**-4% was used for DMFCs because of high proton conductivity, and it showed the OCV of 0.75 V with the maximum power density of 15.4 $mWcm^{-2}$ at 80°C.

Li *et al.* reported the imidazole, amino, and sulfonic acid groups possessing Schiff-based MOF, **38**, formed by the covalent connection of the residual uncoordinated amino groups with the aldehyde groups of imidazole-2-carboxaldehyde [45]. The value for the proton conduction of this compound was calculated to be 1.54×10^{-1} S/cm (80°C, 98% RH) and, using its PVDF-PVP composite material, a high proton conduction of 1.19×10^{-2} S/cm was observed under the same conditions. XRD, SEM,

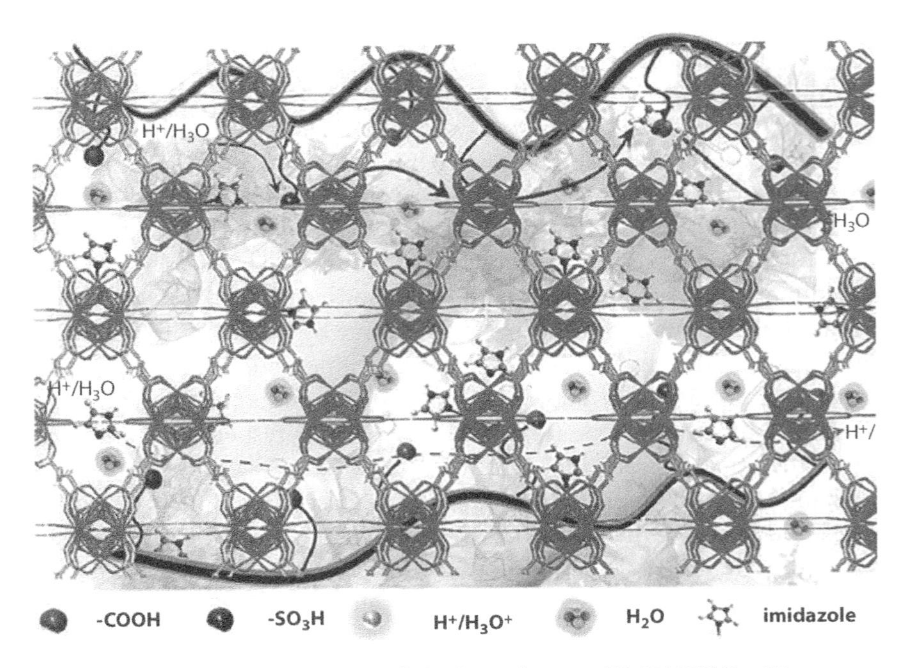

Figure 8.12 The proton conduction in hybrid membranes of C-SPAEKS/Im-**37** (reproduced from J. Membrane Sci., 607 (2020) 118194, [44]).

TEM, FT-IR, and EDX techniques were employed for studying morphologies and characterization of the compounds. In H_2/O_2 fuel cells, the maximum OCV was found to be 0.78 V at 80°C and 98% RH with the maximum power density of 17.5mWcm^{-2}.

Zhang *et al.* reported PEMs synthesized by blending MOF-801 (**39**) with PVDF-PVP matrix, which demonstrated the proton conductivity of 1.84×10^{-3} S cm^{-1} at 52°C and 98% RH [46]. The MOF, **39**, was stable toward hydrochloric acid, dilute sodium hydroxide aqueous solutions, and boiling water. The phase purity of samples was studied by PXRD. Characterization was done using FT-IR, XPS. The membrane was incorporated into H_2/O_2 fuel cell for tests, depicting OCV of 0.95 V at 30°C and 100% RH, with maximum power density of 2.2 mW cm^{-2}.

Gui *et al.* reported a MOF, **40**, synthesized via ionothermal method with stable anhydrous proton conductivity value of 1.45×10^{-3} S cm^{-1} at 180°C [47]. NPD, SCXRD, and PXRD were used for structural analysis. TGA was done to study thermal stability. **40** with PVDF was further employed as the membrane into a H_2/O_2 fuel cell, showing an electrical power density of 12 mW cm^{-2} at 180°C with OCV of 0.72 V. For DMFC applications, maximum OCV was found to be 0.44 V with the maximum power density of 0.13 mW cm^{-2}.

Lo and co-workers reported a MOF (**41**), within which two different loading amounts of imidazole (HIm) were anchored to form $HIm_9 \subset$**41** and $HIm_{11} \subset$**41** and were further utilized for studying the proton transfer process [48]. SCXRD, PXRD, and TGA were employed for structural analysis and thermal stability. The proton conductivity values at 85% RH and 70°C were calculated to be 5.9×10^{-3} and 0.153×10^{-3} S/cm, respectively, for $HIm_{11} \subset$**41** and $HIm_9 \subset$**41**, displaying better performance of $HIm_{11} \subset$**41** than $HIm_9 \subset$**41**. In contrast, the proton conductivity for parent **41** was found to be 6.65×10^{-6} S/cm at 98% RH and 70°C. The *Ea* were found to be 0.27, 0.44, and 0.47 eV for $HIm_{11} \subset$**41**, $HIm_9 \subset$**41**, and **41**, respectively, indicating the Grotthuss-type mechanism for proton conduction in $Him_{11} \subset$**41**.

Rao and group synthesized composite membranes by single doping and co-doping of the two functionalized MOFs: UiO-66-SO_3H (**42a**) and UiO-66-NH_2 (**42b**) in Nafion membrane [49]. The structural analysis and morphologies were studied using XRD, TGA, and FE-SEM. It was found that the proton conductivity after co-doping of these two MOFs, **42a** + **42b**/Nafion-0.6, reached up to 0.256 S cm^{-1}at 90°C and RH of 95%, which was ~ 1.17 times higher than that of the recast Nafion (0.118 S cm^{-1}). This enhanced proton conductivity could be attributed to the synergistic effect of $-SO_3H$ and $-NH_2$ groups present in **42a** and **42b**, that led to the generation of more consecutive hydration channels in the composite PEM

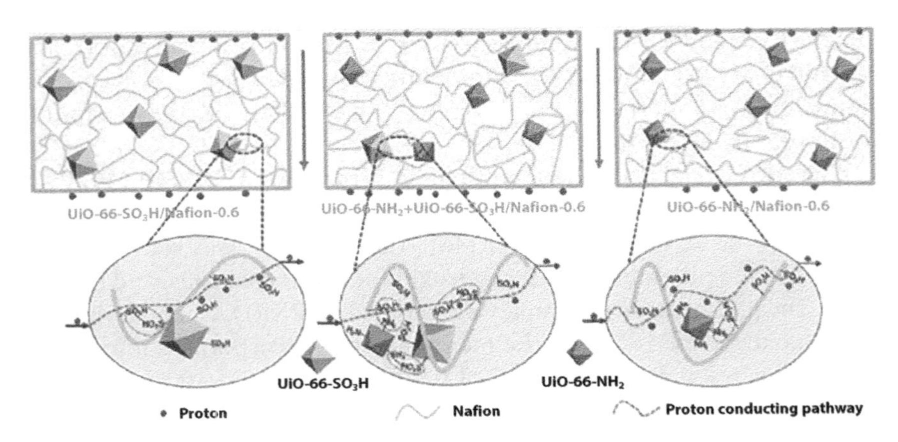

Figure 8.13 The proton conduction in **42a**/Nafion-0.6, **42a+42b**/Nafion-0.6, and **42b**/Nafion-0.6 (reproduced from ACS Appl. Mater. Interfaces, 9 (2017) 22597–22603, [49]).

(Figure 8.13). The proton conductivity of the co-doped PEM at temperature of 90°C and 95% RH remained unaltered throughout 50 h of testing. The methanol crossover of co-doped PEM also decreased as compared to Nafion.

Rao and co-workers attached UiO-66-NH_2(**43**) onto graphene oxide (GO) surfaces to obtain GO@**43** and characterized the morphology and structure of prepared composite membrane using FE-SEM, TGA, XRD, and FT-IR techniques [50]. The proton conductivity values of the membrane were found to be 0.303 and 3.403×10^{-3} S cm^{-1} under hydrated (90°C and 95% RH) and anhydrous conditions, respectively, which were about 1.57 and 1.88 times higher than that of the recast Nafion (0.118 and 1.182 $\times 10^{-3}$ S cm^{-1}, respectively) (Figure 8.14). GO@**43** was then incorporated into Nafion to obtain GO@**43**/Nafion-x. Out of all composite membranes, GO@**43**/Nafion-0.6 showed the maximum proton conduction of 303 S cm^{-1} at temperature of 90°C and RH of 95%, which was nearly 1.57 times higher than that of the recast Nafion membrane (0.118 S cm^{-1}). In addition, methanol permeability of GO@**43** reduced significantly, which could be explained by the trapping of methanol by **43** pores and barrier effect caused by the two-dimensional GO.

Luo *et al.* synthesized MOF-808 (**44**) and studied its proton conductance by varying temperature, where the maximum value of proton conduction was measured to be 7.58×10^{-3} S·cm^{-1} at 42°C and 99% RH [51]. Further composite membrane was fabricated using the MOF, **44**, along with poly(vinylidene fluoride) (PVDF) labeled as **44**@PVDF-X [where X represented the mass percentage of **44** (as X%) in **44**@PVDF-X and

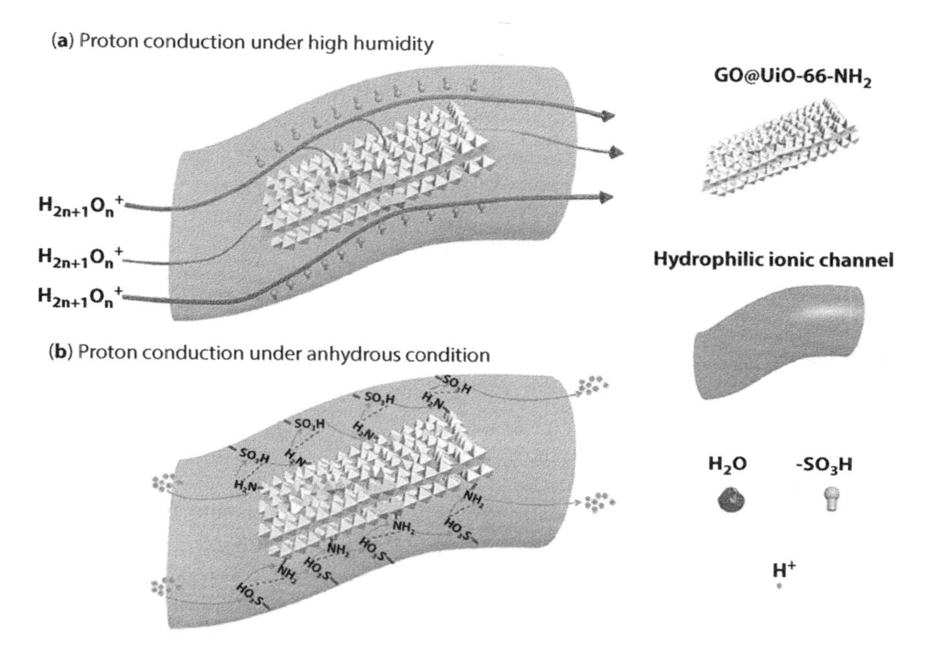

Figure 8.14 Schematic representation of proton conduction under (a) high humidity and (b) anhydrous conditions (reproduced from J. Membrane Sci., 533 (2017) 160–170, [50]).

X = 10%, 25%, 40%, and 55%]. Out of all the compositions, the highest proton conductivity was found for **44**@PVDF-55, which was 1.56×10^{-4} S cm^{-1} in deionized water at 65°C. In addition, the activation energy (*Ea*) values for **44**@PVDF-10, **44**@PVDF-25, **44**@PVDF-40, and **44**@PVDF-55 were measured to be 0.167, 0.145, 0.172, and 0.167 eV, respectively, suggesting occurrence of proton conduction via Grotthuss mechanism in all of the composite membranes.

8.11 Conclusions and Future Prospects

Fuel cell technology needs to be developed to replace non-renewable energy sources. The number of applications of fuel cells can be seen in diverse filed such as data processing; telecommunication; high-technology manufacturing facilities; minimization or elimination of emissions from urban areas, industrial facilities, airports, vehicles, etc.; and the management of biological waste gases. The important component of fuel cells is the PEM that is responsible for proton transfer across it and separates the fuel from the oxidant. Both these factors affect the performance and life of the battery. The MOFs are emerging as promising substitutes of the commercial Nafion membrane, which has certain limitations like reduced proton conductivity in anhydrous conditions and the cost factor, thus limiting its usage. This chapter focuses upon the variety of MOFs that have been designed for the use as PEMFCs. The knowledge gained from this recent progress will pave a way to design more sophisticated MOFs possessing promising proton conduction properties.

References

1. Petreanu, I., Dragan, M., Badea, S.L., pp. 1–21, Thermodynamics and energy engineering, InTech Open, 2020.
2. Pollet, B.G., Franco, A.A., Su, H., Liang, H., Pasupathi, S., F. Barbir, A. Basile, T.N. Veziroğlu (Eds.), Proton exchange membrane fuel cells, in: Compendium of Hydrogen Energy, pp. 3–56, Elsevier, 2016.
3. Zhao, Y., Li, X., Wang, S., Li, W., Wang, X., Chen, S., Chen, J., Xie, X., Proton exchange membranes prepared via atom transfer radical polymerization for proton exchange membrane fuel cell: Recent advances and perspectives. *Int. J. Hydrogen Energy*, 42, 30013, 2017.
4. Liu, Q., Li, Z., Wang, D., Li, Z., Peng, X., Liu, C., Zheng, P., Metal organic frameworks modified proton exchange membranes for fuel cells. *Front. Chem.*, 8, 1, 2020.
5. Alaswad, A., Palumbo, A., Dassisti, M., Olabi, A.G., Fuel cell technologies, applications, and state of the art. A Reference Guide. *Mater. Sci. Eng.*, 1, 2016.
6. Alberti, G., Membranes, in: *Encyclopaedia of Electrochemical Power Source*, J. Garche, C. Dyer, P. Moseley, Z. Ogumi, D. Rand, B. Scrosati, (Eds.), pp. 650–666, Elsevier, 2009.
7. Ioroi, T., Siroma, Z., Yamazaki, S.-I., Yasuda, K., Electrocatalysts for PEM fuel cells. *Adv. Energy Mater.*, 9, 1801284, 2019.
8. Feng, L., Hou, H.B., Zhou, H., UiO-66 derivatives and their composite membranes for effective proton conduction. *Dalton Trans.*, 49, 17130, 2020.
9. Wang, L., Deng, N., Wang, G., Ju, J., Cheng, B., Kang, W., Constructing amino-functionalized flower-like metal–organic framework nanofibers in sulfonated poly(ether sulfone) proton exchange membrane for simultaneously enhancing interface compatibility and proton conduction. *ACS Appl. Mater. Interfaces*, 11, 39979, 2019.
10. Ahmadian-Alam, L. and Mahdavi, H., A novel polysulfone-based ternary nanocomposite membrane consisting of metal-organic framework and silica nanoparticles: as proton exchange membrane for polymer electrolyte fuel cells. *Renew. Energy*, 126, 630, 2018.

11. Tsai, C.H., Wang, C.-C., Chang, C.-Y., Lin, C.-H., Chen-Yang, Y.W., Enhancing performance of Nafion®-based PEMFC by 1-D channel metal-organic frameworks as PEM filler. *Int. J. Hydrogen Energy*, 39, 15696, 2014.
12. Liu, S.S., Han, Z., Yang, J.S., Huang, S.Z., Dong, X.Y., Zang, S.Q., Sulfonic groups lined along channels of metal–organic frameworks (MOFs) for super-proton Conductor. *Inorg. Chem.*, 59, 396, 2020.
13. Anahidzade, N., Abdolmaleki, A., Dinari, M., Tadavani, K.F., Zhiani, M., Metal-organic framework anchored sulfonatedpoly(ether sulfone) as a high temperature proton exchange membrane for fuel cells. *J. Membr. Sci.*, 565, 281, 2018.
14. Yang, F., Xu, G., Dou, Y., Wang, B., Zhang, H., Wu, H., Zhou, W., Li, J.-R., Chen, B., A flexible metal–organic framework with a high density of sulfonic acid sites for proton conduction. *Nat. Energy*, 2, 877, 2017.
15. Chen, H., Han, S.-Y., Liu, R.-H., Chen, T.-F., Bi, K.-L., Liang, J.-B., Deng, Y.-H., Wan, C.-Q., High conductive, long-term durable, anhydrous proton conductive solid state electrolyte based on a metal-organic framework impregnated with binary ionic liquids: Synthesis, characteristic and effect of anion. *J. Power Sources*, 376, 168, 2018.
16. Li, X.M., Dong, L.-Z., Li, S.-L., Xu, G., Liu, J., Zhang, F.M., Lu, L.S., Lan, Y.Q., Synergistic Conductivity Effect in a Proton Sources-Coupled Metal–Organic Framework. *ACS Energy Lett.*, 2, 2313, 2017.
17. Zhang, B., Cao, Y., Li, Z., Wu, H., Yin, Y., Cao, L., He, X., Jiang, Z., Proton exchange nanohybrid membranes with high phosphotungstic acid loading within metal-organic frameworks for PEMFC applications. *Electrochim. Acta*, 240, 186, 2017.
18. Dong, X.Y., Li, J.J., Han, Z., Duan, P.G., Li, L.K., Zang, S.Q., Tuning the functional substituent group and guest of metal–organic frameworks in hybrid membranes for improved interface compatibility and proton conduction. *J. Mater. Chem. A*, 5, 3464, 2017.
19. Li, Z., He, G., Zhang, B., Cao, Y., Wu, H., Jiang, Z., Tiantian, Z., Enhanced proton conductivity of nafion hybrid membraneunder different humidities by incorporating metal-organic frameworks with high phytic acid loading. *ACS Appl. Mater. Interfaces*, 6, 9799, 2014.
20. Wu, B., Ge, L., Lin, X., Wu, L., Luo, J., Xu, T., Immobilization of N-(3-aminopropyl)-imidazole through MOFs in proton conductive membrane for elevated temperature anhydrous applications. *J. Membr. Sci.*, 458, 86, 2014.
21. Dybtsev, D.N., Ponomareva, V.G., Aliev, S.B., Chupakhin, A.P., Gallyamov, M.R., Moroz, N.K., Kolesov, B.A., Kovalenko, K.A., Shutova, E.S., Fedin, V.P., High proton conductivity and spectroscopic investigations of metal–organic framework materials impregnated by strong acids. *ACS Appl. Mater. Interfaces*, 6, 5161, 2014.
22. Ponomareva, V.G., Kovalenko, K.A., Chupakhin, A.P., Dybtsev, D.N., Shutova, E.S., Fedin, V.P., Imparting high proton conductivity to a metal–organic

FrameworkMaterial by controlled acid Impregnation. *J. Am. Chem. Soc.*, 134, 15640, 2012.

23. Tran, V.M.H. and Aguey-Zinsou, K.F., Encapsulation of silicotungstic acid into chromium (III) terephthalate metal–organic framework for high proton conductivity membranes. *Res. Chem. Intermediat.*, 47, 61, 2021.
24. Moi, R., Ghorai, A., Banerjee, S., Biradha, K., Amino- and sulfonate-functionalized metal–organic framework for fabrication of proton exchange membranes with improved proton conductivity. *Crys. Growth Des.*, 20, 5557, 2020.
25. Niluroutua, N., Pichaimuthu, K., Sarmaha, S., Dhanasekaran, P., Shuklaa, A., Unnia, S.M., Bhat, S.D., Copper-trimesic acid metal-organic framework incorporated sulfonatedpoly(ether ether ketone) based polymer electrolyte membrane for directmethanol fuel cells. *New J. Chem.*, 42, 16758, 2018.
26. Wu, M., Xing, Y., Zeng, L., Guo, W., Pan, M., Salt-recrystallization preparation of metal organic framework derived porous carbon support for highly-efficient proton exchange membrane fuel cell. *Int. J. Energy Res.*, 45, 2334, 2021.
27. Liang, X., Cao, T., Wang, L., Zheng, C., Zhao, Y., Zhang, F., Wen, C., Feng, L., Wan, C., From organic ligand to metal-organic coordination polymer, andto metal-organic coordination polymer-cocrystal composite: acontinuous promotion of proton conductivity of crystalline materials. *Cryst. Eng. Comm.*, 22, 1414, 2020.
28. Wang, X.X., Cullen, D.A., Pan, Y.T., Hwang, S., Wang, M., Feng, Z., Wang, J., Engelhard, M.H., Zhang, H., He, Y., Shao, Y., Su, D., More, K.L., Spendelow, J.S., Wu, G., Nitrogen-coordinated single cobalt atom catalysts for oxygen reduction in proton exchange membrane fuel cells. *Adv. Mater*, 30, 1706758, 2018.
29. Ru, C., Gu, Y., Na, H., Li, H., Zhao, C., Preparation of a cross-linked sulfonated poly(arylene ether ketone)proton exchange membrane with enhanced proton conductivityand methanol resistance by introducing an ionic liquid-impregnated metal organic framework. *ACS Appl. Mater. Interfaces*, 11, 31899, 2019.
30. Wang, H., Yin, F.-X., Liu, N., Kou, R.-H., He, X.-B., Sun, C.-J., Chen, B.-H., Liu, D.-J., Yin, H.-Q., Engineering Fe–Fe_3C@Fe–N–C active sites and hybrid-structures from dual metal–organic frameworks for oxygen reduction reaction in H_2–O_2 fuel cell and Li–O_2 battery. *Adv. Funct. Mater.*, 29, 1901531, 2019.
31. Bunzen, H., Javed, A., Klawinski, D., Lamp, A., Grzywa, M., Kalytta-Mewes, A., Tiemann, M., Krug von Nidda, H.-A., Wagner, T., Volkmer, D., Anisotropic water-mediated proton conductivity in large iron(II) metal–organic framework single crystals for proton-exchange membrane fuel Cells. *ACS Appl. Nano Mater.*, 2, 291, 2019.
32. Zhang, F.M., Dong, L.-Z., Qin, J.-S., Guan, W., Liu, J., Li, S.-L., Lu, M., Lan, Y.-Q., Su, Z.-M., Zhou, H.-C., Effect of imidazole arrangements on

proton-conductivity in metal–organic frameworks. *J. Am. Chem. Soc.*, 139, 6183, 2017.

33. Yang, K., Jiang, P., Chen, J., Chen, Q., nanoporous PtFe nanoparticles supported on NDoped porous carbon sheets derived from metal–organic frameworks as highly efficient and durable oxygen reduction reaction catalysts. *ACS Appl. Mater. Interfaces*, 9, 32106, 2017.
34. Afsahi, F., Kaliaguine, S., Non-precious electrocatalysts synthesized from metal–organic frameworks. *J. Mater. Chem. A*, 2, 12270, 2014.
35. Kadirov, M.K., Minzanova, S.T., Nizameev, I.R., Khrizanforov, M.N., Mironova, L.G., Kholin, K.V., Kadirov, D.M., Nefedev, E.S., Morozov, M.V., Gubaidullin, A.T., Budnikova, Y.H., Sinyashin, O.G., A nickel-based pectin metal-organic framework as a hydrogen oxidation reaction catalyst for proton-exchange-membrane fuel cells. *ChemistrySelect*, 4, 4731, 2019.
36. Phang, W.J., Lee, W.R., Yoo, K., Ryu, D.W., Kim, B., Hong, C.S., pH-dependent proton conducting behavior in a metal–organic framework material. *Angew. Chem. Int. Ed.*, 53, 8383, 2014.
37. Afsahi, F., Vinh-Thang, H., Mikhailenko, S., Kaliaguine, S., Electrocatalyst synthesized from metal organic frameworks. *J. Power Sources*, 239, 415, 2013.
38. Deng, R., Han, W., Yeung, K.L., Confined PFSA/MOF composite membranes in fuel cells for promoted water management and performance. *Catal. Today*, 331, 12, 2019.
39. Cai, K., Sun, F., Liang, X., Liu, C., Zhao, N., Zou, X., Zhu, G., An acid-stable hexaphosphate ester based metal–organic framework and its polymer composite as proton exchange membrane. *J. Mater. Chem. A*, 5, 12943, 2017.
40. Ye, Y., Wu, X., Yao, Z., Wu, L., Cai, Z., Wang, L., Ma, X., Chen, Q.-H., Zhang, Z., Xiang, S., Metal–organic frameworks with a large breathing effect to host hydroxyl compounds for high anhydrous proton conductivity over a wide temperature range from subzero to 125°C. *J. Mater. Chem. A*, 4, 4062, 2016.
41. Lai, Q., Zheng, L., Liang, Y., He, J., Zhao, J., Chen, J., metal–organic-framework-derived Fe-N/C electrocatalyst with five-coordinated Fe-N_x sites for advanced oxygen reduction in acid media. *ACS Catal.*, 7, 1655, 2017.
42. Wang, S., Luo, H., Li, X., Shi, L., Cheng, B., Zhuang, X., Li, Z., Amino acid-functionalized metal organic framework with excellent proton conductivity for proton exchange membranes. *Int. J. Hydrogen Energy*, 46, 1163, 2021.
43. Mukhopadhyay, S., Das, A., Jana, T., Das, S.K., Fabricating a MOF material with polybenzimidazole into an efficient proton exchange membrane. *ACS Appl. Energy Mater.*, 3, 7964, 2020.
44. Zhang, Z., Ren, J., Xu, J., Wang, Z., He, W., Wang, S., Yang, X., Du, X., Meng, L., Zhao, P., Adjust the arrangement of imidazole on the metal-organic framework to obtain hybrid proton exchange membrane with long-term stable high proton conductivity. *J. Membr. Sci.*, 607, 118194, 2020.

45. Li, X.-M., Liu, J., Zhao, C., Zhou, J.-L., Zhao, L., Lia, S.-L., Lan, Y.-Q., Strategic hierarchical improvement of superprotonic conductivity in a stable metal–organic framework system. *J. Mater. Chem. A*, 7, 25165, 2019.
46. Zhang, J., Bai, H.J., Ren, Q., Luo, H.B., Ren, X.M., Tian, Z.F., Lu, S., Extra water- and acid-stable MOF-801 with high proton conductivity and its composite membrane for proton-exchange membrane. *ACS Appl. Mater. Interfaces*, 10, 28656, 2018.
47. Gui, D., Dai, X., Tao, Z., Zheng, T., Wang, X., Silver, M.A., Shu, J., Chen, L., Wang, Y., Zhang, T., Xie, J., Zou, L., Xia, Y., Zhang, J., Zhang, J., Zhao, L., Diwu, J., Zhou, R., Chai, Z., Wang, S., Unique proton transportation pathway in a robust inorganic coordination polymer leading to intrinsically high and sustainable anhydrous proton conductivity. *J. Am. Chem. Soc.*, 140, 6146, 2018.
48. Lo, T.H.N., Nguyen, M.V., Tu, T.N., An anchoring strategy leads to enhanced proton conductivity in a new metal–organic framework. *Inorg. Chem. Front.*, 4, 1509, 2017.
49. Rao, Z., Tang, B., Wu, P., Proton conductivity of proton exchange membrane synergistically promoted by different functionalized metal–organic frameworks. *ACS Appl. Mater. Interfaces*, 9, 22597, 2017.
50. Rao, Z., Feng, K., Tang, B., Wu, P., Construction of well interconnected metal-organic framework structure for effectively promoting proton conductivity of proton exchange membrane. *J. Membr. Sci.*, 533, 160, 2017.
51. Luo, H.B., Wang, M., Liu, S.X., Xue, C., Tian, Z.F., Zou, Y., Ren, X.-M., Proton conductance of a superior water-stable metal–organic framework and its composite membrane with poly (vinylidene fluoride). *Inorg. Chem.*, 56, 4169, 2017.

9

Fluorinated Membrane Materials for Proton Exchange Membrane Fuel Cells

Pavitra Rajendran[1], Valmiki Aruna[1], Gangadhara Angajala[1]* and Pulikanti Guruprasad Reddy[2,3]

[1]Department of Chemistry, Kalasalingam Academy of Research and Education, Anand nagar, Krishnan Koil, Tamil Nadu, India
[2]School of Basic Sciences, Indian Institute of Technology Mandi, Kamand, Himachal Pradesh, India
[3]Institute for Drug Research, School of Pharmacy, Faculty of Medicine, The Hebrew University of Jerusalem, Jerusalem, Israel

Abstract

Over the past several years, proton exchange membrane fuel cells (PEMFCs) have become popular as one of the most crucial energy sources amongst the various fuel cells (FCs) in view of energy and environmental sustainability. Proton exchange membranes (PEMs) are the main component of PEMFCs; primarily, fluoromembrane polymer was used as membrane material in PEMFCs. It has a good ionic conductivity, low permeability, mechanical stability, low costs, long service life, thermal stability, and easy construction of membrane electrode. In contrast to other FCs, the assembly and handling of PEMFCs is less complex compared to other FCs and operates at low temperature (60°C–80°C). Membrane materials are used in the FC system: fluorinated ionomers, perfluorinated ionomers, nonfluorinated ionomers, sulfonated poly(arylenes), and an acid-base complex. Perfluorosulfonic acid (PFSA) and poly(arylene ethers) with fluorinated and sulfonated polymers have high H^+ conductivity and good physicochemical characteristics, which make polymers a good PEM in PEMFCs. This chapter focuses on research in PEM material based on fluorinated polymers. It focuses on the preparation process and properties of the membrane material to achieve effective FC compatibility. The most reliable polymers with potential for PEMFCs are also reviewed.

**Corresponding author*: gangadharaangajala@gmail.com

Inamuddin, Omid Moradi and Mohd Imran Ahamed (eds.) Proton Exchange Membrane Fuel Cells: Electrochemical Methods and Computational Fluid Dynamics, (245–270) © 2023 Scrivener Publishing LLC

***Keywords*:** Fuel cells, PEMFCs, proton exchange membrane materials, PFSA ionomers, fluorination, proton conductivity, permeability

Abbreviations

PEMFC	Proton exchange membrane fuel cell
FC	Fuel cell
PEM	Proton exchange membrane
PFSA	Perfluorosulfonic acid
S-FPAE	Sulfonated-fluorinated poly(arylene ether)
PBI	Poly(bibenzimidazole)
HFA	2,2′-bis(4-carboxyphenyl)-hexafluoropropane
BTMDS	1,3-bis(carboxypropyl)tetramethyldisiloxane
HFDP	4,4-(hexafluoroisopropylidene) diphenol
DFBP	Decafluorobiphenyl
BPA	Bisphenol-A
SPPSU	Sulfonated (polyphenylsulfone)
PVDF-HFP	Poly(vinylidene fluoride-co-hexafluoro propylene)
PAEK	Poly(arylene ether ketones)
SPE	Sulfonated polyarylene ether sulfone
FBPAQSH	Fluorinated sulfonated poly(arylene ether sulfone)
c-SPFAES	Fluorinated poly(aryl ether sulfone) membranes cross-linked sulfonated oligomer
SPABES-PAE	Sulfonated poly(arylene biphenylether sulfone)-poly(arylene ether)
FBB	1,4-bis (4-fluorobenzoyl)benzene
SIPAES	Sulfonimide functionalized poly(arylene ether sulfones)
DCDPS	Dichlorodiphenylsulfone
SDCDPS	3,3′-disulfonate-4,4′-dichlorodiphenylsulfone
IEC	Ion exchange capacity
GO	Graphene oxide

9.1 Introduction

In 21st century, the requirement of alternative and green energy sources is causing a major concern on energy demand because of energy necessity, increase of global energy depletion, and environmental pollution. The researchers put forward a possible solution for developing a fuel cell (FC) system, and it is expected that performance will increase in the next few years [1–5]. In general, the FCs are an electrochemical device, like batteries. An individual FC is comprised of a number of smaller cells that are stacked together in order to form a FC. The FC process involves converting chemical power into electricity. The electrochemical reaction like oxidation and reduction has produced energy when hydrogen (H_2) combines with oxygen (O_2) in a specific environment. FCs are made up of anodes and cathodes, with the anode oxidizing the fuel and the cathode reducing the oxidant. Two electrodes were separated with an electrolyte for transferring ions instead of electrons. As the electrons are released at the anode, they move to the cathode via the external circuit, where they participate in the cathodic semi-reaction, generating electricity [6]. Figure 9.1 shows the operating structure for FCs. The FC device has been known for over 150 years. The PEMFCs were applied to submarines in the 1980s by the Royal British Navy. NASA used PEMFCs in spacecraft in the 1950s and in the mid-1960s, known as Gemini 5 and Gemini 7. In the early 2000s, AeroVironment operated long-term solar aircraft using PEMFCs. FCs are among the most alternative forms of electricity generation. The fuel to the electrical efficiency of the simple heat engine with diesel is at around 50 %, but the FC can reach around 60%. Whereas the theoretical maximum efficiency is 83% at room temperature. Although FC is simple, the practical

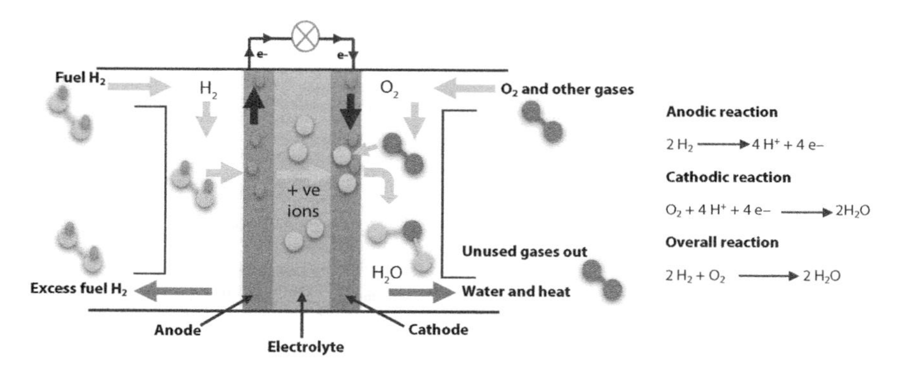

Figure 9.1 Diagram of fuel cell operation.

application of FCs is not still succeeded. Research continues to improve FCs into practical applications such as portable batteries [2, 6–10]. Wind and photovoltaic energy production are reliant upon the supply of primary energy (solar or wind), while FCs are independent of energy sources other than H_2 [4]. The FC is considered an unregulated power source. Factors such as H_2 and O_2 input flow rate, temperature, and pressure affect FC electrical outputs [11, 12].

Proton exchange membrane FCs or polymer electrolytic membrane FCs (PEMFCs) have received significant attention due to energy and environmental sustainability [13, 14]. The advantages of PEMFCs are high power density, easy to handle, minimal operating temperature, low emissions, low corrosion, and fast startup and shutdown operation [15–20]. The polymer exchange membrane (PEM) is the major part of PEMFCs (Figure 9.2). Usually, protons' conductivity is derived by solid electrolytes, typically polymers; however, inorganic and ceramic materials can also be used [21, 22]. To reduce the corrosion of FC systems, a solid-phase polymer membrane is used as an electrolyte in PEMFCs. To reduce the corrosion of FCs, PEMFCs use a solid-phase polymer membrane as their electrolyte. The PEM (electrolyte) is a semipermeable membrane manufactured from ionomers and must be (i) electrical insulation that that cannot conduct electrons, only allowing ion, (ii) the reactant separator, (iii) electrical insulation between anode and cathode, (iv) good ionic conductivity, (v) low permeability, (vi) low costs, (vii) mechanical stability, (viii) thin material membrane (thickness

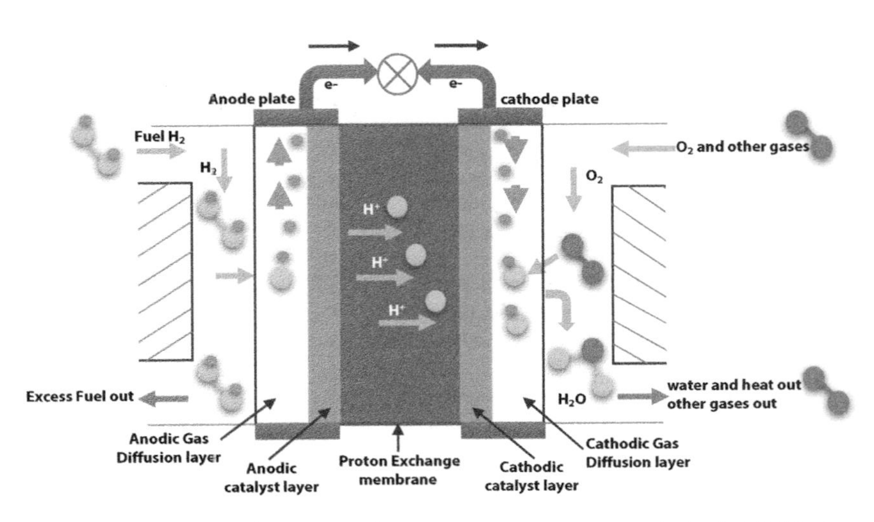

Figure 9.2 The simplified structure of proton exchange membrane fuel cell (PEMFC) operation.

10–250 μm), (ix) long shelf life, (x) easy fabrication of membrane electrode assemblies, (xi) thermal stability, (xii) oxidative stability, and (xiii) low electro-osmosis drag coefficient [9, 23]. Unlike other FCs, assembling and handling PEMFCs is less complex because PEMFCs operate at low temperature (60°C–80°C) and the cell separator is a polymer film [22]. Membrane materials are used in the FC system: fluorinated ionomers, perfluorinated ionomers, nonfluorinated ionomers, sulfonated poly(arylenes) and an acid-base complex [24–26]. There are number of membrane materials used in fabrication of PEM. The fabrication of membrane material depends on the polymer backbone and the presence of functional groups. Fluoropolymers are suitable polymeric materials for the proton exchange membrane (PEM) in the FC [27, 28]. The PEM used in PEMFCs are perfluorosulfonic acid (PFSA) polymers as it has a high proton conductivity with good physical and chemical properties. The nature of fluorine electronegativity and low polarizability are the most important features of the membrane material for FCs. Stability of the chemical compound against bases, reducing and oxidizing agents, mechanical stability, formation of thin non-porous films, and prominent ion exchange capacity (IEC) are the key characteristics of PFSA membranes relevant to FC system [9, 22, 29]. The drawbacks of PFSA in large scale are high costs of material, high temperatures result in low conductivity, and high permeability to methanol [30]. The conductivity of the PFSA polymer membrane like Nafion®, Gore®, Forblue®, Fumion®, Aquivion®, or the 3M® membrane appears to have declined above 100°C. However, an extensive range of fluoride-based polymers has been developed for wide-ranging applications.

Three methods have been reported such as synthesis/grafting/blending of proton conductivities groups on polymer backbone, incorporation of fillers into a proton conductive copolymer, and the dispersion of Proton conductive fillers in non-proton conductive polymer. For example, grafting of sulfonated moieties on to the copolymer poly(vinylidene fluoride-co-hexafluoropropylene) [31, 32]. FPAEs [fluorinated poly(arylene ethers)], polymers are the basic building blocks of a PEM and exhibit good physical properties, hydrophobicity, thermal stability, mechanical properties, and economical affordability. In addition, S-FPAEs (sulfonated-FPAEs) possess PEM properties, like enhanced proton conductance and water uptake. S-FPAEs can be treated easily and more economically than Dow and Nafion [33]. Other PEM materials include polytriazole with good solubility; thermal, dimensional, and chemical stability; mechanical properties; and low water absorption [34]. These developments have not been completed because of some issues that have not been resolved. Research aimed at solving challenges related to FC sustainability, energy, energy efficiency, and waste management. This section is intended to provide a comprehensive overview of PEMFCs. It deliberates preparation, properties of PEMFCs, and output of PEMFCs.

9.2 Fluorinated Polymeric Materials for PEMFCs

A portion of the process of preparing fluorinated membrane materials and performances of PEMFCs has been the subject of discussion in this section. Figure 9.3 represents schematic diagram of fluorinated polymer nanocomposite PEM for FC practicability.

9.3 Poly(Bibenzimidazole)/Silica Hybrid Membrane

The poly(bibenzimidazole) (PBI)/silica hybrid membrane containing fluorine is used as a PEM in high-temperature PEMFC, since it has shown significant thermooxidative stability, the coefficients of thermal expansion, mechanical properties, methanol permeability, and conductivities. The solubility of fluorinated copolymer containing PBI that contains hydroxyl

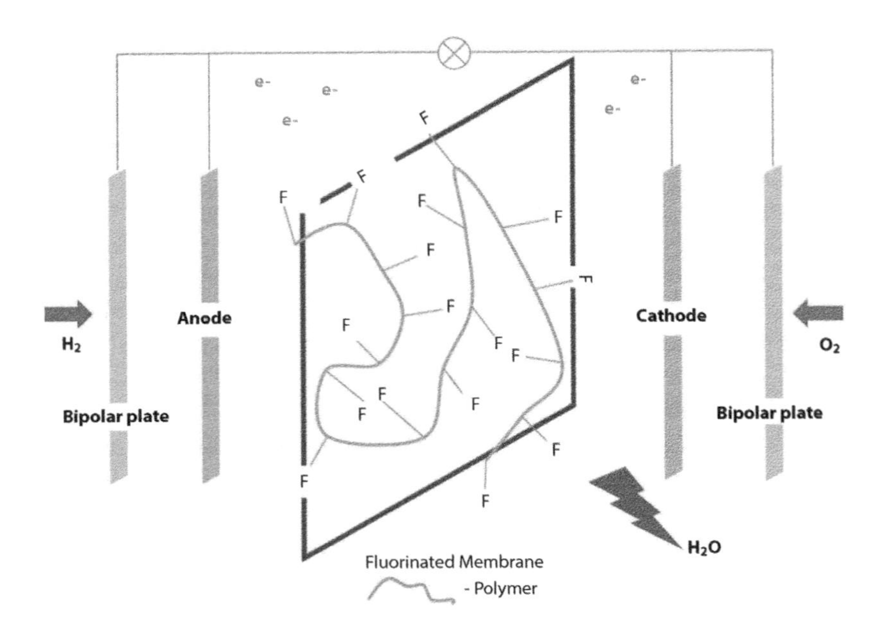

Figure 9.3 Fluorinated polymer PEM for fuel cell applications.

groups is the important factor for reaching exceptional applications. The increases in methanol crossover of acid-doped PBI/silica nanocomposite happen due to protonation of PBI that could reduce the intermolecular interaction force; however, it was much lower than that of Nafion 117 membrane 30×10^{-8} cm^2/s. The proton conductivities of nanocomposites films show up to 160°C, whereas the comparison with the Nafion 117 membrane

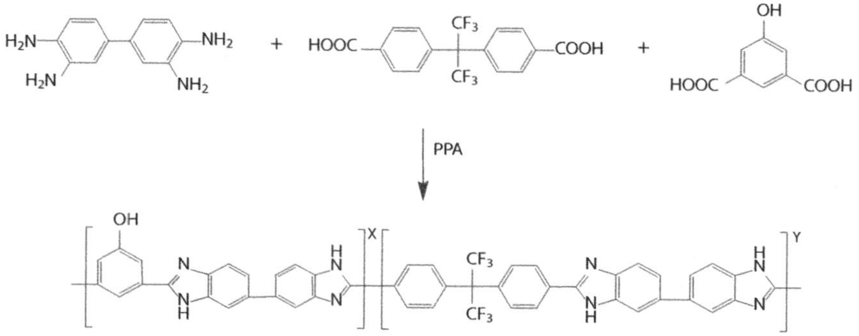

Scheme 9.1 (a) Synthesis of fluorine containing PBI polymer [15].

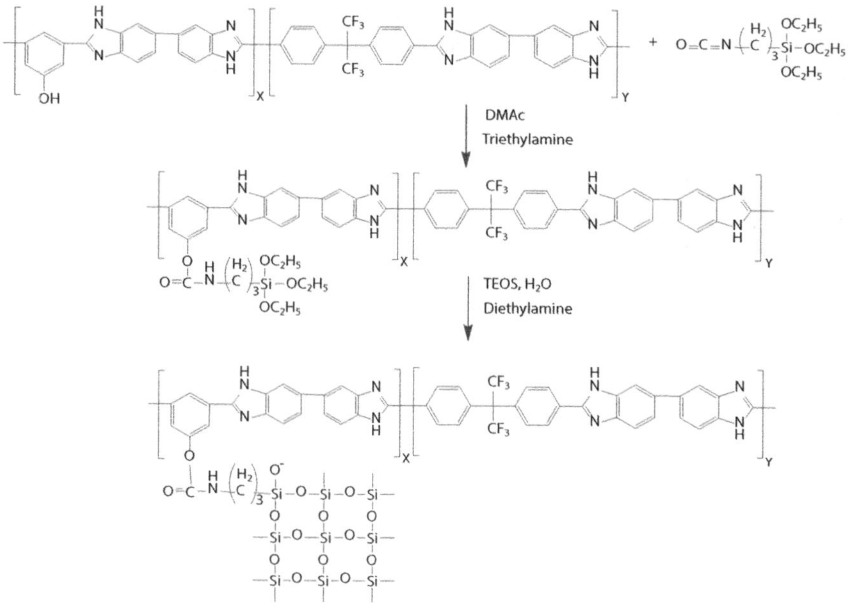

Scheme 9.1 (b) Preparation process of PBI/silica hybrid membrane material [15].

decreased sharply above 100°C due to dehydration. Furthermore, adding silica to a system can be achieved through a combination of organic polymers and inorganic compounds [15]. Schemes 9.1a and 9.1b elaborate the preparation process of the PBI/silica hybrid membrane.

9.4 Poly(Bibenzimidazole) Copolymers Containing Fluorine-Siloxane Membrane

The synthesis of PBI copolymers containing fluorine-siloxane can be fabricated by copolymerizing 2,2′-bis(4-carboxyphenyl)-hexafluoropropane (HFA), 1,3-bis(carboxypropyl)tetramethyldisiloxane (BTMDS), and 3,3′-diaminobenzidine (Scheme 9.2). Polymer backbones with bulky HFA groups or flexible BTMDS groups are significantly more soluble than PBI without these groups. Despite the copolymer's thermal and mechanical stability, the polymers exhibited good physical properties. In comparison to PBI without the siloxane group, the polymer backbone displayed hydrophobic siloxane groups, which led to decreased permeability for methanol at high temperatures and increased proton conductivity [35].

The PBI copolymers containing fluorine-siloxane membrane mainly constitutes the sulfonated and fluorinated multiblock co polymers, which was made form nucleophilic aromatic substitution of highly activated fluorine terminated telecheli polymer and hydroxyl-terminated telecheli (Schemes 9.3a and 9.3b). As expected, the proton conductivities of multiblock polymer decreased exponentially as the relative humidity (RH) decreased. However, the proton conductivities of multiblock polymer increased with lower RH. The reason could be related to a morphologically distinctive structure consisting of sulfonated hydrophilic units surrounding fluorinated hydrophobic units. An increase in sulfonated and fluorinated units in the multiblock polymer led to improved membrane

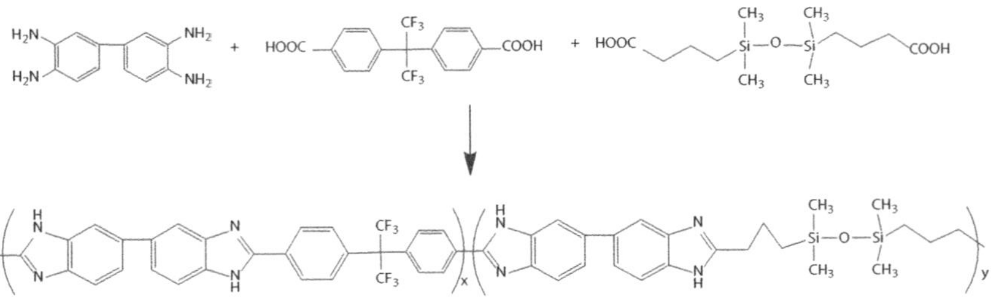

Scheme 9.2 Synthesis of PBI copolymers containing fluorine-siloxane membrane [35].

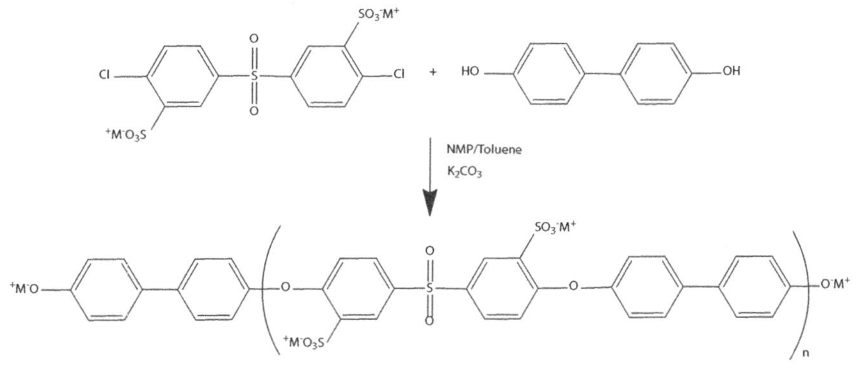

Scheme 9.3 (a) Synthesis of hydroxy-terminated sulfonated poly(arylene ether sulfone) [36].

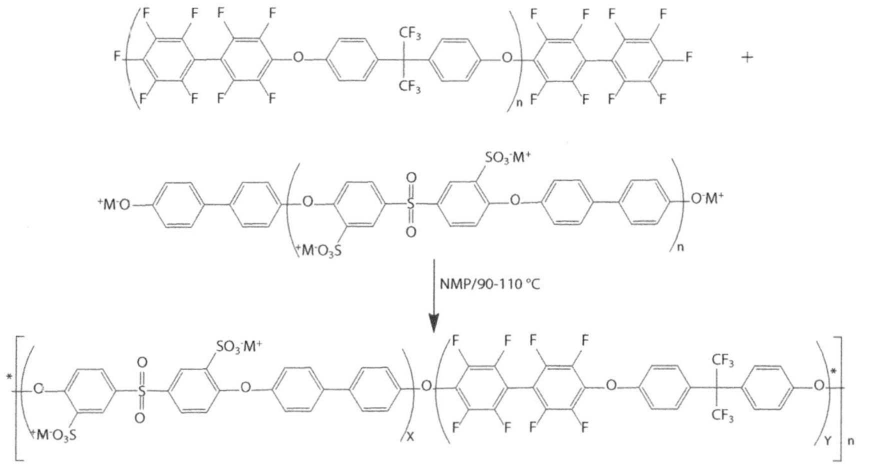

Scheme 9.3 (b) Sulfonated and fluorinated multiblock polymer by copolymerization with decafluorobiphenyl-terminated poly(arylene ethers) and hydroxy-terminated sulfonated poly(arylene ether sulfone) [36].

IEC (2.2 mequiv. g^{-1}), water uptake (40 to more than 400%), proton conductivity (0.32 S cm^{-1}), and mechanical stability [36].

9.5 Sulfonated Fluorinated Poly(Arylene Ethers)

Other membrane material, like S-FPAE, possessed high thermal properties and a tensile strength superior to that of Nafion 117 with an appropriate level of sulfonation and with acceptable conductivity. In Scheme 9.4a, S-FPAEs were developed using 4,4-(hexafluoroisopropylidene) diphenol (HFDP) and decafluorobiphenyl (DFBP). S-FPAEs also made from bisphenol-A (BPA) are represented in Scheme 9.4b. The sulfonation of FPAEs was achieved by sulfuric acid (30% SO_3). Higher sulfonated polymers exhibited a higher exchange capacity of ions, water absorption, and proton conductance.

A higher level of sulfonation occurred in S-DFBP-BPA than in S-DFBP-HFDP in similar reaction conditions, owing to the existence of the electro-donor arylene substituent, the methyl group, and the assisted electrophilic sulfonation reaction. Both S-DFBP-BPA and S-DFBP-HFDP (S-FPAEs) exhibited higher water absorption than Nafion at the similar

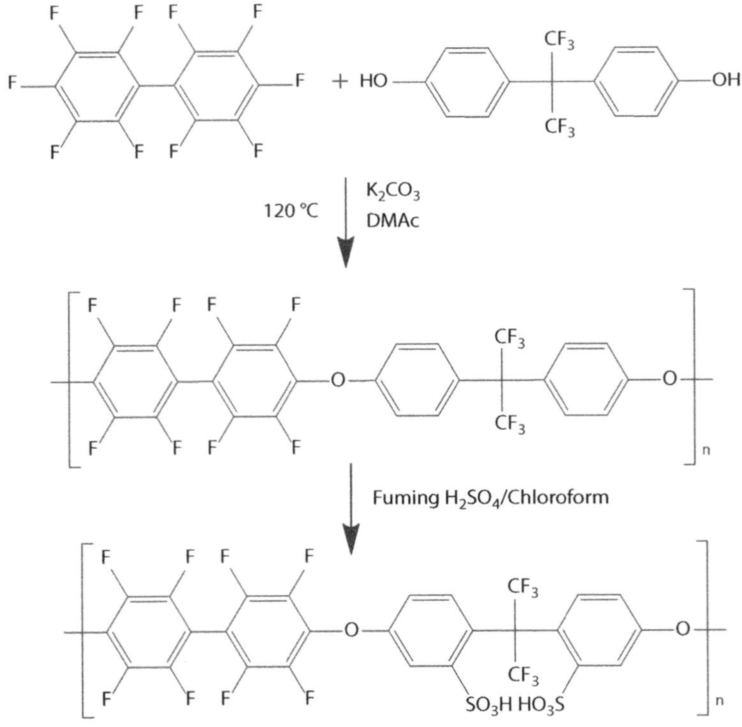

Scheme 9.4 (a) S-DFBP-HFDP (S-FPAEs) was developed by 4,4-(hexafluoroisopropylidene)diphenol (HFDP) and decafluorobiphenyl (DFBP) [37].

ionic exchange capacity as their aromatic group spine is less hydrophobic than that of the perfluorinated vertebral network. The PEM, having conductivity greater than.10^2 S/cm^2, was unsuitable for the use of PEMFC due to its low mechanical strength. The power density of Nafion 117 exhibited 368 mW/cm^2, whereas 425.5 mW/cm^2 was observed for S-DFBP-HFDP at 800 mA/cm^2 and 1,150 mA/cm^2, respectively. S-DFBP-HFDP displayed a voltage lower than Nafion 117 at low currents. In a study, S-FPAEs have been identified as promising materials. Their cost-effective nature, thermal stability, and electrochemical properties make them perfect for applications with high-power density, high-stability, and mechanical properties [33]. The excess density of the sulfonic acid group (SO_3H) influences the functioning of the PEMFC through swelling or hydrogel formation. Sulfonated hydrophilic oligomers and fluorinated hydrophobic oligomers were combined to enhance proton conductance, radical resistance, dimensional stability, and permeability in membrane materials. Since, C-F bonds in the hydrophobic regions offered high thermo-mechanical strength. In addition, microphase separation between hydrophobic and hydrophilic segments enhances proton conductivity without deteriorating mechanical and chemical performance [37].

9.6 Fluorinated Sulfonated Polytriazoles

Polytriazoles become one of the polymeric materials used in PEMFCs membrane material. Since fluorinated sulfonated polytriazoles have been found to be thermally stable with a decomposition temperature ranging from 190°C to 320°C. The tensile strength of fluorinated sulfonated polytriazoles polymers ranges from 51 to 65 MPa compared to Nafion 117 (22 MPa) and other sulfonated membranes. It was a result of the presence of a rigid quadriphenyl moiety and acid-base interaction of $-SO_3H$ group with triazole in the polymer core. The low water uptake, good oxidative properties, and excellent dimensional stability were attributed by the presence of $-CF_3$ hydrophobic group and triazole and quadriphenyl rings together. Moreover, fluorinated sulfonated polytriazole exhibited high molecular weights, solubility, mechanical stability, dimensional stability, and low water uptake. Fluorinated sulfonated polytriazoles were synthesized in the process given in Schemes 9.4b and 9.5a [30].

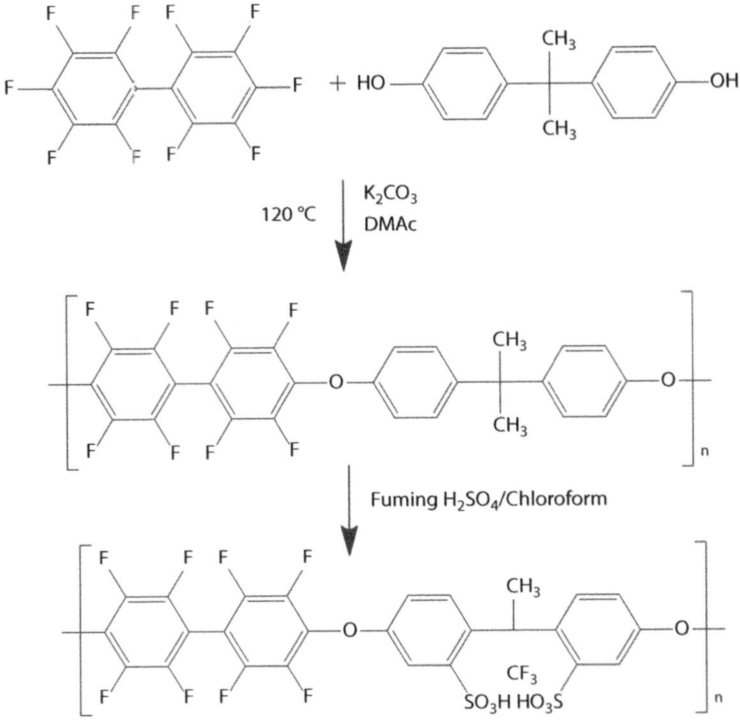

Scheme 9.4 (b) S-DFBP-BPA [sulfonated-fluorinated poly(arylene ethers)] was developed by combining decafluorobiphenyl (DFBP) with bisphenol-A (BPA) [37].

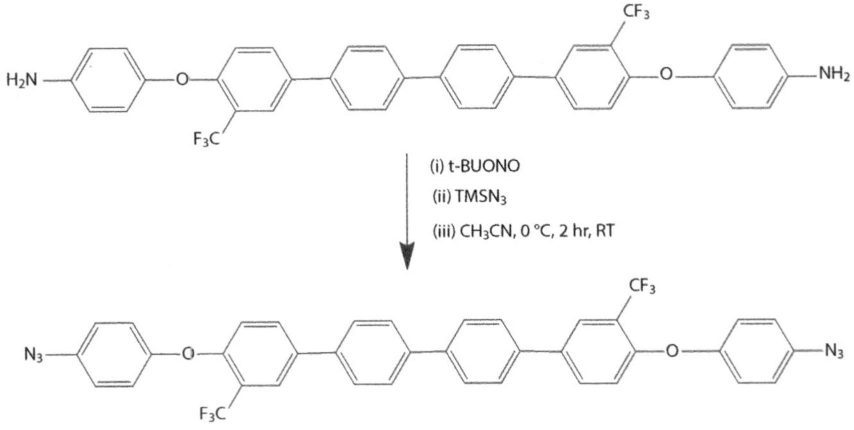

Scheme 9.5 (a) Synthesis of 4,4-bis[3-trifluoromethyl-4(4-azidophenoxy)phenyl]biphenyl monomer [30].

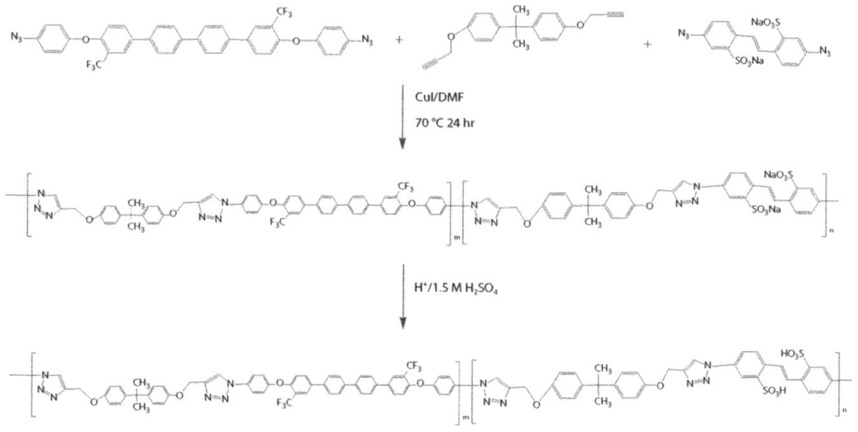

Scheme 9.5 (b) Synthesis of sulfonated polytriazoles polymer [30].

9.7 Fluorinated Polybenzoxazole (6F-PBO)

Sulfonated (polyphenylsulfone) (SPPSU) is one of the polymer exchange membranes for alternate Nafion. SPPSU has been incorporated into fluorinated polybenzoxazole nanofiber membranes to improve the drawbacks of SPPSU such as poor proton conductivity, high water uptake, and large dimensionality change at dehumidify conditions. Essentially, fluorinated polybenzoxazole membranes serve as mechanical networks that enhance mechanical strength and dimensional stability of composite membranes. The preparation process of fluorinated polybenzoxazole was enlightened in Scheme 9.6. SPPSU/fluorinated polybenzoxazole composite membranes with H_3PO_4 showed the tensile strength and modulus 84.61 MPa and 4.17 GPa, which was higher compared than SPPSU and Nafion. The elongation of the composite membrane decreased 33%–51% than SPPSU, representing excellent dimensional stability, since dimension stability is an important factor in determining the performance of FCs. Whereas the proton conductivity of SPPSU/fluorinated polybenzoxazole composite showed reduced proton conductivity even the temperature and acid dopant increased [38].

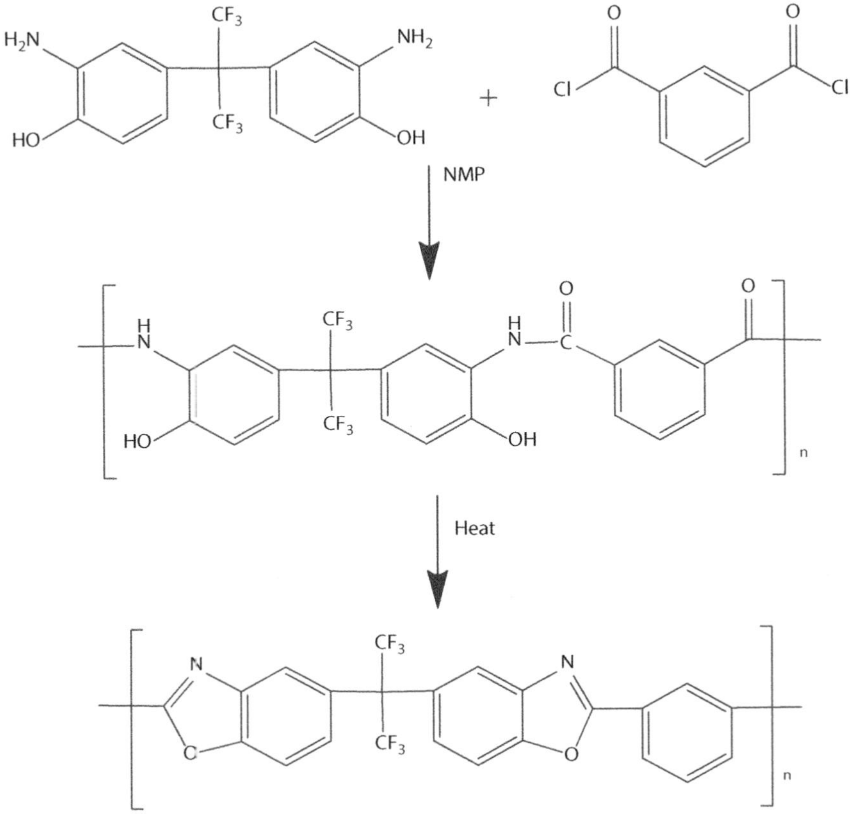

Scheme 9.6 Synthesis of fluorinated polybenzoxazole membranes by poly(hydroxyamide) [38].

9.8 Poly(Bibenzimidazole) With Poly(Vinylidene Fluoride-Co-Hexafluoro Propylene)

PBI with PVDF-HFP have been achieved for the development of stable mechanically and thermochemically PEMs that are capable of high proton conductivity for use in FCs at high temperatures. The proton conductivity of PBI/PVDF-HFP was higher than PBI and PVDF-HFP. This was attributed by the presence of large electronegative F atoms, since it stimulates H+ of the N-H ring of an imidazole, thereby triggering proton transport. Besides, the electronegative fluorine moiety increases the hydrophobicity of the PBI/PVDF-HFP membranes that the membrane exhibited lower water uptake. Since PEM FC system exposed to oxidative environment, oxidative stability of the membrane should be accounted. The PBI/PVDF-HFP membranes showed high oxidative stability in the oxidizing environment containing Fenton's reagent (2 ppm $FeSO_4$, 30% H_2O_2). The oxidative stability increased as content of PVDF-HFP increased. The mechanical and dimensional stability increased for PBI/PVDF-HFP membranes compared to PBI [39].

9.9 Fluorinated Poly(Arylene Ether Ketones)

Hexasulfophenyl pendants on fluorinated poly(arylene ether ketones) (6F-PAEK-SPx) showed excellent proton conductivities (148 mS/cm at 100°C) and low swelling ratio due to the presence of sulfonic acid pendant

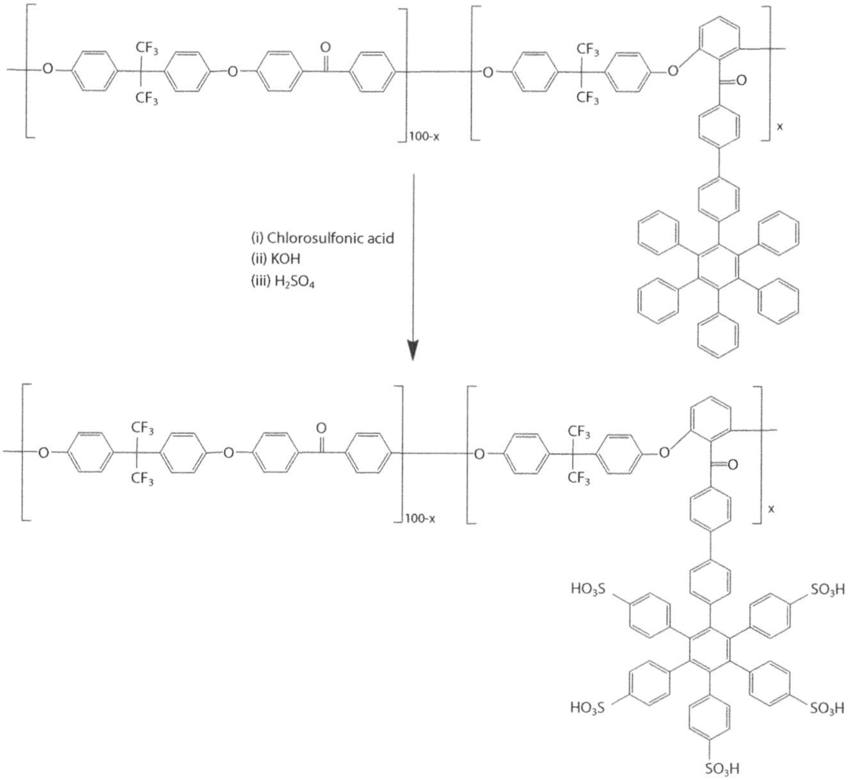

Scheme 9.7 Synthesis of hexasulfophenyl pendants containing fluorinated poly(arylene ether ketones) [39].

and fluorinated backbone. The tensile stress of membrane showed maximum load of 29.5–43.4 MPa and elongation at 110.5%–160.3%. The advantages of 6F-PAEK-SPx such as high proton conductivity, high thermal stability, and dimensional stability have received attention as PEM material for FC applications. Synthesis of 6F-PAEK-SPx is presented in Scheme 9.7 [39].

9.10 Fluorinated Sulfonated Poly(Arylene Ether Sulfone) (6FBPAQSH-XX)

Polymer containing a fluorenyl group sulfonated polyarylene ether sulfone (SPE) used in PEMFCs. The conductivity of the protons of the block copolymer membrane was 140 mS/cm to 800°C, which is greater than Nafion [40]. Fluorinated sulfonated poly(arylene ether sulfone) (6FBPAQSH-XX) polymer found that tensile strength and proton conductivity of membrane varied with different degree of sulfonation. 6FBPAQSH-60 membranes have

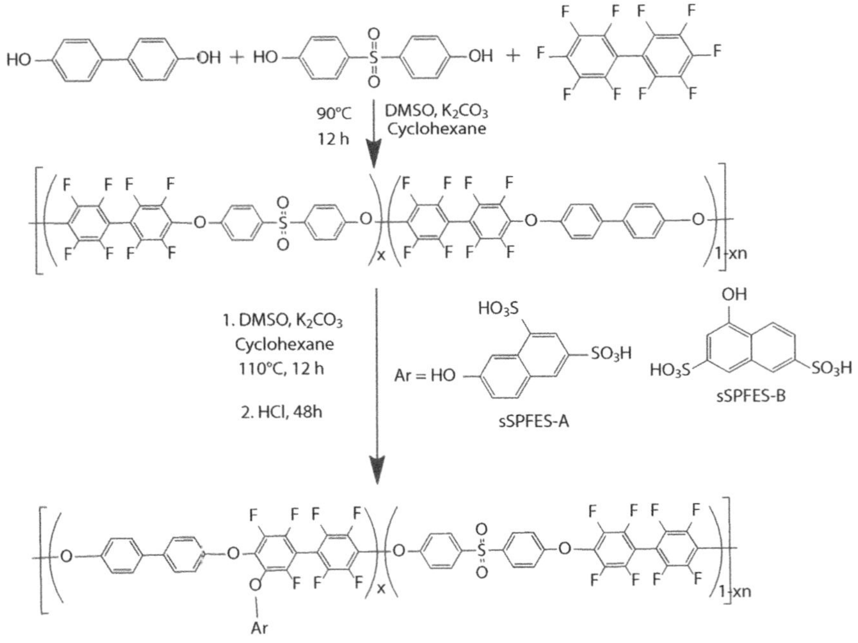

Scheme 9.8 Synthetic route of fluorinated poly(ether sulfones) with disulfonated naphthyl groups [42].

a factor of 1.31 smaller tensile strength compared to 6FBPAQSH-20 membranes. With respect to 6FBPAQSH-20, 6FBPAQSH-60 has a higher proton conductivity by a factor of 8.92 [41]. Fluorinated poly(ether sulfones) with disulfonated naphthyl groups had proton conductivity 0.17–0.28 S/cm and demonstrated thermal stability, mechanical ductility, dimensional aging of below 17%, and water uptake less than 70%. Power output for a membrane between 593 and 658 mW/cm^2 was observed humidity ranged from 60% to 80% in a H_2/O_2 test at 80°C, which suggests they are promising candidates for FC applications. Scheme 9.8 illustrates the procedure for manufacturing fluorinated poly(ether sulfones) with disulfonated naphthyl groups [42].

9.11 Fluorinated Poly(Aryl Ether Sulfone) Membranes Cross-Linked Sulfonated Oligomer (c-SPFAES)

In addition, IEC of a c-SPFAES is 1.04 to 1.78 mmol g^{-1} with excellent mechanical toughness, Young's modulus is greater than 1.3GPa, and tensile strength is greater than 35.8MPa with heat-resistant properties. In terms of power density, the c-SPFAES-1.61 membrane is rated at 587 mW cm^{-2} at 80°C and proton conductivity of 162 mS cm^{-1} under fully hydrated conditions in a H_2/O_2 single FC test [43].

9.12 Sulfonated Poly(Arylene Biphenylether Sulfone)-Poly(Arylene Ether) (SPABES-PAE)

Moreover, SPABES-PAE copolymer including fluorinated units provided membranes with exceptional oxidative and dimensional stabilities. Therefore, The SPABES has water retention capability, high proton conductance, reasonable dimensional stability, and the largest peak power density of the entire SPABES-PAE was 333.29 $mW{\cdot}cm^{-2}$ at 60°C under 100% RH [51]. Compared to Nafion 212, 6F-SPAESKK displayed high thermal stability and similar IEC ranges between 0.888 and 1.532 meq g^{-1}. The synthesis of 6F-SPAESKK were achieved using monomer [1,4-bis (4-fluorobenzoyl)benzene: 1,4 FBB] as presented in Scheme 9.9. As high current-density region of 6F-SPAESKK60 showed superior performance compared to Nafion 212, water uptake and dimensional swelling on the membrane were lower compared to Nafion 212 for the other membranes.

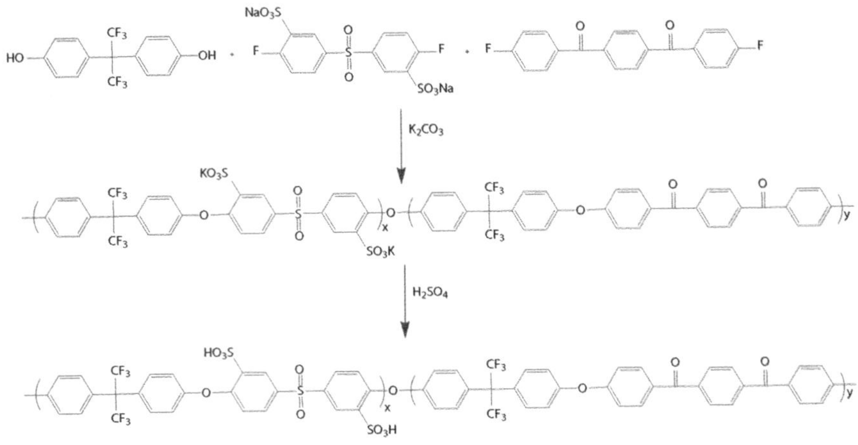

Scheme 9.9 The synthetic route of membranes of 6F-sulfonated poly(arylene ether sulfone ketone) [44].

Researchers found that a 6F-BPA unit led to a very stable mechanical and operational behavior over the long run [44].

The ever-increasing demand for PEMFCs has been to develop proton-conducting materials with superacidity and low cost. Among the gas-phase superacids, sulfonimide groups exhibit the greatest stability, both thermally and chemically. It was found that the partially fluorinated sulfonimide functionalized poly(arylene ether sulfones) (SIPAES) had a similar proton conductance compared to Nafion® 117. The SPAESs are prepared by copolymerizing 4,4′-dichlorodiphenylsulfone (DCDPS), 3,3′-disulfonate-4,4′-dichlorodiphenylsulfone (SDCDPS), and bisphenol (Scheme 9.10). A partial fluorosulfonyl imide monomer was used to convert arylsulfonic acid groups into sulfonimide acid groups. As a result, IEC measured 0.78–1.41 mequiv. g^{-1} uptake of approximately 6.7% to 40.6% for 20%–40% ionic groups. A further advantage of using polymers containing fluorine in these applications is the ability of the polymer chain to resist oxidative degradation because it is protected from oxidizing radical [45].

It can be seen in Figure 9.4 that the fluorinated PEM has some unique characteristics. Results of the overall study confirmed that PEMs made of fluorinated polymers have high proton conductivity, low methane crossover, and high ion exchange capacities. Fluorine greatly increases the hydrophobic nature of the polymer nanocomposite, which results in minimal water absorption.

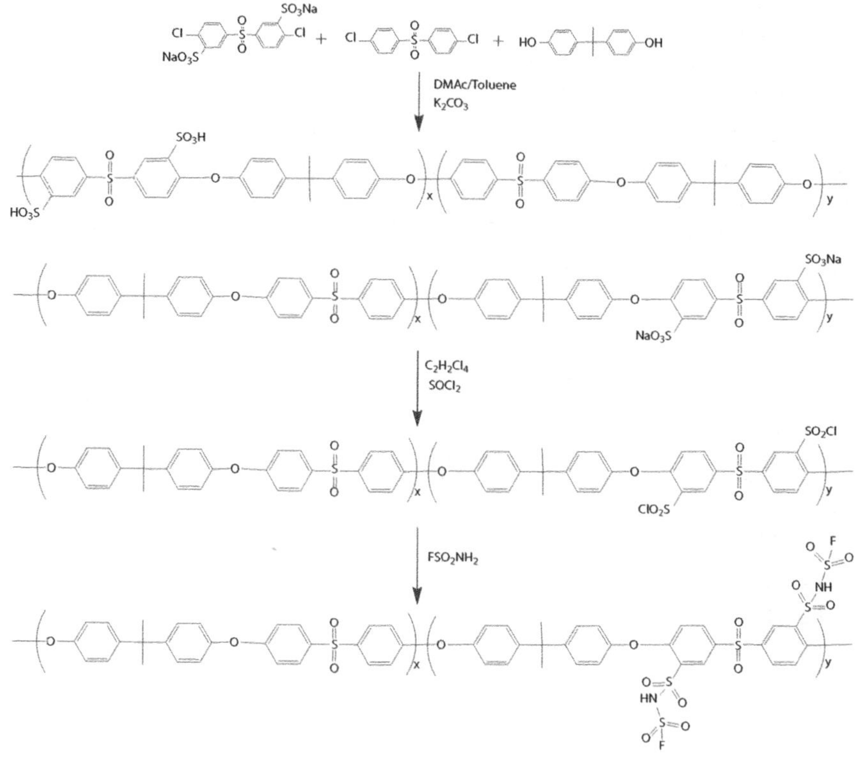

Scheme 9.10 Preparation process of SIPAES [45].

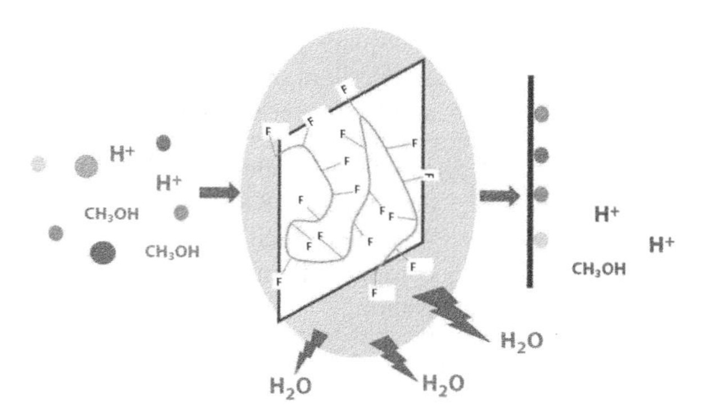

Figure 9.4 Diagram illustrating characteristics of fluorinated polymer proton exchange membranes.

A description of PEM material is given in Table 9.1, which shows that fluorinated and sulfonated multiblock polymers possess advantages on proton transport, ion exchange, and overcome challenges associated with other PEM materials. Although the fluorinated multi-block polymer has a significant conductivity of protons and others, the mechanical module and oxidative stability of the multi-block polymer did not meet the requirements of the PEM. The needs of PEM were improved through chemical cross-linking, long chain polymer linking, mixing inorganic additives, and varying membrane reinforcing process and resulted in

Table 9.1 Membrane material used in PEMFCs.

Membrane materials for PEMFCs	Descriptions	References
Poly(bibenzimidazole)/ silica hybrid membrane	Proton conductivities show up to 160°C	[15]
Polybenzimidazole/ fluorine-siloxane membrane	Increased membrane ion exchange capacity (2.2 meq g^{-1})	[36]
Sulfonated fluorinated poly(arylene ethers)	Enhanced proton conductance, radical resistance, dimensional stability, and permeability in membrane materials	[37]
Fluorinated sulfonated polytriazoles	Due to the presence of a $-CF_3$ hydrophobic group and triazole and quadriphenyl rings, there is an excellent oxidative stability and low water uptake	[30]
Fluorinated polybenzoxazole	Showed the tensile strength and modulus 84.61 MPa and 4.17 GPa which was higher compared than SPPSU and Nafion	[38]
Poly(benzimidazole) with poly(vinylidene fluoride-co-hexafluoro propylene)	Increased proton transport	[39]

(*Continued*)

Table 9.1 Membrane material used in PEMFCs. (*Continued*)

Membrane materials for PEMFCs	Descriptions	References
Fluorinated poly(arylene ether ketones)	Low swelling ratio due to the presence of sulfonic acid pendant and fluorinated backbone	[39]
Fluorinated sulfonated poly(arylene ether sulfone)	Proton conductivity showed 140 mS/cm, which is greater than Nafion	[41]
Fluorinated poly(aryl ether sulfone) membranes cross-linked sulfonated oligomer	Ion exchange capacity of a c-SPFAES is 1.04 to 1.78 mmol g^{-1}	[43]
Sulfonated poly(arylene biphenylether sulfone)-poly(arylene ether)	Ion exchange capacity ranges between 0.888 to 1.532 meq g^{-1}	[44]

composite PEM material with enhanced electrochemical, thermochemical, and physio-chemical properties. For example, functionalization of carbon nanotubes and GO have widely used in PEM material due to physio-chemical properties, mechanical modulus, and excessive ion density. Several composite-based PEMs were investigated [41, 46]. GO/sulfonated polyimide matrix was achieved through electrostatic interaction, thereby showing maximum water absorption, proton conductivity, and mechanical stability [47]. Hybrid inorganic organic materials dispersed in perfluorinated ionomers represent one of the most promising pathways for improving the drawbacks of the proton-conducting membrane. Nanofillers, ionic liquid doped with inorganic oxoclusters, silane-based fillers, zeolites, Pt-SiO_2, Pt-TiO_2 and heteropolyacids, etc., have been investigated in order to improve the efficiency in PEMFCs. The formation of dynamic cross-links between fluorinated polymer and surface of nano filler improved the mechanical characteristics of hybridized materials up to temperature 200°C [48]. Fluorine-decorated graphene nanoribbons with very stable carbon corrosion have contributed to a more sustainable PEMFC for long-standing operation. In addition, fluorinated graphene nanoribbons have shown an effective cathode as additives [49]. A chemical modification of GO will be performed to assess H_2/O_2-based PEM cell performance using fluorine and sulfonic acid groups. Fluorinated GO is obtained by light oxidation with HF and by oxidation and sulfonation, which results in SO_3 and F functionalization. Considering low temperature and moderate humidity membrane electrode assembly performance, both SO_3 and F participate in reducing H_2 crossover. The fluorine groups also contributed to a higher hydrolytic stability by contributing to an inefficiency against structural degradation, whereas sulfonic acid improved stability via, the preservation of proton conductivity. Membrane re-dispersion was prevented by the F-groups in neutral H_2O, thus showing promising hydrolytic stability and structural stability, although water absorption is similar. By comparing them with appropriate reference materials, it was determined how the fluorine and sulfonic acid groups on the GO might influence the performance of the FC [50].

9.13 Conclusion

The new components will be developed capable of producing high performance, long-lasting, and profitable on the basis of existing technologies. Polymer membrane materials considered as a polymer electrolyte among other solid or liquid electrolyte since it reduces corrosion of FC system and tends to produce high conductivity, long-term use of membrane materials, and low cost. The performance and physical and chemical properties of PEM depend on the incorporation of groups such as fluorine, SO_3, and inorganic fillers in the base polymer. Fluorinated polymers, for example, PFSA, are widely used in PEMFC as they produce high conductivity of protons with good chemical and physical properties. The nature of fluorine electronegativity and hydrophobicity and low polarity are the two important factors that have led to reduced water uptake and improved thermal stability. In addition, the excess of sulfonic acid groups led to the formation of swelling and hydrogel, so it recovered by incorporating sulfonated hydrophilic oligomer with fluorinated hydrophobic oligomer. However, fluorinated and perfluorinated ionomers have drawbacks, and they improved by applying nanofillers, graphene, ionic liquid–doped inorganic oxoclusters, silane-based fillers, zeolites, Pt-SiO_2, Pt-TiO_2, and heteropolyacids to enhance efficiency.

Conflicts of Interest

The authors declare no conflict of interest.

Acknowledgements

The authors sincerely thank the management of Kalasalingam Academy of Research and Education for their constant encouragement and support and providing all the necessary facilities for writing this chapter.

References

1. Calderón, A.J., González, I., Calderón, M., Segura, F., Andújar, J.M., A new, scalable and low cost multi-channel monitoring system for polymer electrolyte fuel cells. *Sensors*, 16, 349, 2016.
2. De las Heras, A., Vivas, F.J., Segura, F., Andújar, J.M., From the cell to the stack. A chronological walk through the techniques to manufacture the PEFCs core. *Renew. Sust. Energ. Rev.*, 96, 29, 2018.
3. Segura, F. and Andújar, J.M., Step by step development of a real fuel cell system. Design, implementation, control and monitoring. *Int. J. Hydrog. Energy*, 40, 5496–5508, 2015.
4. Andújar, J.M. and Segura, F., Fuel cells: History and updating. A walk along two centuries. *Renew. Sust. Energ. Rev.*, 13, 2309, 2009.
5. Casteleiro-Roca, J.-L., Barragán, A.J., Manzano, F.S., Calvo-Rolle, J.L., Andújar, J.M., Fuel cell hybrid model for predicting hydrogen inflow through energy demand. *Electronics*, 8, 1325, 2019.
6. Coralli, A., Sarruf, B.J.M., de Miranda, P.E.V., Osmieri, L., Specchia, S., Minh, N.Q., Chapter 2 - Fuel Cells, in: *Science and Engineering of Hydrogen-Based Energy Technologies*, P.E.V. de Miranda, (Ed.), pp. 39–122, Academic Press, USA, 2019.
7. Breeze, P., *Fuel Cells*, Academic Press, USA, 2017.
8. Barbir, F. and Gómez, T., Efficiency and economics of proton exchange membrane (PEM) fuel cells. *Int. J. Hydrog. Energy*, 21, 891, 1996.
9. Walkowiak-Kulikowska, J., Wolska, J., Koroniak, H., Polymers application in proton exchange membranes for fuel cells (PEMFCs). *Phys. Sci. Rev.*, 2, 20170018, 2017.
10. Ortiz-Rivera, E.I., Reyes-Hernandez, A.L., Febo, R.A., Understanding the history of fuel cells, in: *Presented at the 2007 IEEE Conference on the History of Electric Power*, Newark, NJ, USA, IEEE, pp. 117–122, 2007.
11. Andújar, J.M., Segura, F., Durán, E., Rentería, L.A., Optimal interface based on power electronics in distributed generation systems for fuel cells. *Renew. Energy*, 36, 2759, 2011.
12. Segura, F., Andujar, J.M., Duran, E., Analog current control techniques for power control in PEM fuel-cell hybrid systems: A critical review and a practical application. *IEEE Trans. Ind. Electron.*, 58, 1171, 2011.

13. Carapellucci, R., Cipollone, R., Di Battista, D., Modeling and characterization of molten carbonate fuel cell for electricity generation and carbon dioxide capture. *Energy Procedia, ATI 2017 - 72nd Conference of the Italian Thermal Machines Engineering Association*, vol. 126, p. 477, 2017.
14. Revankar, S.T. and Majumdar, P., *Fuel Cells: Principles, Design, and Analysis*, CRC Press, USA, 2014.
15. Chuang, S.-W., Hsu, S.L.-C., Liu, Y.-H., Synthesis and properties of fluorine-containing polybenzimidazole/silica nanocomposite membranes for proton exchange membrane fuel cells. *J. Membr. Sci.*, 305, 353, 2007.
16. Devrim, Y., Devrim, H., Eroglu, I., Development of 500 W PEM fuel cell stack for portable power generators. *Int. J. Hydrog. Energy*, 40, 7707, 2015.
17. Mehta, V. and Cooper, J.S., Review and analysis of PEM fuel cell design and manufacturing. *J. Power Sources*, 114, 32, 2003.
18. Meyer, Q., Ronaszegi, K., Pei-June, G., Curnick, O., Ashton, S., Reisch, T., Adcock, P., Shearing, P.R., Brett, D.J.L., Optimisation of air cooled, open-cathode fuel cells: Current of lowest resistance and electro-thermal performance mapping. *J. Power Sources*, 291, 261, 2015.
19. Millington, B., Du, S., Pollet, B.G., The effect of materials on proton exchange membrane fuel cell electrode performance. *J. Power Sources Fuel Cells Sci. & Technol.*, 196, 9013, 2011.
20. Kim, D., Robertson, G., Guiver, M., Lee, Y., Synthesis of highly fluorinated poly(arylene ether)s copolymers for proton exchange membrane materials. *J. Membr. Sci.*, 281, 111, 2006.
21. Arcella, V., Merlo, L., Ghielmi, A., 15 - Proton exchange membranes for fuel cells, in: *Advanced Membrane Science and Technology for Sustainable Energy and Environmental Applications*, A. Basile, and S.P. Nunes, (Eds.), pp. 465–495, Woodhead Publishing Series in Energy, Woodhead Publishing, UK (United Kingdom), 20112011.
22. Scott, K., Membrane electrode assemblies for polymer electrolyte membrane fuel cells, in: *Functional Materials for Sustainable Energy Applications, Woodhead Publishing Series in Energy*, J.A. Kilner, S.J. Skinner, S.J.C. Irvine, P.P. Edwards, (Eds.), pp. 279–311, Woodhead Publishing, UK (United Kingdom), 2012.
23. Bae, B., Miyatake, K., Watanabe, M., Effect of the hydrophobic component on the properties of sulfonated Poly(arylene ether sulfone)s. *Macromolecules*, 42, 1873, 2009.
24. Arnett, N.Y., Harrison, W.L., Badami, A.S., Roy, A., Lane, O., Cromer, F., Dong, L., McGrath, J.E., Hydrocarbon and partially fluorinated sulfonated copolymer blends as functional membranes for proton exchange membrane fuel cells. *J. Power Sources*, 172, 20, 2007.
25. Smitha, B., Sridhar, S., Khan, A.A., Solid polymer electrolyte membranes for fuel cell applications—A review. *J. Membr. Sci.*, 259, 10, 2005.
26. Tiwari, A., *Innovative Graphene Technologies: Evaluation and Applications*, vol. 2, Smithers Rapra, UK (United Kingdom), 2013.

27. Smith, D.W., Lacono, S.T., Lyers, S.S., *Handbook of Fluoropolymers, Science, and Technology*, Wiley, New York, 2014.
28. Ameduri, B. and Boutevin, B., *Well-Architectured Fluoropolymers: Synthesis, Properties, and Applications*, Elsevier, Amsterdam, *Science*, 2004.
29. Zatoń, M., Rozière, J., Jones, D.J., Current understanding of chemical degradation mechanisms of perfluorosulfonic acid membranes and their mitigation strategies: A review. *Sustain. Energy Fuels*, 1, 409, 2017.
30. Singh, A., Mukherjee, R., Banerjee, S., Komber, H., Voit, B., Sulfonated polytriazoles from a new fluorinated diazide monomer and investigation of their proton exchange properties. *J. Membr. Sci.*, 469, 225, 2014.
31. Seck, S., Magana, S., Prébé, A., Buvat, P., Bigarré, J., Chauveau, J., Améduri, B., Gérard, J.-F., Bounor-Legaré, V., New fluorinated polymer-based nanocomposites via combination of sol-gel chemistry and reactive extrusion for polymer electrolyte membranes fuel cells (PEMFCs). *Mater. Chem. Phys.*, 252, 123004, 2020.
32. Taguet, A., Ameduri, B., Boutevin, B., Grafting of 4-Hydroxybenzenesulfonic acid onto commercially available Poly(VDF-co-HFP) copolymers for the preparation of membranes. *Fuel Cells*, 6, 331, 2006.
33. Lee, H.C., Hong, H.S., Kim, Y.-M., Choi, S.H., Hong, M.Z., Lee, H.S., Kim, K., Preparation and evaluation of sulfonated-fluorinated poly(arylene ether)s membranes for a proton exchange membrane fuel cell (PEMFC). *Electrochim. Acta*, 49, 2315, 2004.
34. Huang, Y.J., Ye, Y.S., Yen, Y.C., Tsai, L.D., Hwang, B.J., Chang, F.C., Synthesis and characterization of new sulfonated polytriazole proton exchange membrane by click reaction for direct methanol fuel cells (DMFCs). *Int. J. Hydrog. Energy*, 36, 15333, 2011.
35. Hsu, S., L.C., Lin, Y.C., Tasi, T.Y., Jheng, L.C., Shen, C.H., Synthesis and properties of fluorine- and siloxane-containing polybenzimidazoles for high temperature proton exchange membrane fuel cells. *J. Appl. Polym. Sci.*, 130, 4107, 2013.
36. Ghassemi, H., McGrath, J.E., Zawodzinski, T.A., Multiblock sulfonated–fluorinated poly(arylene ether)s for a proton exchange membrane fuel cell. *Polymer*, 47, 4132, 2006.
37. Kim, A.R., Vinothkannan, M., Yoo, D.J., Sulfonated fluorinated multi-block copolymer hybrid containing sulfonated(poly ether ether ketone) and graphene oxide: A ternary hybrid membrane architecture for electrolyte applications in proton exchange membrane fuel cells. *J. Energy Chem.*, 27, 1247, 2018.
38. Jin, J., Hao, R., He, X., Li, G., Sulfonated poly(phenylsulfone)/fluorinated polybenzoxazole nanofiber composite membranes for proton exchange membrane fuel cells. *Int. J. Hydrog. Energy*, 40, 14421, 2015.
39. Hazarika, M. and Jana, T., Novel proton exchange membrane for fuel cell developed from blends of polybenzimidazole with fluorinated polymer. *Eur. Polym. J.*, 49, 1564, 2013.

40. Pang, J., Jin, X., Wang, Y., Feng, S., Shen, K., Wang, G., Fluorinated poly(arylene ether ketone) containing pendent hexasulfophenyl for proton exchange membrane. *J. Membr. Sci.*, 492, 67–76, 2015.
41. Mohanty, A.K., Mistri, E.A., Banerjee, S., Komber, H., Voit, B., Highly fluorinated sulfonated poly(arylene ether sulfone) copolymers: Synthesis and evaluation of proton exchange membrane properties *Ind. Eng. Chem. Res.*, 723, 2772, 2013.
42. Hu, Z., Lu, Y., Zhang, X., Yan, X., Li, N., Chen, S., Fluorinated poly(ether sulfone) ionomers with disulfonated naphthyl pendants for proton exchange membrane applications. *Front. Mater. Sci.*, 12, 156, 2018.
43. Zhang, X., Lu, Y., Yan, X., Hu, Z., Chen, S., Sulfonated oligomer-crosslinked fluorinated poly(aryl ether sulfone)-based proton exchange membranes for fuel cells. *Fuel Cells*, 18, 397, 2018.
44. Kim, D.-H., Park, I.K., Lee, D.-H., Fluorinated sulfonated poly (arylene ether)s bearing semi-crystalline structures for highly conducting and stable proton exchange membranes. *Int. J. Hydrog. Energy*, 45, 23469, 2020.
45. Chandra Sutradhar, S., Ahmed, F., Ryu, T., Yoon, S., Lee, S., Rahman, M.M., Kim, J., Lee, Y., Kim, W., Jin, Y., A novel synthesis approach to partially fluorinated sulfonimide based poly (arylene ether sulfone) s for proton exchange membrane. *Int. J. Hydrog. Energy*, 44, 11321, 2019.
46. Kumar, R., Xu, C., Scott, K., Graphite oxide/Nafion composite membranes for polymer electrolyte fuel cells. *RSC Adv.*, 2, 8777, 2012.
47. He, Y., Tong, C., Geng, L., Liu, L., Lü, C., Enhanced performance of the sulfonated polyimide proton exchange membranes by graphene oxide: Size effect of graphene oxide. *J. Membr. Sci.*, 458, 36, 2014.
48. Di Noto, V., Bettiol, M., Bassetto, F., Boaretto, N., Negro, E., Lavina, S., Bertasi, F., Hybrid inorganic-organic nanocomposite polymer electrolytes based on Nafion and fluorinated TiO_2 for PEMFCs. *Int. J. Hydrog. Energy*, 37, 6169, 2012.
49. Jin, S., Yang, S.Y., Lee, J.M., Kang, M.S., Choi, S.M., Ahn, W., Fuku, X., Modibedi, R.M., Han, B., Seo, M.H., Fluorine-decorated graphene nanoribbons for an anticorrosive polymer electrolyte membrane fuel cell. *ACS Appl. Mater. Interfaces*, 13, 26936, 2021.
50. Sandström, R., Annamalai, A., Boulanger, N., Ekspong, J., Talyzin, A., Mühlbacher, I., Wågberg, T., Evaluation of fluorine and sulfonic acid co-functionalized graphene oxide membranes under hydrogen proton exchange membrane fuel cell conditions. *Sustain. Energy Fuels*, 3, 1790, 2019.
51. Lee, K.H., Chu, J.Y., Kim, A.R., Yoo, D.J., Facile fabrication and characterization of improved proton conducting sulfonated poly(arylene biphenylether sulfone) blocks containing fluorinated hydrophobic units for proton exchange membrane fuel cell applications. *Polymers*, 10, 1790–1798, 2018.

10

Membrane Materials in Proton Exchange Membrane Fuel Cells (PEMFCs)

Foad Monemian and Ali Kargari*

Membrane Processes Research Laboratory (MPRL), Department of Chemical Engineering, Amirkabir University of Technology (Tehran Polytechnic), Tehran, Iran

Abstract

The need for energy for the growing population and consuming fossil fuels has made clean energy more popular. One way to generate clean energy is to use fuel cells. The basis of energy production in these systems is based on an electrochemical reaction. In this chapter, fuel cells and their history are described, and then, the applications and energy production mechanisms are expressed. The principal part of this chapter is to get acquainted with the materials used in the proton exchange membranes of the fuel cell and the ways of characterizing these membranes. Various materials have been used for PEM, most of which are sulfonic acid–based membranes. Different methods of PEM preparation have led to varying classifications of these materials, which can be referred to direct polymerization, post-sulfonation, and chemical and radiation grafting methods.

***Keywords*:** Fuel cell, post-sulfonation, PEM, PEMFC, block copolymer, graft copolymer, durability, proton conductivity

10.1 Introduction

Nowadays, energy demand and thus fossil fuels consumption has caused greenhouse gas emissions and other pollutants. These reservoirs have been expensive due to their non-renewability. In 2008, developing countries' CO_2 emission per capita level was about 20% of quantity for major

**Corresponding author*: Ali_kargari@yahoo.com

Inamuddin, Omid Moradi and Mohd Imran Ahamed (eds.) Proton Exchange Membrane Fuel Cells: Electrochemical Methods and Computational Fluid Dynamics, (271–298) © 2023 Scrivener Publishing LLC

industrial countries. Estimations are shown that the CO_2 emission of developing nations is 50% more than the quantity for total CO_2 emission nations [1].

The human lifestyle is made more comfortable by technology development. For this purpose, energy management using hybrid systems [e.g., fuel cells (FCs)] has been investigated [2]. One way to reduce GHG emission is using FCs as an environment-friendly power resource because of their economic benefits and efficiency. Their potential to diminish environmental impacts made this method an alternative to combustion engines [3, 4].

10.2 Fuel Cell: Definition and Classification

A FC is known as an electrochemical device that converts chemical energy to electrical energy by external reactants. FC specifications include the following:

1. It is a converting energy device.
2. Converting energy is happened by electrochemical reaction.
3. Converting energy from chemical to electrical has occurred in one step.
4. Reactants are not stored inside the reactor (stack) [5].

The chemical energy of fuel and oxidants is directly changed to electricity and heat within the FC as an electrochemical device with high efficiency. Unlike the typical heat engines, the Carnot cycle is not followed by electrochemical processes of FCs; hence, their performance is simple and more efficient than internal combustion engines (ICEs) [6]. All FCs contain two electrodes [anode (where fuel enters) and cathode (where oxidant is supplied)] and one electrolyte (where ions transfer from anode to cathode) [3].

According to their type of electrolyte, FCs are classified into two standard models containing:

1) Solid electrolytes: These consist of Polymer Electrolyte Membrane FCs (PEMFCs), Direct Methanol FCs (DMFCs), and Solid Oxide FCs (SOFCs).
2) Liquid electrolytes: These include Alkaline FCs (AFCs), Phosphoric Acid FCs (PAFCs), and Molten Carbonate FCs (MCFCs).

Table 10.1 Type of fuel cells and their operating temperature [3, 8].

Fuel cell type	Alkaline (AFC)	Polymer (PEMFC)	Phosphoric acid (PAFC)	Molten carbonate (MOFC)	Solid oxide (SOFC)
Operating temperature (°C)	60–250	80–120	150–220	600–700	500–2,000
Efficiency (%)	50–70	40–50	40–45	50–60	50–60

For example, DMFCs are similar to PEMFCs, but their fuel is different. In DMFCs, its fuel is methanol, whereas PEMFCs fuel is pure hydrogen or hydrogen-rich gas [1, 3, 7]. Type of FCs and their operating temperature are given in Table 10.1 [8].

In PEMFC, the solid electrolyte is the main part of these FCs, made by polymeric membrane sheet. This polymeric membrane, which is more straightforward, less hazardous, and has higher energy density than other electrolyte types, allows protons (not electrons) to pass through it to reach from anode to cathode [2]. Furthermore, the separation between two reactant gases is another task of this polymeric membrane, which is 30% of PEMFC cost that is related to its membrane [3]. In common PEMFC, H_2 and O_2 (or air) are employed as the fuel and oxidant, respectively. For hydrogen supply, a reformer is required, whereas it is not needed in DMFC. Low operating temperature and pressure with the acceptable price of methanol cause DMFC as a promising energy source [3].

10.3 Historical Background of Fuel Cell

More than 150 years have passed since the FCs developed. In the early 19th century, a British scientist, Humphry Davy, described the FC concept. After that, Christian Friedrich Schönbein published a paper about this field in 1838. The design of the earliest FC as named "gas voltaic battery" is done by William Robert Grove in 1842. In this FC, diluted sulfuric acid was used as the electrolyte, and electrodes were made of platinum. Initially, FCs were an attractive concept for power generation due to the low efficiency of other methods). In 1882, power generation stations converted coal (as fuel) to power with 2.5% efficiency. W. Ostwald published the paper in 1894 that shown that the next century will be the century of electrochemical combustion. In the 1920s, efficiency for steam turbine

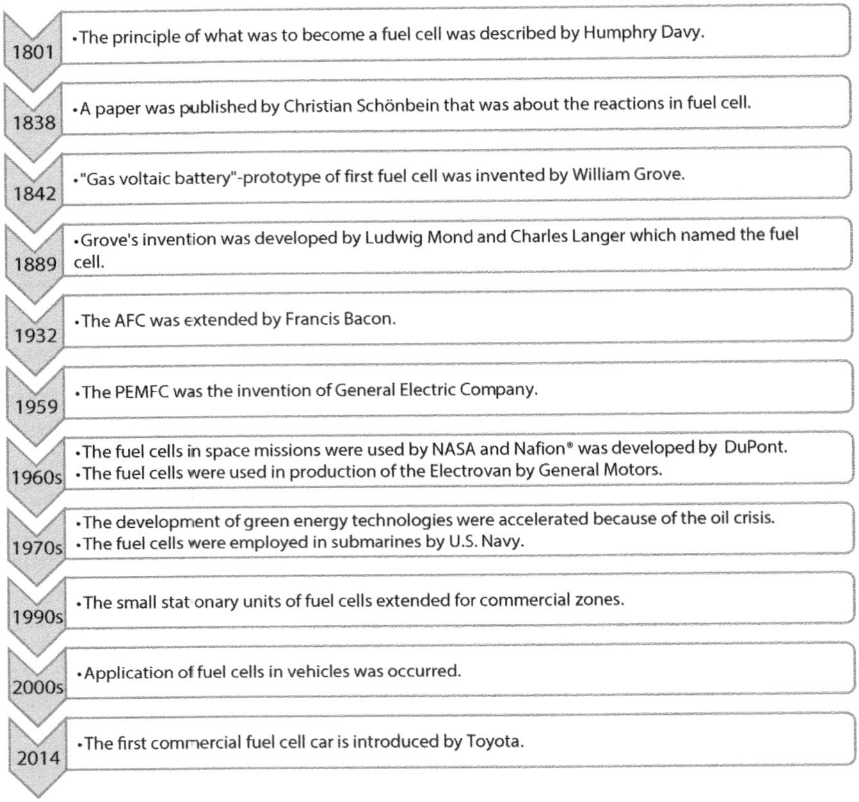

1801
- The principle of what was to become a fuel cell was described by Humphry Davy.

1838
- A paper was published by Christian Schönbein that was about the reactions in fuel cell.

1842
- "Gas voltaic battery"-prototype of first fuel cell was invented by William Grove.

1889
- Grove's invention was developed by Ludwig Mond and Charles Langer which named the fuel cell.

1932
- The AFC was extended by Francis Bacon.

1959
- The PEMFC was the invention of General Electric Company.

1960s
- The fuel cells in space missions were used by NASA and Nafion® was developed by DuPont.
- The fuel cells were used in production of the Electrovan by General Motors.

1970s
- The development of green energy technologies were accelerated because of the oil crisis.
- The fuel cells were employed in submarines by U.S. Navy.

1990s
- The small stat onary units of fuel cells extended for commercial zones.

2000s
- Application of fuel cells in vehicles was occurred.

2014
- The first commercial fuel cell car is introduced by Toyota.

Figure 10.1 The historical background of fuel cell technology [9].

and reciprocating steam engines were 13%–14% and 20%, respectively, and these low conversions led to FC advancement. Coal was the major fuel for power generation in that century, and gasification was a new process of H2 production, developed by Ludwig Mond and Charles Langer. However, the gasification catalyst was poisoned by sulfur and other impurities in the gas; hence, FCs were weakened. In the 1950s, simultaneously with the peak of space competition, the application of FCs became stronger [1]. The historical background of FCs is shown in Figure 10.1 [9].

10.4 Fuel Cell Applications

High efficiency has led to their application for various applications consist of transportation, electronic devices (portable and stationary), and power generation (because of the depletion of fossil fuels). Some of these usages are explained below [3].

10.4.1 Transportation

Limited space and rapid response time (or faster startup) in the cars make the application of FCs difficult. AFC systems in hybrid vehicles are favorable for this purpose. Pure hydrogen is used as fuel in this type of FC. AFC vehicles are restricted to a few types of applications (e.g., fleet buses and other centralized vehicles) due to limited pure hydrogen distribution centers. In addition to AFC, PEMFC is also used for transportation applications, but it still needs to be tested to overcome usage concerns [10]. For medium- to heavy-duty vehicles, PEMFC advantages such as low operating temperature, zero greenhouse gas emissions, high energy density, and lightweight structure are caused to use them instead of electric vehicles [11]. Degradation of the membrane (e.g., pinholes formation and thickness reduction) negatively affects PEMFC performance; therefore, durability is essential in PEMFC. In this way, a particular protocol is established by the US Department of Energy for durability measuring of PEMFCs; for example, power loss must be under 10% after 8,000 h operation for PEMFC cars and <25,000 h for heavy-duty vehicles [12].

10.4.2 Stationary Power

The possibility of distributed power generation with high efficiency is an essential point of FCs. For stationary applications, FCs that operate in high-temperature and low-temperature are employed, and each of these has its unique advantages. For instance, low-temperature FCs have a faster startup time, while high-temperature FCs (e.g., MCFC and SOFC) produces powerful heat (high-grade), which can be used in heat cycle lonely or integrating with FCs and increasing efficiency [10].

10.4.3 Portable Applications

FC requirements in small power applications (e.g., laptops or cell phones) are more specific than vehicle applications. The requirements concentrate on the dimensions and weight of the system as well as on the temperature. Rapid startup, high efficiency and energy density, low operating temperature, and other properties of PEMFCs have led to use them for this purpose [10, 11]. Applications of PEMFCs in different industries are given in Table 10.2 [13].

Table 10.2 PEMFCs in different industries: Applications, prospect, and the main reason for usage [3, 13].

Application	Prospect*	Main reason	Competition	Comments
Transportation				
Lightweight vehicles, Bus and RV	4*	More space for the fuel processing equipment	None	A hybrid system combined with PEMFC is desired
Passenger car	3*	Healthy for people (zero GHG emissions)	A hybrid system combined with the ICE without PEMFC	PEMFC is desired
Powered bicycle	2*	Hydrogen supplies are not convenient	Battery	Batteries or hybrid systems are desired
Recreation purposes (Sailing yacht)	3*	Bottled LPG is more common than PEMFC	DBFC and DMFC	Used as an APU
Stationary				
Stationary middle-scale power generation	2*	Hydrogen supplies with high purity are the major challenge	SOFC and MCFC	SOFC and MCFC are desired
Continuous small-scale power supply	3*	Long blackout periods are possible	Battery	A hybrid system is desired

(*Continued*)

Table 10.2 PEMFCs in different industries: Applications, prospect, and the main reason for usage [3, 13]. (*Continued*)

Application	Prospect*	Main reason	Competition	Comments
Portable				
Laptops	2	Hydrogen supplies in liquid form are not possible	DBFC, DMFC, and Batteries	DBFC or DMFC is desired
Overall Comment: PEMFC is suitable for transportation, small-scale power generation, and domestic power considering PEMFC operating temperature and other parameters.				

* 1 = least positive, 2 = less positive, 3 = positive, and 4 = most positive.

10.5 Comparison between Fuel Cells and Other Methods

A FC and a battery have similarities, such as converting chemical energy to electrical energy, high efficiency, and not following the Carnot cycle limit. Still, they have a difference in the location of reactants. Reactants store within the battery, while the reactants are outside of the FC. The mechanisms of the consumption of the reactants in the battery, FC, and heat engine are explained as follows [5]:

- When the reactants are finished in the primary battery, the battery is unused and by totally consuming the reactants in the secondary battery, electricity charging is needed. Two parameters are essential in the electrical energy quantity of the battery: amount of the reactants and rate of discharge. The opposite of discharging process is the recharging process. In this process, reactants are converted to their initial state by external electrical energy; therefore, in a secondary battery, two cycles of discharging (chemical energy converts to electrical power) and charging (electrical energy converts to chemical energy) have happened [5].
- In contrast, reactants are not located in the FC's reactor (named as stack). They are outside of the stack and only flow through the stack when electrical energy is needed. By terminating the power generating, entering the reactants into the stack is stopped. The amount of electrical energy of a FC depends on the reactant quantity stored in the storage tank. The refueling process is needed by finishing the reactants and takes just a few minutes [5].
- An ICE has similarities to a FC. The quantity of electrical energy is related to the amount of reactants placed outside the reactors in both of them. However, the efficiency of the heat engine is lower than a FC due to obeying the Carnot limit. In addition, power generating has happened in several stages for a heat engine instead of a single step for a FC. In Table 10.3, comparisons of battery, FC, and heat engine have been brought [5].

Table 10.3 Comparisons of battery, heat engine, and fuel cell [5].

Parameter	Battery	Fuel cell	Heat engine
Reaction type	Electrochemical	Electrochemical	Combustion
Efficiency	High	High	Low
Carnot limit	No	No	Yes
Fuel location	Inside	Outside	Outside
Refueling	Electrical charging	Add fuel	Add fuel
Refueling time	Long (hours)	Short (minutes)	Short (minutes)
Running time per refueling	Short	Long	Long
Reaction noise	No	Low	High
Conclusions	It can be said summarily that the fuel cell is a battery whose refueling is done by adding the fuel (placed outside the fuel cell), refueling time is short, and running time per refueling is long compared to the battery.		

In a FC, chemical energy converts to electrical energy by an electrochemical process within the electrochemical reactor. This reactor contains two electrodes: an anode for oxidation of fuel and a cathode for oxidant reduction. The electrolyte is put between the electrodes to prevent direct contact and allows ionic particles to move through the electrodes. Power generation in a FC is done in this procedure; electrons are lost from the oxidant at the anode surface and electrons are reached to the cathode by the external circuit. This movement generates electrical energy [5]. Schematic of a PEM FC and procedure of power generation in a PEM FC are given in Figure 10.2 [14].

As shown in Figure 10.2, usually, in a PEMFC, H_2, air (or pure O_2), and electrolyte are used as fuel, oxidant, and proton conductor, respectively. At the anode, H_2 is decomposed to H^+ (protons) and e^- (electrons) by the **(R. 10.1)**. Protons are transferred from anode to cathode via electrolyte and electrons move through the external circuit, generate power, and reach the cathode. By achieving to the cathode, ionic particles react with oxygen **(R. 10.2)** and water produces. **(R. 10.3)** displays the overall reaction [5, 15]:

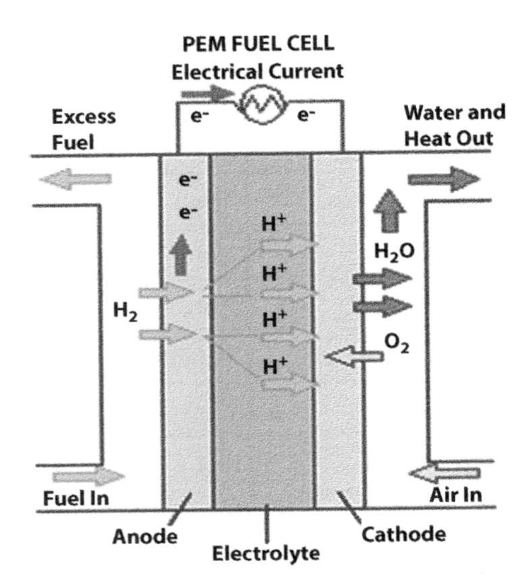

Figure 10.2 Schematic of a PEM fuel cell. H_2 (fuel) and air (oxidant) [14].

Anode: $$H_2 \rightarrow 2H^+ + 2e^-: \quad \text{(R. 10.1)}$$

Cathode: $$0.5O_2 + 2H^+ + 2e^- \rightarrow H_2O \quad \text{(R. 10.2)}$$

Overall: $$H_2 + 0.5O_2 \rightarrow H_2O \quad \text{(R. 10.3)}$$

10.6 PEMFCs: Description and Characterization

Material selection can be improved the performance of the membrane in a PEMFC. Membrane usage in high temperatures without loss in quality can increase its applications; however, choosing an appropriate material is needed [16]. An electrolyte polymer membrane and two electrodes with an applied catalyst layer are the main components of a PEMFC, known as "membrane electrode assembly (MEA)". These electrodes are electrically insulated and separated by electrolyte [17]. The electrolyte in a PEMFC is a polymeric membrane commonly known as proton exchange membrane (PEM) that has three primary duties: 1. proton transportation from anode to cathode; 2. impermeability to gases; and 3. electrical insulation [18].

Since the invention of ion exchange membranes as the ideal membrane for FCs in the 1950s, finding a suitable membrane for harsh conditions is still a challenge for researchers. There are some desired properties for the membrane acting as a PEM:

1. Stability in cell operating conditions such as resistance to oxidation, reduction, and hydrolysis.
2. Insignificantly permeable to reactive gas (minimize Columbic inefficiency).
3. High proton conductivity for establishing high current as well as zero electrical conductivity.
4. High water transfer to create a uniform volume of water and prevent local drying.
5. Reasonable production cost for FC application [3].

Characterization is one of the essential factors for examining FCs. It helps to better compare the membranes with each other, and, finally, researchers can decide to select the optimal membrane for a specific purpose. The most critical parameters that are considered in the FCs are as follows:

10.6.1 Ion Exchange Capacity–Conductivity

One of the critical parameters for characterizing the PEM is ion exchange capacity (IEC). It is obtained by dividing dry mass equivalent to one mole of acid groups [reciprocal of equivalent weight (EW)] and its unit is g/mol H^+. Characterizing a polymer with the feature of proton exchange conductivity is done by IEC and EW values. However, IEC may be misleading, because each polymer backbone has different masses and different IEC is generated. Thus, it is better to use a volume-based IEC (mole/cm^3) instead of a mass-based IEC while it is not common [19]. EW, relative humidity, and temperature are significant parameters that affect conductivity. In addition, other factors such as mechanical annealing, using various solvents for casting, and thermal treatments impact on proton conductivity and morphology in PEM [20].

10.6.2 Durability

Maintaining chemical and mechanical properties during the membrane lifetime is an essential requirement. Gradually reduction in membrane conductivity results from loss of ion exchange groups associated with decreased mechanical strength (by forming pinhole) and finally rupture. All of this is not desired for a PEMFC [19].

10.6.3 Water Management

Water management in PEM is one of the main factors in determining the applicability and operability of FCs. Water plays a crucial role in proton conductivity, because it can accelerate the proton separation from the membrane and provides hydrated mobile protons [21]. Therefore, it can be said that proton movement through the PEM is a function of water content and water sorption of PEM. However, the transport properties of the membrane cannot be ignored [19]. Different parameters affecting permeability that water quantity in membrane and ionomer structure are the most important [20]. Water transport in membrane contains several stages such as diffusion (because of water content gradient), hydraulic permeation (due to pressure gradient between anode and cathode), and electroosmotic drag [19]. Other leading factors like thermal and mechanical annealing and casting process affect permeability [20].

10.6.4 Cost

The biggest obstacle in the commercialization of membranes is the cost of materials. Many researchers have tried to optimize the low-cost materials to achieve highest structural properties with the lowest price. In this way, various membranes such as Nafion® (DuPont, USA), Flemion® (Asahi Glass, Japan), and Aciplex® (Asahi Kasei, Japan) are commercialized by the different companies that are still expensive [19].

10.7 Membrane Materials for PEMFC

The choice of suitable material for a PEMFC relies on operating temperature. For instance, in a FC that works at low temperature (<100°C), sulfonated polymers and other ion exchange materials are commonly used. Water is needed for proton separation from sulfonic acid groups in the main chain of the polymer. Suggested materials for 100°C–200°C operating temperature are amphoteric proton self-conductive materials (e.g., acidic phosphates, phosphonic acid/phosphoric acid, or heterocycles capable of getting or giving protons without the need for a mobile molecule (used as the PEM separated the electrodes). These FCs are known as "intermediate temperature FCs" (ITFCs) [22].

PEM materials are divided into the following three overall categories based on their chemical structure:

1. Statistical copolymers.
2. Block and graft copolymers.
3. Polymer Blending and other PEM compounds.

Each division has some subdivisions, which are explained further.

10.7.1 Statistical Copolymer PEMs

Statistical copolymers are a group of copolymers in which the distribution of monomers through their backbones follows statistical rules; for example, the monomer sequence distribution may obey Markovian statistics of first, second, or other statistical laws [23]. Many existing PEMs are made up of statistical copolymers. In addition, sulfonic acid–based copolymers are a significant part of this copolymers' type. Direct polymerization and postsulfonation are two ways for their preparation. In general, several general categories can be expressed for statistical copolymers [24]:

- **Perfluorinated and Partially Fluorinated**
  PEM on perfluorocarbon polymers is one of the most favored FC systems, which has succeeded in commercial applications such as transportation, stationary, and portable usages. In addition to high performance, the durability of polymers based on perflourosulfonic acid (PFSA) is outstanding [24]. A PFSA contains a nonpolar polytetrafluoroethylene (PTFE) in the backbone and the polar side chain terminated with sulfonic acid [25]. Sulfonic acid groups in contact with water are disassociated and protons are transported through the membrane. At the same time, remaining SO_3^- groups become hydrated and form a neutralizing atmosphere in the presence of water and protons [26]. These materials have high proton conductivity along with mechanical, chemical, and thermal stability. However, PFSA polymers are mechanically unstable at temperatures above 100°C [25, 27]. In high temperatures, water dehydration has happened in this kind of ionomer, which reduces protons' conduction.

Nevertheless, many efforts have been made to overcome this issue. Therefore, at this moment, these materials are suitable for low temperatures.

Nafion®, Fumion®, Flemion®, Aquivion®, and 3M™ are the commercial polymers that have been widely used in various industries used as electrolytes due to their electrochemical stability, proton conductivity, and lifetimes [28, 29]. The range of different structures of PFSA is shown in Figure 10.3.

Nafion® **(1a)** is a PFSA polymer, which is commercialized in the 1960s by DuPont. Aliphatic perfluorinated and ether-linked end to sulfonic acid are placed as backbone and side groups in Nafion, respectively. Nafion has some advantages, such as high mechanical strength, proton conductivity, and long-term durability. However, some drawbacks (e.g., high methanol permeation, high price, and low proton conductivity in high temperature) caused hesitation in its use [30].

Different EWs of polymers make various type of polymers. For example, similar structures to Nafion (EW ~ 1,100 g/mol) have been developed. Flemion® (EW ~ 1,000 g/mol), Aciplex® **(1b)** (EW ~ 1,000–1,200 g/mol), and Nafion have a difference in EW. Dow Chemical invented a polymer with a simpler structure than Nafion **(2a)**, which has higher performance due to its shorter chain length. However, its monomer is more expansive than Nafion and this issue stopped the project [24].

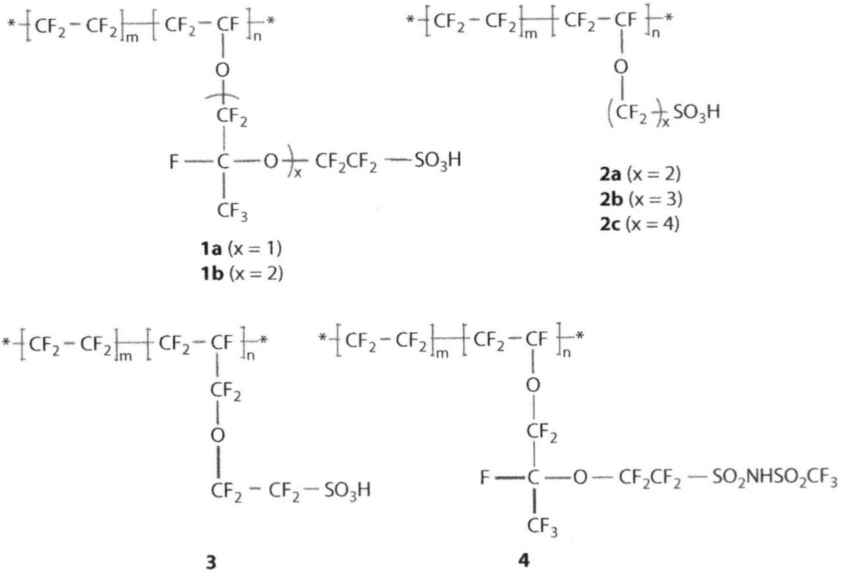

$*\!-\![CF_2-CF_2]_m-[CF_2-CF(O-[CF_2-CF(CF_3)-O]_x-CF_2CF_2-SO_3H)]_n-*$

1a (x = 1)
1b (x = 2)

$*\!-\![CF_2-CF_2]_m-[CF_2-CF(O-(CF_2)_x-SO_3H)]_n-*$

2a (x = 2)
2b (x = 3)
2c (x = 4)

$*\!-\![CF_2-CF_2]_m-[CF_2-CF(CF_2-O-CF_2-CF_2-SO_3H)]_n-*$

3

$*\!-\![CF_2-CF_2]_m-[CF_2-CF(O-CF_2-CF(CF_3)-O-CF_2CF_2-SO_2NHSO_2CF_3)]_n-*$

4

Figure 10.3 The range of different structures of PFSA: Nafion™ and Flemion™ (1a); Aciplex™ (1b); DOW, Hyflon-ion (2a); 3M™ (2b); Asahi Kasei (2c); Asahi Glass (3) and *bis* [perfluoroalkyl sulfonyl] groups (4) [24].

A novel sample similar to the Dow structure is synthesized by 3M Company **(2b)** with one extra CF_2 side group compares to **(2a)**. Asahi Kasei designed a new material **(2c)** that is structurally similar to **(2a-b)**, with the difference that its carbon chain is longer. Structure **(3)** (produced by Asahi Glass) has a CF_2 group that separates the ethereal oxygen and remaining sulfonic acid side group from each other. Its durability owes to the quinone group in this polymer [24].

In other perfluorocarbon PEM **(4)**, sulfonamide groups are used instead of sulfonic acid that causes this material to have significant stability against thermal, electrochemical, and chemical situations. In addition, they are not too sensitive to oxidation decomposition and dehydration compared to PEMs with fluorosulfonic acid groups and assumed as a nanophase-separated structure. This low sensitivity leads to better conductivity for samples **(4)** (EW ~ 1,200–1,400 g/mol) than Nafion® (EW ~ 1,100 g/mol) in RH < 70% [24].

For optimal performance, membrane hydration is necessary for membrane swelling and forming the ion transfer bridge. For this aim, one or some reactant gas flows must be hydrated, or water products provide humidity in integrated systems [21]. As considering the benefits mentioned for the perfluorinated PEM, using them faces challenges that can be pointed out by the following:

1. Unusable in high temperature.
2. Swelling and shrinking during the application due to change water uptake by changing temperature and humidity.
3. Expansive raw materials (~$700/m^2).
4. The complex production process as well as not environment-friendly.
5. The proton conductivity depends on water content (humidifier is required for maintaining the humidity in the membrane) [31].

To solve synthesis problems of suitable perfluorocarbon-based monomers, partially fluorinated systems with advantages (e.g., reasonable price, sustaining durability, and performance) have been suggested. Sulfonated copolymer incorporating with α,β,β-trifluorostyrene (BAM3G) (synthesized by Ballard) is one of the most popular perfluorinated membranes. In addition, its cost is lower than PFSA. Two examples of these membranes are given in Figure 10.4 [24, 29].

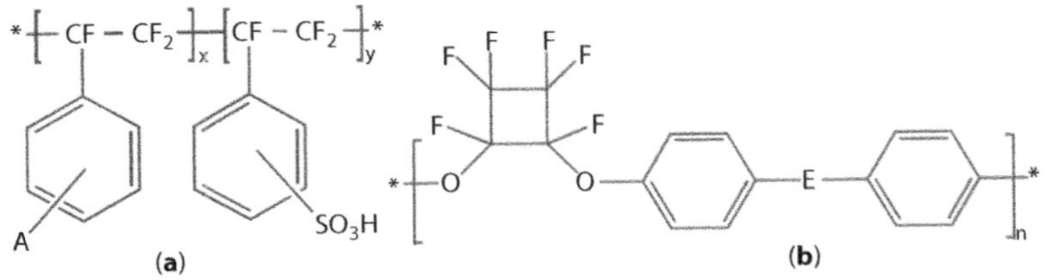

Figure 10.4 Examples of partially fluorinated systems: trifluorostyrene-based PEM (a) and a trifluorovinyl ether-based PEM (b) [24].

- **Polyarylenes**
  One of the most versatile hydrocarbon-based PEMs is polyarylene-based PEMs. Polyarylene is a compound with aryl or heteroaryl ring in its backbone. Recent research into using these materials as PEMs has shown that these materials have better oxidation, mechanical and thermal stability than polystyrene-based systems. Initial samples of this kind of material, which were also used in the space project, are phenol-formaldehyde sulfonated resins (shown in Figure 10.5). In space projects, these materials have a convenient synthesis and sulfonation as well as low cost, but their low oxidation stability causes limitations in their usage [24].
- **Other Statistical Copolymer PEM**
  Various compounds are taken in this category. For instance, polyphosphazene (a combination of phosphorus and nitrogen) is an inorganic polymer used as a PEM due to its excellent chemical, thermal, and oxidative stability and ease of functionalization with other functional groups [24].

10.7.2 Block and Graft Copolymers

Block copolymer is known as a copolymer with a linear arrangement of monomers blocks in the main chain of the copolymer. This chain is propagated by different monomer blocks connection. In a graft copolymer, one

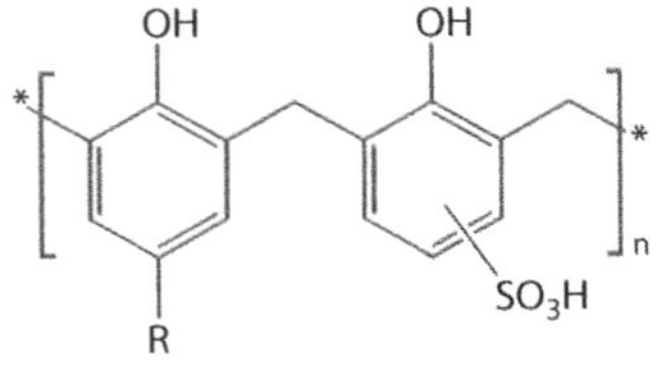

Figure 10.5 Phenol-formaldehyde sulfonated resin: It is used as PEM [24].

or more kinds of monomers are linked to the backbone as a side chain, and they are commonly arranged non-statistical. The simplest model of a graft copolymer is A_n-*g*-B_m (A and B are monomers). In the following, each of these categories is explained better to understand them and their application as a PEM [23]:

- **Block Copolymers**
  The first sample of a block copolymer that is used as an inexpensive PEM is sulfonated polystyrene-*b*-poly(ethylene-*r*-butylene)-*b*-polystyrene (S-SEBS) (designed by Dais-Analytic Corp.). This copolymer is operated in the hydrogen PEMFC at ambient conditions and low current densities. By testing these types of copolymers, researchers found that proton conductivity has independence of IEC in some range of IECs (IEC = 0.94–1.71 meq/g) [24]. Many copolymers are employed as a PEM in a FC that some are brought by Table 10.4.
- **Graft Copolymers**
  Synthesis of graft copolymer PEMs typically occurs by irradiation grafting monomers' reactions into a dense base substrate (a commercial polymer film such as styrene) with further sulfonation of the reactive species. Its main benefit is using a base substrate, which is commonly a commercial polymer. It can provide suitable mechanical strength and

Table 10.4 Various types of block copolymers were examined as a membrane in a fuel cell [24].

Name of block copolymers	Synthesis or examined by
S-SEBS	Dais-Analytic Corp
Sulfonated poly(styrene-*b*-isobutylene-*b*-styrene), S-SIBS	Elabd *et al.*
Poly(arylene ether sulfone)-*b*-poly(arylene ether ketone)	Ghassemi, Ndip, and McGrath.
PVDF-co-hexafluoropropylene) (PVDF-*co*-HFP))	Holdcroft and Shi
Sulfonated polysulfone-b-PVDF (PSF-*b*-PVDF)	Rubtat *et al.*, Norsten *et al.*, and Yang

excellent proton conductivity of grafted chains based on appropriate choices. Irradiation can be done in two ways:

1. *In situ* irradiation is done in the presence of monomer and base substrate.
2. The substrate is irradiated before monomer irradiation.

The reaction between monomers and radicals, which are produced in the base substrate, is happened in both cases and leads to the propagation and structure formation of polymer chains attached to the film (irradiated one). In Figure 10.6, some samples of grafted PEMs are given [24].

Poly(ethylene-alt-tetrafluoroethylene)-graft-poly (styrene sulfonic acid) (ETFE-g-PSSA (Figure 10.6b) is developed by Rouiley *et al.* Its substrate is FEP, and its performance has a similarity to Nafion under 60°C and fully humidified conditions. However, its lifetime is short, but it can be improved to 1,000 h by adding divinyl benzene (DVB) as a cross-linker meanwhile the graft copolymerization with styrene. The time can be increased by

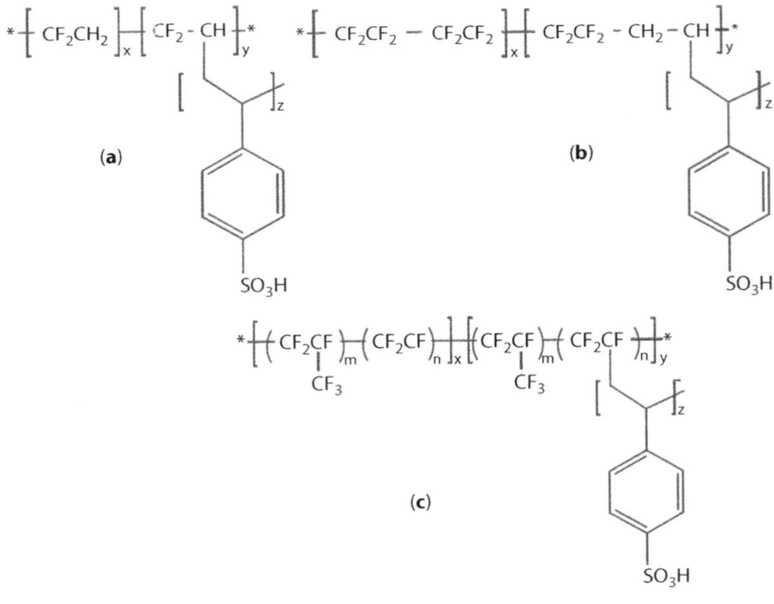

Figure 10.6 Samples of PEMs grafted by irradiation: PVDF-g-PSSA (a); poly(ethylene-alt-tetrafluoroethylene)-graft-poly (styrene sulfonic acid) () (b); and poly(tetrafluoroethylene-cohexafluoropropylene)-g-PSSA (FEP-g-PSSA) (c) [24].

using DVB and triallyl cyanurate altogether. Other kinds of substrates have been used in other research. For example, Horsfall, Lovell, and Shen *et al.* utilized ethylene tetrafluoroethylene (ETFE) and polyvinylidene fluoride (PVDF) as a substrate. PEMs with materials (Figures 10.6a and b) have been shown significant FC performance in PEMFC and DMFC in comparison to material (Figure 10.6c). The research results show that the important parameters in FC performance are the grafting conditions (type of cross-linker, operating conditions, etc.) and overall sulfonic acid content. The substrate composition has less importance on FC performance than late parameters [24].

10.7.3 Polymer Blending and Other PEM Compounds

PEMs without any additives are mentioned in the previous two categories. In this section, a group of PEMs combined with additives is described. Polymer blends, composite PEMs, ionomer-filled porous substrates, and reinforced PEMs are three general parts of these materials that each of them is expressed below:

- **Polymer Blends**

  Polymer blending is referred to mixing two or more polymers altogether to integrate the favorable properties of each component with each other. High IEC polymers mix with low IEC polymers to improve FC performance and mechanical strength simultaneously. An example of efforts in this area is mixing Nafion with other polymers, especially PVDF and its copolymer, such as PVDF-co-HFP, to achieve higher mechanical strength due to the semi-crystalline nature and thermal and chemical stability of PVDF [24]. However, this blending harms proton conductivity and the conductivity of the resulting blend is remarkably lower than pure Nafion because of the hydrophobic nature of PVDF and lower water uptake as a result [24].

  Other attempts were made on other polymers; for example, poly(ether sulfone) (PES) or sulfonated poly(ether sulfone) (SPES) with various contents (typically 20–60 wt%) are used for improving the mechanical strength of SPEEK. FC testing results showed a 40% increase in conductivity for modified SPEEK compared to pure SPEEK. Another blending polymer for reinforcing SPEEK is PVDF [24].

- **Composite PEMs**
  In the last decade, a group of materials known as composite PEMs for operating above 100°C has attracted the attention of researchers, because high-temperature operation (>120°C) has the following benefits:

  1. Increment in FC efficiency
  2. Improvement in the kinetics of fuel oxidation
  3. Reduction in CO poisoning effect of the anode catalyst
  4. Advancement in gas transport (absence of liquid water) [24]

Despite the benefits, commercial polymers (e.g., pure Nafion) are inappropriate for high-temperature operation. This condition challenges other PEMFC components such as seals, catalysts, and diffusion gas layer. At high temperatures, the tendency of dehydration (and consequently reduction in proton conductivity) is increasing. In addition, loss in mechanical strength and increase in gas permeability are happened.

There are several ways to dominate this problem. Preventing water loss through proton transport pathways is one of the ways; therefore, for this purpose, the proton conduction is maintained above the boiling point of water by adding hydrophilic mineral components to the membrane. These particles also can conduct protons [24].

Various factors affect composite PEMs features, such as the amount of dispersed materials, homogeneous dispersion, size and the orientation of the solid compounds dispersed in the polymer matrix, and ionomer nature [27].

Zeolites, alumina, titania, silica, zirconia, and clays are various mineral compounds employed in composite PEMs. In most composite PEMs, Nafion is used as the primary polymer but SPEEK, or other polymers have also been used in some studies. Inorganic materials in PEMs are used in two methods: the first one is done by adding compounds to an ionomer solution before casting and the other one is mixing to an ionomer solution during the particles' impregnation through a preformed PEM. Application of preexisting material before casting increases the possibility of forming a hydrophilic zone, inhomogeneous dispersion, and particles aggregation [24].

- **Ionomer-Filled Porous Substrates and Reinforced PEMs**
  Reinforcing materials are usually used for mechanical stability and proton conductivity improvement. One of their

uses is to mix these materials to polymers into thin membranes. Although thin membranes have some benefits (e.g., low resistance to proton transportation, inexpensive cost, and good water management in PEMFC applications), their preparation is more complicated than thicker membranes because of a reduction in mechanical strength. Porous polymers (e.g., PP, PTFE, and PC), expanded PTFE, PSF, and glass microfiber fleece are the most common reinforcing materials [24].

In addition, properties of reinforced PEMs are affected by porosity and thickness of substrate; for instance, based on H_2/O_2 FC testing at 60°C/0.2 MPa for Nafion, the current density is increased from 500 to 800 mA cm^{-2} by increment in substrate pore size from 0.3 to 0.5 μm. Furthermore, current densities for Nafion 115, 45-μm Nafion/PTFE, and 25-μm Nafion/PTFE are 800, 800, and 950 mA cm^{-2}, respectively. These results show that current density has a reverse relationship with substrate thickness [24].

There are several methods for PEM preparation, which are mentioned in Figure 10.7. In addition, in Table 10.5, PEM preparation methods, their advantages and disadvantages, and some PEM characterizations are given.

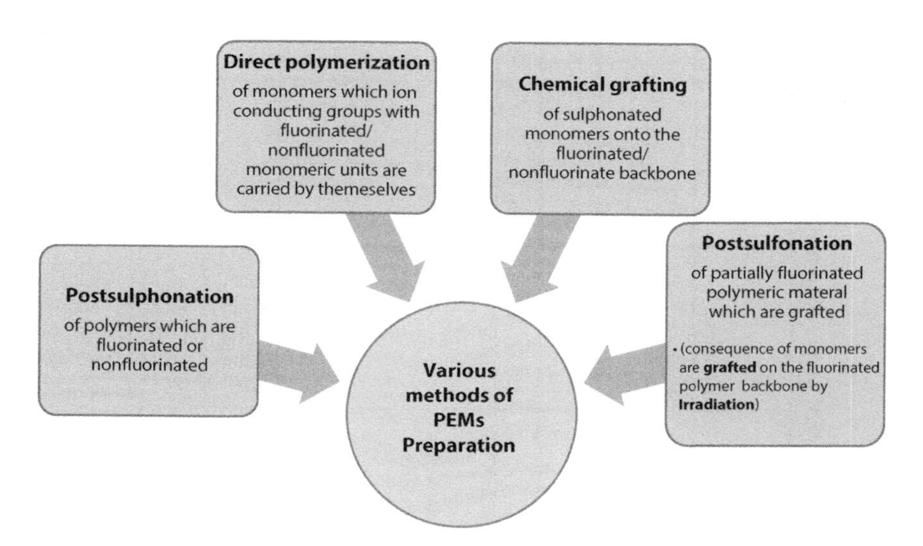

Figure 10.7 Various methods of PEMs preparation [9].

Table 10.5 PEM preparation methods, their advantages and disadvantages, and some PEM characterizations.

Modified methods	Samples	EW (IEC mequiv. g^{-1})	Water content (%)	Proton conductivity (S cm^{-1})	Single-cell performance	Overall advantage	Overall disadvantage
Use commercial products	Nafion 112 [3]	1100	12 (25°C)	0.033 (25°C)	(1) excellent performance at low temperature	(1) High mechanical strength and proton conductivity	(1) High methanol permeation, (2) High price
	Nafion 117 [3]	1100	13 (25°C)	0.075 (25°C)			
	Nafion 115 [25]	N.A.*	N.A.	0.095 (50°C)			
Post-sulfonation of commercial products	SPAEK (DS** = 1.16) [3]	555 (1.8)	37 (25°C)	0.063 (25°C)	(1) At high temperatures (80°C–100°C), their performance is excellent, but their conductivity is low [30]	(1) Simplicity and feasibility [30]	(1) For real application, DS = 40%–70% is needed (2) Their proton conductivity is lower than Nafion® [30]
	SPEEK (DS = 0.68) [3]	N.A.	17.4	0.0463 (25°C)			
	SPPEK (DS = 1.09) [3]	N.A.	25	$>10^{-2}$			

(Continued)

Table 10.5 PEM preparation methods, their advantages and disadvantages, and some PEM characterizations. (*Continued*)

Modified methods	Samples	EW (IEC mequiv. g^{-1})	Water content (%)	Proton conductivity (S cm^{-1})	Single-cell performance	Overall advantage	Overall disadvantage
Direct synthesis from sulfonated monomers	AL-SPI-5 [32]	(1.79)	0.106 (30°C)	0.106 (30°C)	(1) Their performance is better than sulfonated commercial products [30]	(1) Different types of membranes are introduced (2) Higher DS (> 100%) might be obtained in the presence of more than one sulfonic acid group in sulfonated monomers [30]	(1) Complexity of the process (2) Their proton conductivity is lower than Nafion® [30]
	SPSF (DS = 0.33) [29]	(1.24)	20.93 (25°C)	0.071 (25°C)			
Blocked or branched membranes	Graft: Dense PVDF (80 μm)/Styrene (monomer)/ sulfuric acid (activating agent) [3]	(0.68–1.70)	N.A.	0.02–0.03	(1) Proton conductivity improves due to forming well non-phase separation but increased methanol crossover [30].	(1) Better proton transfer channel is happened due to forming well non-phase separation [30]	(1) Complexity of the process (2) Methanol permeability is high [30]
	Porous PVDF (125 μm)/ Styrene [3]	N.A.	Variable	Up to 0.13			
	PVDF-*g*-PSSA [19]	(2.78)	N.A.	0.13			

(*Continued*)

Table 10.5 PEM preparation methods, their advantages and disadvantages, and some PEM characterizations. (*Continued*)

Modified methods	**Samples**	**EW (IEC mequiv. g^{-1})**	**Water content (%)**	**Proton conductivity (S cm^{-1})**	**Single-cell performance**	**Overall advantage**	**Overall disadvantage**
Composite membranes with other polymers	PVDF (10%)/ SPPESK [33]	N.A.	N.A.	0.0085	(1) Determined by choice of the second polymers, and whether a strong interaction be formed [30]	(1) Outstanding performance (2) forming several composite structures might be occurred [30]	(1) Commonly, their proton conductivity is lower than neat polymer membranes [30]
	SPEEK (80%)/ (20%) SPVDF-HFP [34]	(1.76)	25.8 (25°C)	0.0015 (25°C)			
	SPEEK (62%)/ (38%) Aquivion® [35]	N.A.	96	0.023 (80°C, RH 50%)			

(*Continued*)

Table 10.5 PEM preparation methods, their advantages and disadvantages, and some PEM characterizations. (*Continued*)

Modified methods	Samples	EW (IEC mequiv. g^{-1})	Water content (%)	Proton conductivity (S cm^{-1})	Single-cell performance	Overall advantage	Overall disadvantage
Organic-inorganic composite membranes	SPEEK (DS = 0.72)/BPO_4 (20%) [3]	N.A.	66 (25°C)	0.0032 (25°C)	(1) The inorganic fillers selection to form a strong interaction is vital in this type (2) At high temperatures (>100°C), their performance is much better [30]	(1) The proton conductivity is high. However, the methanol permeability is low. (2) Their chemical and thermal stability is high, and their dimensions don't change during the process [30]	(1) Inorganic filler should be functionalized for better dispersion and interaction with polymer matrix [30]
	Nafion/SiO_2 [27]	N.A.	N.A.	0.023 (25°C)			
	SPEEK (DS = 0.72)/ OMMT (10%) [3]	N.A.	N.A.	0.0088 (100°C)			

* Not Applicable.
** Degree of Sulfonation.

10.8 Conclusions

FC membranes play an essential role in PEMFC. These membranes are made of different materials, the most important of which are perfluorinated membranes. The material choice depends on the operating conditions. The durability and cost of these materials are significant challenges for their commercialization. However, researchers are working to reduce the cost of the membrane and improve its FC properties. Therefore, different methods are invented to prepare PEM, shown in Figure 10.7; for instance, hydrocarbon-based systems have lower costs than other materials and do not have concerns about fluorocarbon membranes applications. Because of this, block and graft copolymers are the combination of different types of monomers and their properties are integrated to improve overall properties. They have been very effective in enhancing proton conductivity and reducing prices. Therefore, composite membranes and polymer blends have also played a favorable effect on the proton conduction of PEMs. By developing the knowledge of structure-property, more organized efforts will be made in PEM preparation to achieve high-performance, durable, and low-cost PEM.

References

1. Gasik, M., Introduction: materials challenges in fuel cells, in: *Materials for Fuel Cells*, pp. 1–5, Woodhead Publishing, Cambridge, United Kingdom, 2008.
2. Khatib, F.N., Wilberforce, T., Thompson, J., Olabi, A.G., A comparison on the dynamical performance of a proton exchange membrane fuel cell (PEMFC) with traditional serpentine and an open pore cellular foam material flow channel. *Int. J. Hydrog. Energy Int. J.*, 46, 5984, 2021.
3. Zaidi, J. and Matsuura, T., *Polymer membranes for fuel cells*, pp. 1–25, Springer Science & Business Media, Berlin/Heidelberg, Germany, 2008.
4. Yang, Y., Zhou, X., Li, B., Zhang, C., Recent progress of the gas diffusion layer in proton exchange membrane fuel cells: Material and structure designs of microporous layer. *Int. J. Hydrog. Energy Int. J.*, 46, 4259, 2020.
5. Qi, Z., *Proton Exchange Membrane Fuel Cells*, pp. 1–56, CRC Press, Florida, United States, 2013.
6. Hamrock, S.J. and Yandrasits, M.A., Proton exchange membranes for fuel cell applications. *J. Macromol. Sci., Part C: Polym. Rev.*, 46, 219, 2006.
7. Odeh, A.O., Osifo, P., Noemagus, H., Chitosan: A low cost material for the production of membrane for use in PEMFC-A revi. *Energ. Sources, Part A: recovery, utilization, Environ. effects*, 35, 152, 2013.

8. Brandon, N.P. and Parkes, M.A., *Fuel Cells: Materials*, pp. 1–6, Elsevier, Amsterdam, Netherlands, 2016.
9. Walkowiak-Kulikowska, J., Wolska, J., Koroniak, H., 10. Polymers application in proton exchange membranes for fuel cells (PEMFCs). *Phys. Sci. Rev.*, 2, 293, 2017.
10. Pu, H., *Polymers for PEM fuel cells*, pp. 1–49, John Wiley and Sons, New Jersey, United States, 2014.
11. Saadat, N., Dhakal, H.N., Tjong, J., Jaffer, S., Yang, W., Sain, M., Recent advances and future perspectives of carbon materials for fuel cell, *Renew. Sustain. Energy Rev.*, 138, 110535, 2020.
12. Dickinson, E.J. and Smith, G., Modelling the proton-conductive membrane in practical polymer electrolyte membrane fuel cell (PEMFC) simulation: A review. *Membranes*, 10, 310, 2020.
13. Wee, J.H., Applications of proton exchange membrane fuel cell systems, Renew. *Sustain. Energy Rev.*, 11, 1720, 2007.
14. Zhang, J., Zhang, H., Wu, J., Zhang, J., Chapter 1-PEM fuel cell fundamentals, in: *PEM Fuel Cell Testing and Diagnosis*, pp. 1–42, Elsevier, Amsterdam, Netherlands, 2013.
15. Srinivasan, S., *Fuel cells: from fundamentals to applications*, pp. 189–233, Springer Science and Business Media, Berlin/Heidelberg, Germany, 2006.
16. Okonkwo, P.C., Belgacem, I.B., Emori, W., Uzoma, P.C., Nafion degradation mechanisms in proton exchange membrane fuel cell (PEMFC) system: A review. *Int. J. Hydrog. Energy Int. J.*, 46, 27956, 2021.
17. Stropnik, R., Lotrič, A., Bernad Montenegro, A., Sekavčnik, M., Mori, M., Critical materials in PEMFC systems and a LCA analysis for the potential reduction of environmental impacts with EoL strategies. *Energy Sci. Eng.*, 7, 2519, 2019.
18. Branco, C.M., El-kharouf, A., Du, S., Materials for polymer electrolyte membrane fuel cells (PEMFCs): Electrolyte membrane, gas diffusion layers, and bipolar plates. *Reference Module Mater. Sci. Mater. Eng.*, 1–11, 2017.
19. Gubler, L. and Scherer, G.G., A proton-conducting polymer membrane as solid electrolyte–function and required properties, in: *Fuel Cells I*, pp. 1–14, Springer, Berlin/Heidelberg, Germany, 2008.
20. Kopasz, J. and Mittelsteadt, C., Polymer electrolytes, in: *Fuel Cells: Data, Facts, and Figures*, pp. 101–109, John Wiley and Sons, New Jersey, United States, 2016.
21. Yang, C., Costamagna, P., Srinivasan, S., Benziger, J., Bocarsly, A.B., Approaches and technical challenges to high temperature operation of proton exchange membrane fuel cells. *J. Power Sources*, 103, 1–9, 2001.
22. Hartnig, C.H., Jörissen, L., Kerres, J., Lehnert, W., Scholta, J., Polymer electrolyte membrane fuel cells, in: *Materials for Fuel Cells*, pp. 101–184, Woodhead Publishing, Cambridge, United Kingdom, 2008.

23. Smith, W.L., Borgeat, P., Hamberg, M., Jackson Roberts, L., Willis, A.L., Yamamoto, S., Ramwell, P.W., Rokach, J., Samuelsson, B., Corey, E.J., Pace-Asciak, C.R., Nomenclature. *Methods Enzymol.*, 187, 1–9, 1990.
24. Peckham, T.J., Yang, Y., Holdcroft, S., Proton exchange membrane, in: *Proton Exchange Membrane Fuel Cells: Materials Properties and Performance*, pp. 107–189, CRC Press, Florida, United States, 2009.
25. Kwon, S.H., Kang, H., Lee, J.H., Shim, S., Lee, J., Lee, D.S., Lee, S.G., Investigating the influence of the side-chain pendants of perfluorosulfonic acid membranes in a PEMFC by molecular dynamics simulations. *Mater. Today Commun.*, 21, 100625, 2019.
26. Kornyshev, A.A. and Spohr, E. S. J. K. S., 10 Proton transport in polymer electrolyte membranes using theory and classical molecular dynamics, in: *Device and Materials Modeling in PEM Fuel Cells*, pp. 349–361, Springer Science & Business Media, Berlin/Heidelberg, Germany, 2008.
27. Rhee, H.W. and Ghil, L.J., Polymer nanocomposites in fuel cells, in: *Advances in Polymer Nanocomposites*, pp. 433–471, Woodhead Publishing, 2012.
28. Khoo, K.S., Chia, W.Y., Wang, K., Chang, C.K., Leong, H.Y., Maaris, M.N.B., Show, P.L., Development of proton-exchange membrane fuel cell with ionic liquid technology. *Sci. Total Environ. Sci. Total Environ.*, 793, 148705, 2021.
29. Li, X., Wang, S., Zhang, H., Lin, C., Xie, X., Hu, C., Tian, R., Sulfonated poly (arylene ether sulfone) s membranes with distinct microphase-separated morphology for PEMFCs. *Int. J. Hydrog. Energy Int. J.*, 46, 33978, 2021.
30. Xu, M., Xue, H., Wang, Q., Jia, L., Sulfonated poly (arylene ether) s based proton exchange membranes for fuel cells. *Int. J. Hydrog. Energy Int. J.*, 46, 31727, 2021.
31. Baroutaji, A., Carton, J.G., Sajjia, M., Olabi, A.G., *Materials in PEM fuel cells*, Elsevier, 2015.
32. Yao, Z., Zhang, Z., Hu, M., Hou, J., Wu, L., Xu, T., Perylene-based sulfonated aliphatic polyimides for fuel cell applications: Performance enhancement by stacking of polymer chains. *J. Membr. Sci.*, 547, 43, 2018.
33. Gu, S., He, G., Wu, X., Hu, Z., Wang, L., Xiao, G., Peng, L., Preparation and characterization of poly (vinylidene fluoride)/sulfonated poly (phthalazinone ether sulfone ketone) blends for proton exchange membrane. *J. Appl. Polym. Sci.*, 116, 852, 2010.
34. Martina, P., Gayathri, R., Pugalenthi, M.R., Cao, G., Liu, C., Prabhu, M.R., Nanosulfonated silica incorporated SPEEK/SPVdF-HFP polymer blend membrane for PEM fuel cell application. *Ionics*, 26, 3447, 2020.
35. Boaretti, C., Pasquini, L., Sood, R., Giancola, S., Donnadio, A., Roso, M., Cavaliere, S., Mechanically stable nanofibrous sPEEK/Aquivion® composite membranes for fuel cell applications. *J. Membr. Sci.*, 545, 66, 2018.

11

Nafion-Based Membranes for Proton Exchange Membrane Fuel Cells

Santiago Pablo Fernandez Bordín[1,2], Janet de los Angeles Chinellato Díaz[3,4] and Marcelo Ricardo Romero[3,4]*

[1]*Instituto de Física Enrique Gaviola (IFEG), Consejo Nacional de Investigaciones Científicas y Técnicas CONICET, Córdoba, Argentina*
[2]*Universidad Nacional de Córdoba, Facultad de Matemática, Astronomía, Física y Computación (FAMAF), Córdoba, Argentina*
[3]*Universidad Nacional de Córdoba, Facultad de Ciencias Químicas, Departamento de Química Orgánica, Córdoba, Argentina*
[4]*Instituto de Investigación y Desarrollo en Ingeniería de Procesos y Química Aplicada (IPQA), Consejo Nacional de Investigaciones Científicas y Técnicas (CONICET), Córdoba, Argentina*

Abstract

Proton exchange membrane fuel cells are devices that directly convert the chemical energy of a chemical reaction into electrical energy in the form of direct current. This device has characteristics that position itself as a promising source of clean energy. An essential component that directly affects the performance of fuel cells is the membrane used as electrolyte. In order to improve its efficiency, the membrane should be designed to comply with specific requirements such as high proton transport, good electrical insulation, low fuel permeability, and excellent thermal and chemical stability, to name only the most relevant elements. One of the most widely used and researched materials is the Nafion membrane, a hydrophobic fluoropolymer that contains hydrophilic side chains or ramifications ended in sulfonic acid group ($-SO_3H$). Throughout this chapter, the properties of Nafion and specifically its relevant structural and transport models were analyzed in detail. In this sense, small-angle X-ray/neutron spectroscopy (SAXS-SANS) are mandatory supplements and powerful tools for characterizing Nafion membranes and allow us to understand the fuel cell performance and how they are modified when subjected to different stimuli. In addition, results from different studies and authors are summarized.

**Corresponding author*: marceloricardoromero@gmail.com

Inamuddin, Omid Moradi and Mohd Imran Ahamed (eds.) Proton Exchange Membrane Fuel Cells: Electrochemical Methods and Computational Fluid Dynamics, (299–330) © 2023 Scrivener Publishing LLC

Keywords: Nafion membrane, proton exchange membrane, Nafion properties, Nafion characterizations

11.1 Introduction: Background

In the past decades, global energy consumption has considerably increased, and, as a consequence, energy production has become a topic of interest for research. This urgent need has driven the development of renewable energies and the improvement of old ones, which also seek to slow down climate change and solve other environmental problems. In this context, fuel cells (FC) have positioned themselves as an alternative energy source. FCs are electrochemical devices that directly convert the energy of a chemical reaction into electrical energy in the form of direct current. In particular, proton exchange membrane FCs (PEMFCs) are positioned as an energy source with great potential to replace power generation from fossil fuels [1].

One of the key components of PEMFCs, as the name suggests, is the membrane used as the electrolyte. This membrane is based on polymeric molecules and is designed to selectively allow the passage of protons from the anode to the cathode. In addition to acting as an impermeable electron barrier or for reactant liquids, gases, and anions [2, 3]. The main characteristics of an optimal polymeric membrane for use in PEMFCs are as follows [4, 5]:

- High proton conductivity: directly related with cell performance.
- High chemical, mechanical, and thermal stability: FCs generally operate in hostile chemical environments, are subjected to high-temperature variations, and operate under pressure.
- Low fuel permeability: fuel transport through the membrane should be avoided because it can cause short circuits, affecting cell performance.
- Low electronic conductivity: electrons must flow through the external circuit of the cell and not through the membrane.

Nafion membrane was developed in the 1960s by Walther Grot of the DuPont company [6] and is the most widely used and researched proton exchange membrane (PEM). The general chemical structure of Nafion is represented in Figure 11.1. The polymeric structure presents high similarity

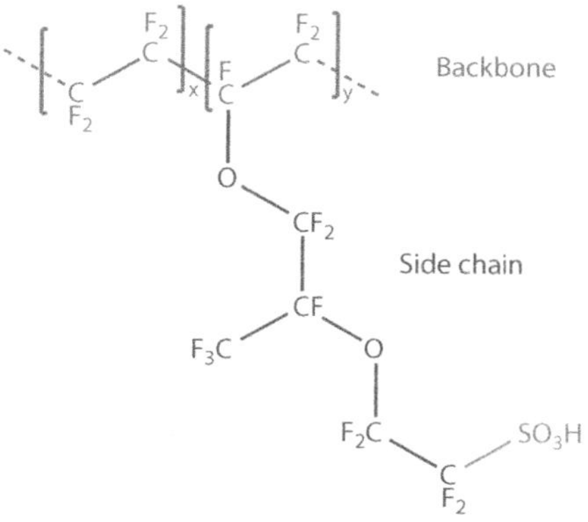

Figure 11.1 Chemical structure of Nafion (from the authors).

with teflon, since it is a hydrophobic fluoropolymer (polytetrafluoroethylene, PTFE), with hydrophilic side chains or branches ending in sulfonic acid group $-SO_3H$. These groups not only provide a certain concentration of acid groups to Nafion but also the number of acid groups is proportional to the proton conductivity of the membrane.

The initial idea of Nafion applications was as a separating membrane for chlorine and caustic soda production in chlor-alkali cells. However, the economic interests of that time period did not fulfill the conditions necessary to the development of the chlor-alkali industry [7]. Fortunately, in the 1960s, General Electric Company (GE) was working on the development of PEMFCs for the NASA space program in the United States. In these investigations, GE found that Nafion satisfied the requirements for proton conduction and oxidative stability for implementation in FC.

Despite these important discoveries, the commercial development of PEMFCs was relegated to space and military industries for about 20 years. Subsequently, in the late 1970s and early 1980s, DuPont focused its Nafion membrane business for use in the chlor-alkali industry, which had begun to increase its development. It was only until the 1990s that the company focused its production on Nafion membranes applied to the PEMFC industry. Thus, FCs began to gain relevance as an alternative energy source, especially to replace combustion engines in cars. However, 30 years later, PEMFCs are still under development and have not been applied on a large scale. Nevertheless, over time, the applications of Nafion have diversified, among which we can mention some developments such as vanadium redox flow battery [8], dehumidification systems [9], and in the generation of artificial solar fuels (artificial photosynthesis) [10] and biosensors [11], among others.

Since its discovery, many studies have been carried out on the composition, properties, and performance of Nafion membranes in a wide variety of applications, and the large number of results position this material as the most widely used PTFE membrane on the market. However, there are other fluorinated membranes with very similar characteristics, such as those manufactured by the companies Asashi and Dow Chemical, which consist of fluorinated membranes with short side chains, and those produced by the company 3M, which manufactures analogous membranes but with medium-length side chains. In addition, the FuMATech company also markets fluorinated membranes with similar characteristics, such as Fumapem F-1850 and Fumapem F-14100 [12].

Nafion properties can depend on specific factors such as the chemical identity of the membrane, the preparation method used, the thermal and chemical history to which it has been subjected, and activation of Nafion in I-V measurements, among others [13]. These factors must be taken into account when making comparisons between different studies.

11.2 Physical Properties

In the market, there are a broad availability of Nafion membranes, among which two main types can be distinguished from preparation method, as follows:

a) Nafion™ Solution Cast Membranes SCM (NR211 and NR212)
b) Nafion™ Extrusion Cast Membranes ECM (N115, N117 and N110).

SCMs are usually obtained from Nafion dispersions, generally have thin thickness, and are selected when the control of low ion transport resistance is a crucial requirement. On the other hand, ECMs present opposite properties with an elevated resistance and low permeability. Typically, these membranes have an equivalent weight (EW) around 1,100 Da.

The typical thickness of SCM is in the range of 25–50 μm with a basis weight of 50–100 g/m^2. On the other, ECM are thicker and heavier, with a thickness around 120 μm and 250 μm, and a basis weight in the range of 250–500 g/m^2. Table 11.1 displays the relevant physical properties provided by the manufactured.

Another fundamental physical property relevant for Nafion membrane applications is related to its thermal stability. In this sense, FCs can operate most of the time at temperatures close to 100°C. In order to get a better picture of this characteristic, Figure 11.2 shows a typical schematic of thermogravimetric analysis (TGA) and the respective graph of

Table 11.1 Physical properties of ECM.

Properties	**ECM**
Tensile Modulus (MPa)	249 [1], 114 [2] and 64 [3]
Tensile Strength, Max. (MPa)	43 [1MD], 32 [1TD] 34 [2MD], 26 [2TD] 25 [3MD], 24 [3TD]
Elongation at Break (%)	225 [1MD], 310 [1TD] 200 [2MD], 275 [2TD] 180 [3MD], 240 [3TD]
Tear Resistance-Initial (g/mm)	6,000 [1], 3,500 [2], and 3,000 [3]
Tear Resistance-Propagating (g/mm)	>100 [1MD], >150 [1TD] 92 [2MD], 104 [2TD] 74 [3MD], 85 [3TD]
Conductivity (S/cm)	0.10 min
Available Acid Capacity (meq/g)	0.90 min
Total Acid Capacity (meq/g)	0.95–1.01
Thickness Change (%)	10 (from [1] to [2]) 14 (from [1] to [3])
Linear Expansion (%)	10 (from [1] to [2]) 15 (from [1] to [3])

[1]50% RH, 23°C
[2]water soaked, 23°C
[3]water soaked, 100°C
[MD]machine direction
[TD]transverse direction

its derivative (DTG) performed on a Nafion membrane. It can be seen that the weight of the membrane starts to decrease from 100°C as a consequence of the evaporation of water absorbed by the sulfonic groups. However, these functional groups will start to degrade at a temperature of ~380°C [14]. Finally, a significant weight loss corresponds to the decomposition and subsequent degradation of the fluorocarbon chain of Nafion in the range between 400°C and 600°C [15]. It should be noted that Nafion membranes composed of polymers and other elements present similar curves evidencing shifts in the peaks of the derivative weight, depending on their particular composition.

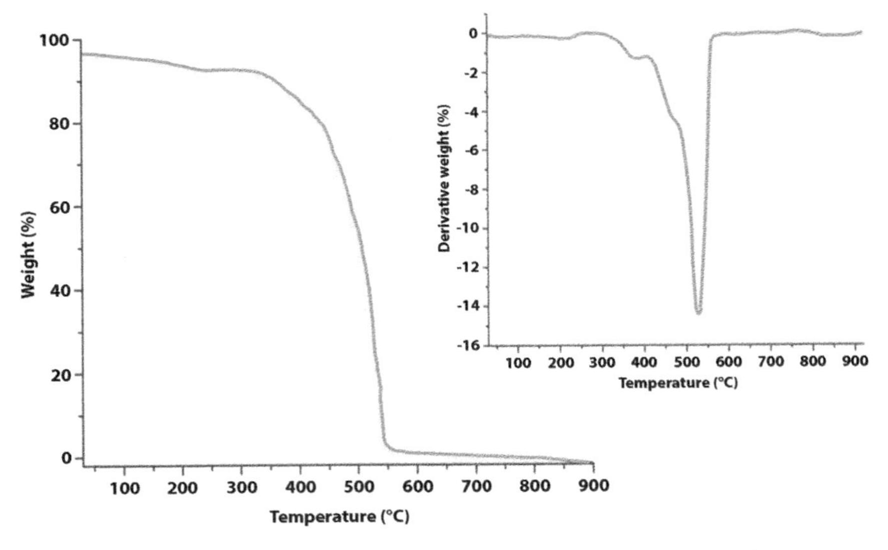

Figure 11.2 The TGA and DTG of Nafion 117. Reprinted from [14], Copyright 2019, with permission from Elsevier.

11.3 Nafion Structure

The morphology of Nafion membranes is a current subject of study and discussion since it directly influences their mechanical, thermal, and oxidative stability. However, a definitive model to describe their morphology has not yet been developed and instead in the literature can be found a variety of theoretical models proposed. Thus, various existing models are useful to explain certain properties of fluorinated membranes, such as the conformational behavior of hydrophilic channels, the spatial distribution of water molecules, ions, and counterions within them, and the mechanisms of proton transport as a function of the degree of hydration.

One of the first widely accepted structural models was proposed in 1981 by Gierke [16]. The so-called "cluster-channel" model consisted of 4-nm-diameter clusters within sulfonate ions ($-SO_3^-$) distributed in a continuous fluorocarbon network (CF_2). These clusters were connected by 1-nm-diameter channels, which served to explain the different transport mechanisms.

Subsequent research proposed other theoretical structural models, such as the sphere model describing the ion-poor core surrounded by a layer rich in sulfonic groups [17]. Later, the rod model was proposed in which the sulfonic groups are arranged in crystalline rods [18]. Then, a "sandwich" model was developed, consisting of two polymer layers whose sulfonic groups interact attractively through a water layer forming the ion channel [19].

Recently, Schmidt-Rohr and Chen proposed a model based on long, parallel, cylindrical water channels forming inverted micelles of 2.4-nm average diameter and PTFE crystals within an amorphous PTFE matrix [20]. This model was improved by Fernandez Bordín *et al.* [21] who proposed a detailed description of the channels structure and their grouping into hexagonal clusters. In addition, the mentioned authors performed an analysis under various humidity and temperature conditions, likewise the crystalline part was deeply analyzed. Both works involved experimental measurements that complemented the simulations of the Nafion structure to reinforce the proposed model.

Despite the differences between the existing models, most of them agree on the presence of an ionic domain, amorphous and crystalline phases, and water, providing a simple description of the membrane structure. In summary, the models fundamentally differ on the distribution of each phase.

The main experimental technique for the study of Nafion structure is small-angle X-ray/neutron scattering (SAXS/SANS). The obtained experimental curve or SAXS/SANS pattern produced by the Nafion membrane shows the scattering intensity, $I(q)$, as a function of the scattering vector, $q = |\boldsymbol{q}| \frac{4\pi sin(\theta)}{\lambda_b}$, (where λ_b and θ are the wavelength of the incident light in

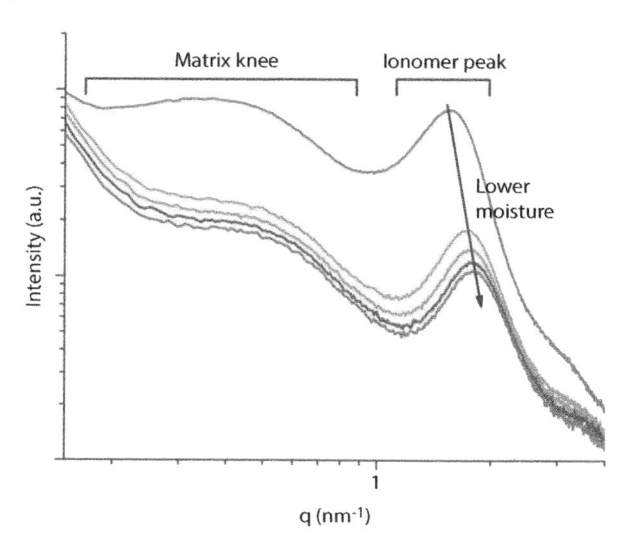

Figure 11.3 Typical SAXS pattern for Nafion membrane at different moisture conditions (from the authors).

the scattering medium and the scattering angle, respectively), has two main features (see Figure 11.3): a broad and sharp peak, called ionomer, around q ~1.6 nm^{-1}, and a matrix region with a shape of knee at $q < 0.7$ nm^{-1}.

The ionomer peak is a consequence of the regular arrangement of the water channels in the membrane. The width of this peak is associated with perturbations in the periodicity of the structure; while the center of the peak is related to the degree of hydration of the membrane, which moves toward higher q as the membrane dehydrates. This means that the correlation distance between the hydrophilic domains of the membrane shrinks [21].

On the other hand, the matrix knee is produced by the crystalline domains of the fluorocarbon polymer, dispersed in the amorphous polymer matrix. Moreover, the degree of crystallinity determines the scattering intensity of the matrix knee [16]. These crystals are important components that influence the mechanical properties. Furthermore, the characterization of the crystalline domains can be complemented by the wide-angle X-ray scattering (WAXS) and Figure 11.4 shows the characteristic spectrum obtained for Nafion membranes by means of this technique. In this spectrum, the first maximum corresponds to the overlap of the amorphous region with high intensity at $q = 11.64$ nm^{-1} superimposed with the crystalline region observed at $q = 12.34$ nm^{-1}. The spectrum is also characterized by another maximum occurring at $q = 27.5$ nm^{-1}, which corresponds to a Bragg distance of 2.28 Å. This distance is related to the inter-chain separation of the Nafion membrane. Therefore, WAXS measurements can

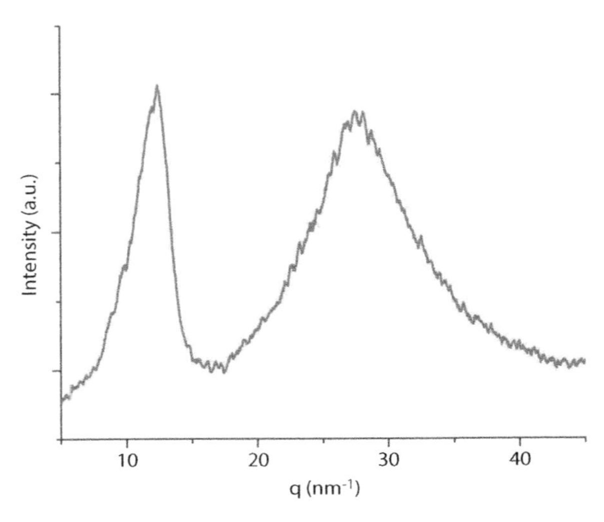

Figure 11.4 Typical WAXS pattern for Nafion membrane (from the authors).

provide a measure of membrane crystallinity using the equation proposed by Heijden *et al.* [22], which relates both integration of the crystalline intensity I_c with the total intensity I_T:

$$\chi_c = \frac{\int q^2 I_c(q)dq}{\int q^2 I_T(q)dq} \tag{11.1}$$

It should be noted that the crystallinity measurement can be interpreted as a tool to improve membrane performance since it can control species transport and affect proton conductivity.

11.4 Water Uptake

The conductivity of the Nafion membrane to hydrogen ions and the cell performance in operation are closely related to the membrane structure and the water content. Adequate control of the sample environment is essential for the proper functioning of the PEM, since deficiency or excess of relative humidity may be accompanied by a decay in the membrane conductivity [23].

The water content in the membrane can be expressed as both weight percent of water (w) and water content (λ). w relates the weights of the dry and hydrated membrane as in Equation (11.2):

$$w[\%] = \frac{W_{wet} - W_{dry}}{W_{dry}} \tag{11.2}$$

While λ is more frequently used and expresses the number of molecules per sulfonic acid group present in the polymer defined as

$$\lambda = \frac{N(H_2O)}{N(SO_3H)} = \frac{w \times EW}{M_{H_2O}} \tag{11.3}$$

where EW is the equivalent weight and M_{H_2O} is the molecular weight of water.

The highest amount of water that can be absorbed by the membrane depends on several factors such as the aggregate state of the water (liquid/

vapor), the environmental temperature, and the different treatments that the membrane can receive. For instance, it has been established that the Nafion membrane absorbs different amounts of water depending on whether the reservoir is in vapor or liquid form, the water contents being $\lambda = 14$ and $\lambda = 22$ at 25°C for these states, respectively [24].

In addition, there are two temperature-dependent properties that must be taken into account, the absorption capacity and the time to reach equilibrium. Figure 11.5 exhibits the variation of λ about five times when temperature rises from 20°C to 140°C [23]. The variation of the reached hydration level is related to the morphological changes that the membrane undergoes at elevated temperatures.

Figure 11.6 shows the membrane hydration dynamics as a function of time, and Table 11.2 shows the time required to reach equilibrium, the λ values at equilibrium after 1 h, and the percentage of total hydration achieved after 1 h. In general, the times for which the Nafion membrane reaches hydration at equilibrium are long, around 150 h at low temperatures and over 200 h at high temperatures. Although this could be considered as a drawback, in practice, after 1 h, the membrane has exceeded

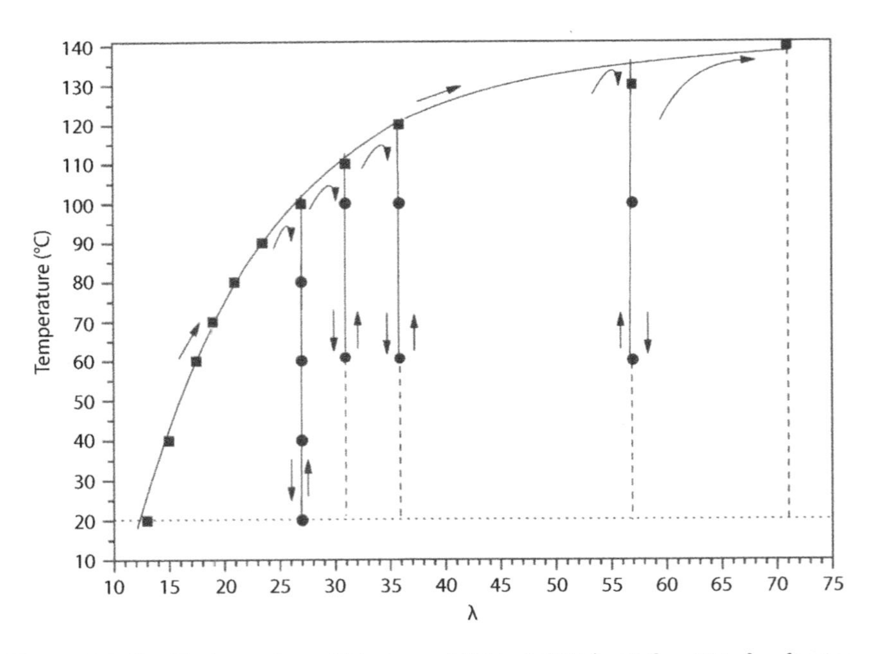

Figure 11.5 Equilibrium values of λ between 20°C to 140°C for Nafion 117 after dipping in water for 20 days. The not reversibility of this process is indicated with the arrows. Reprinted from [23], Copyright 2007, with permission from Elsevier.

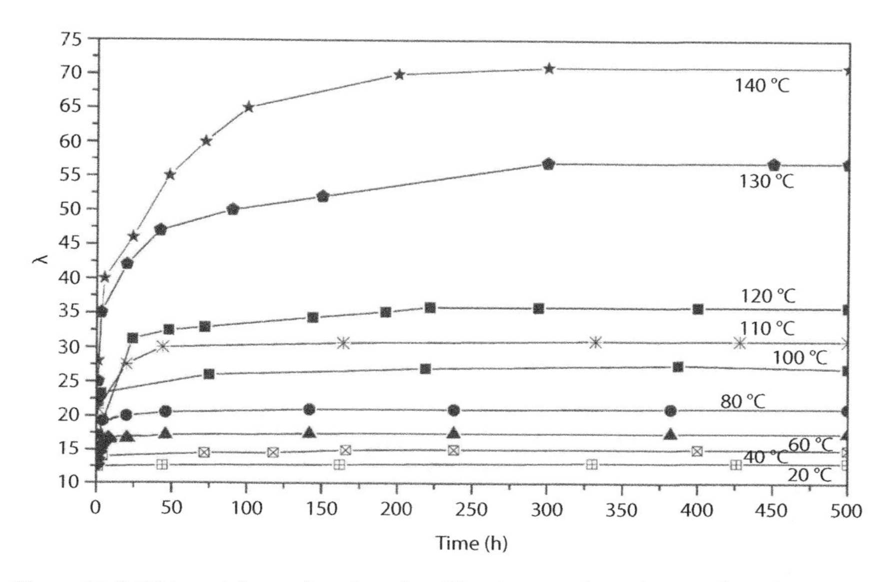

Figure 11.6 Water uptake as a function of equilibration time for Nafion 117 dipped in water at different temperatures. Reprinted from [23], Copyright 2007, with permission from Elsevier.

Table 11.2 Water uptake at different temperatures (°C) and eq. times (h) for Nafion (117) membranes. Reprinted from [23], Copyright 2007, with permission from Elsevier.

T (°C)	λ (Equilibrium values)	Equilibrium time (h)	λ (1 h of equilibration)	Equilibrium % (1 h)
20	13.0	170	12.4	95.4
40	15.0	160	13.2	88
60	17.5	150	14.0	80
70	19.0	200	12.7	67
80	21.0	150	15.4	75
90	23.5	160	15.0	64
100	27.5	150	21.5	78
110	31.0	165	21.0	68
120	35.9	220	17.2	48
130	57.0	225	25.0	44
140	71.0	200	28.0	39

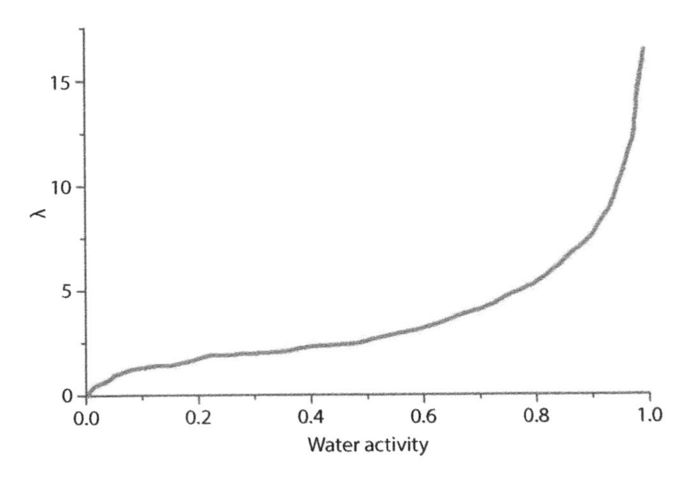

Figure 11.7 λ as a function of water activity (from the authors).

60% of its capacity, considering that the operating temperature of PEMFCs ranges between 50°C and 80°C [25].

Another relevant property is the ability of vapor absorption, which has an influence when the reactant gases in PEMFCs contain a certain percentage of water in the form of vapor. Figure 11.7 shows the characteristic shape of λ as a function of water activity for the Nafion membrane. In the graph, the curve exhibits phases of low activity and a high activity, corresponding to the solvation of the ions and the hydration of the membrane, respectively. Although these curves usually have a characteristic shape, presenting certain variations depending on the history of the Nafion membrane (e.g. treatments and cycles of use).

11.5 Protonic Conductivity

Membranes used as electrolytes in a PEMFC must exhibit good proton conductivity since their specific function is to allow the passage of protons from the anode to the cathode. They must also serve as an impermeable barrier to the passage of electrons, reactant liquids or gases, and anions.

As mentioned earlier, Nafion membranes contain two main regions: a skeleton based on hydrophobic polymeric phase, and a proton conductive area dependent on hydrophilic phase, which is dispersed in the above.

The mobility of protons through the membrane can be carried out by three mechanisms [26, 27]:

- Mass diffusion: this mechanism is based on the displacement of the solvated protons through the water molecules present in the medium from the anode to the cathode.
- Grotthuss: in which the protons are conducted by jumps between the water molecules, producing an agile regrouping of the chemical bonds so that the proton moves from one water molecule to another. Proton transport through this second mechanism is much faster than mass diffusion because the friction produced when the proton makes its path between the water molecules is absent.

In both mechanisms, the presence of water is essential since, under conditions of high humidity, the hydrophilic domains increase in size exhibiting favorable channels for proton conduction.

- Surface: this mechanism originates after a diminution of humidity inside the polymer channels, triggered by the hydrophilic property of the walls. Under these circumstances, the protons are conducted through the $-SO_3^-$ functional groups. However, when the humidity inside is high, the two previous mechanisms predominate.

The different types of proton conduction are schematized in Figure 11.8.

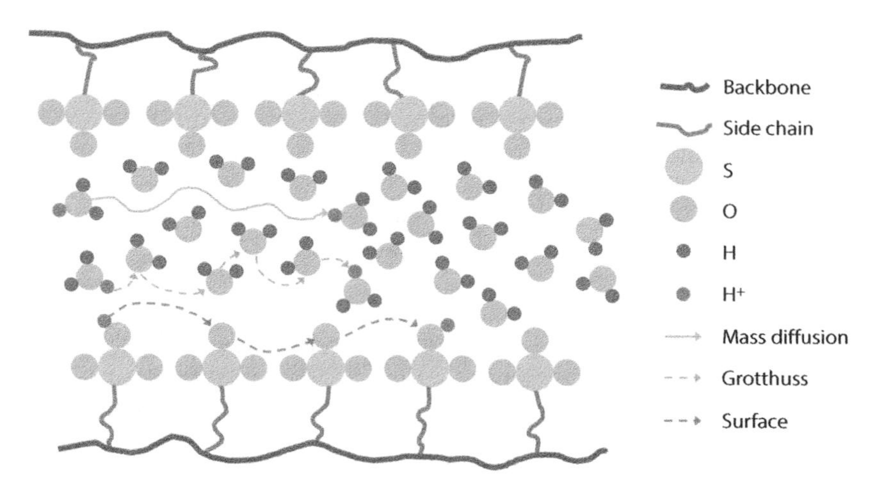

Figure 11.8 Schematic representation of the different proton conduction mechanisms (from the authors).

Since proton conductivity occurs through three mechanisms, the total conductivity in a porous membrane can be expressed as the sum of the contributions of each one [28]:

$$\sigma_p = \sigma^{\Sigma}{}_{H^+} + \sigma^{G}{}_{H^+} + \sigma^{E}{}_{H^+} \tag{11.4}$$

where $\sigma^{\Sigma}{}_{H^+}$, $\sigma^{G}{}_{H^+}$, and y $\sigma^{E}{}_{H^+}$ are the conductivities corresponding to the surface, Grotthuss, and mass diffusion mechanisms, respectively. Thus, the Nernst-Einstein relation allows to express each contribution of the proton conductivity using the diffusion coefficients

$$\sigma^{\alpha}{}_{H^+} = \frac{F^2}{RT} D^{\alpha}{}_{H^+} C^{\alpha}{}_{H^+} \tag{11.5}$$

where F is the Faraday constant, R is the gas constant, T is the absolute temperature, and $D^{\alpha}{}_{H^+}$ y $C^{\alpha}{}_{H^+}$ are the diffusion coefficient and the concentration of protons involved in the α mechanism, respectively.

For the mass diffusion mechanism, the diffusion coefficient can be written as

$$\frac{1}{D^{E}{}_{H^+}} = \frac{x_W}{D^{W}{}_{H^+}} \left(1 + \frac{1 - x_W}{x_W} \frac{D^{W}{}_{H^+}}{D^{M}{}_{H^+}} \right) \tag{11.6}$$

where x_W in the water molar fraction in the membrane, and $D^{W}{}_{H^+}$ and $D^{M}{}_{H^+}$ are the Stefan-Maxwell coefficients for hydronium ions for both bulk water in the pore and polymer matrix M, respectively. However, at low water activity, the mole fraction of it in PEMs is elevated (e.g., $x_W = 0.67$ when $a_i = 0.1$, and rapidly tends to 1); Equation (11.6) can be simplified as

$$\frac{1}{D^{E}{}_{H^+}} \approx \frac{1 + \delta_c}{D^{W}{}_{H^+}} \tag{11.7}$$

where $\delta_c = \left(D^{W}{}_{H^+} / D^{M}{}_{H^+} \right) \left[(1 - x_W) / x_W \right]$. Therefore, the total proton conductivity in a Nafion pore can be written in mathematical terms of the molar concentrations, the diffusion coefficients, and the ratio δ_c:

$$\sigma_p = \frac{F^2}{RT} \left(D^{\Sigma}{}_{H^+} C^{\Sigma}{}_{H^+} + D^{G}{}_{H^+} C_{H^+} + \frac{D^{W}{}_{H^+}}{1 + \delta_c} C_{H^+} \right) \tag{11.8}$$

Thus, the parallel pore model is a useful tool to take into account the pore tortuosity and the reduced cross-sectional area available for transport of hydrogen ions. Then, the product between an individual pore diffusion coefficient times ε_i/τ results in the effective diffusion coefficient for the membrane (ε_i is the membrane porosity and τ is the tortuosity factor). The porosity can be expressed as

$$\varepsilon_i = \lambda_i / (\lambda_i + r) \tag{11.9}$$

where λ_i is the number of moles of water absorbed by each acid site, r is the ratio of the partial molar volumes of the Nafion film and H_2O, respectively. In this way, an expression can be obtained to calculate the conductivity over the whole membrane

$$\sigma_{H^+} = \frac{\varepsilon_i}{\tau}\left[\frac{F^2}{RT}\left(D^{\Sigma}{}_{H^+}C^{\Sigma}{}_{H^+} + D^{G}{}_{H^+}C_{H^+} + \frac{D^{W}{}_{H^+}}{1+\delta_c}C_{H^+}\right)\right] \tag{11.10}$$

Finally, the total conductivity is determined by the structural characteristics δ_c and τ, together with the proton concentration distribution between the external superficial region ($C^{\Sigma}{}_{H^+}$) and the internal region (C_{H^+}) of the Nafion film, which are related to the acidic strength of the sulfonic functional groups.

Over the years, the conductivity of Nafion membranes has been studied by different techniques and under various experimental conditions, which makes it difficult to select a homogeneous criterion for comparing the results obtained. Furthermore, there are clear indications, suggesting that the conductivity determinations could be influenced by factors such as cell geometry, the applied technique, the electrolyte, and the sample preparation [29]. In this sense, Silva *et al.* [29] summarized the results of the experiments performed by different authors for the determination of conductivity in Nafion membranes, which are detailed in Tables 11.3 to 11.5.

In addition, Ochi *et al.* [30] performed relevant advances of the conductivity these membranes as a function of the most important parameters for FC applications: relative humidity (Figure 11.9) and water content (Figure 11.10), at different temperatures. In both cases, the conductivity of Nafion strongly depends on the presence of water in the membrane, mainly in the first stages of hydration where a marked increase in this parameter is observed. On the other hand, as the temperature increases, the proton conductivity of the membrane follows the same tendency until close to 80°C,

Table 11.3 Tangential direction conductivity measurements of Nafion® 1100 EW membranes. Reprinted from [29], Copyright 2004, with permission from Elsevier.

λ	Technique	σ (mS cm^{-1})	R (Ω cm^2)	T (°C)
0–14[1]	AC Impedance (5 kHz)	5–63	3.5–0.29	30
14[1]	AC 4 Electrodes (20 kHz)	15–43	1.2–0.425	45
18–19[1]	AC Impedance (100 kHz)	81	0.225	25
19[1]	AC Impedance (100 kHz)	52	0.350	20
22–23[2]	AC Impedance (5 kHz)	102	0.180	30
22–23[2]	AC Impedance (10 kHz)	91	0.200	25
Wet state	AC Coaxial Probe method	72	0.255	22

[1]controlled relative humidity;
[2]fully immersed in water.

Table 11.4 Normal direction conductivity (NDC-1) measurements of Nafion® 1100 EW membranes in acidic electrolyte. Reprinted from [29], Copyright 2004, with permission from Elsevier.

H_2SO_4 electrolyte	Technique	σ (mS cm^{-1})	R (Ω cm^2)	T (°C)
10^{-5} M	DC (4-point probe)	55–85	0.327–0.215	20
1 M		129	0.142	25
1 M		140	0.130	25
1 M	DC Current pulse	70	0.260	20
1 M	AC Impedance	71	0.259	25

where it remains practically constant even at subsequent warming up [31]. In this sense, Springer *et al.* [32] found an expression for conductivity as a function of temperature and hydration state, for Nafion between 30°C and 80°C:

$$\sigma(T) = (0.005139\lambda - 0.00326)\, exp\left[1268\left(\frac{1}{303} - \frac{1}{T}\right)\right] \tag{11.11}$$

Table 11.5 Normal direction conductivity (NDC-2) measurements of Nafion® 1100 EW membranes in direct contact with electrodes by using AC Impedance technique. Reprinted from [29], Copyright 2004, with permission from Elsevier.

λ	σ (mS cm^{-1})	R (Ω cm^2)	T (°C)
3	3–10	-	25
12–13	49	0.375	25
23	82 (90)	0.224	20
Wet state	34	0.530	25
Wet state	83	0.220	30

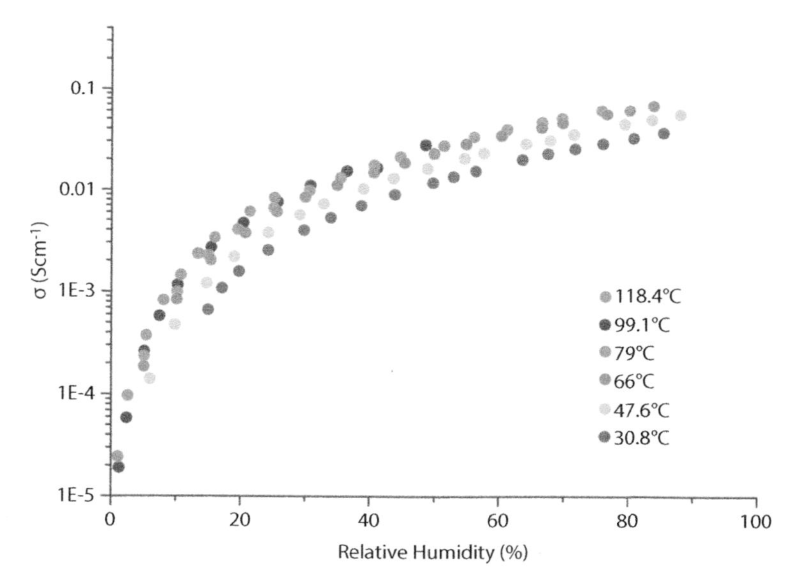

Figure 11.9 Conductivity in Nafion 117 versus percent of relative humidity of this membrane. Reprinted from [30], Copyright 2009, with permission from Elsevier.

Moreover, Kopitzke *et al.* [33] measured conductivity of hydrogen ions as a function of temperature in these membranes dipped in liquid water. These authors found a correlation between conductivity and temperature given by

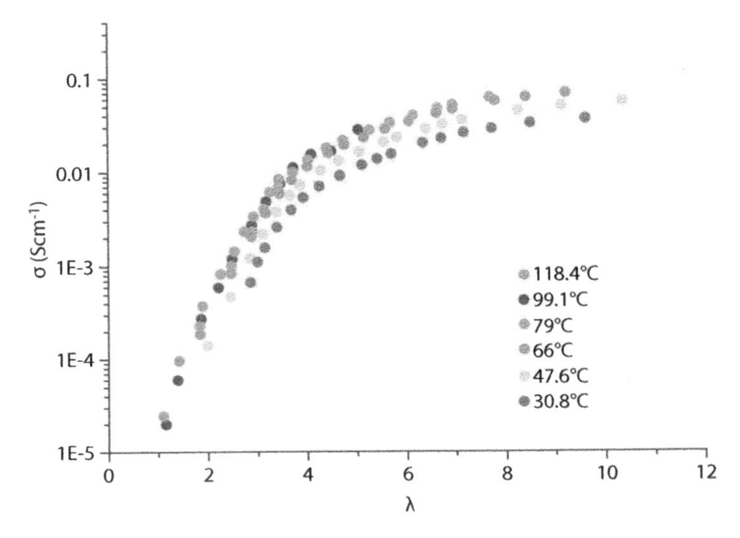

Figure 11.10 Conductivity in Nafion 117 vs. of λ. Reprinted based on reference [30], Copyright 2009, with permission from Elsevier.

$$\sigma(T) = \sigma_o exp\left(-\frac{E_k}{RT}\right) \tag{11.12}$$

where E=7.829kJ mol^{-1} and σ_0=2.29 S cm^{-1}.

11.6 Water Transport

Water transport is a factor that directly influences the performance of the membrane. Membrane dehydration causes an increase in internal resistance, while excess water obstructs the transport of reactant gases and increases polarization, especially in the cathode zone.

In Nafion membranes, this phenomenon involves several processes, the main ones being electro-osmosis (EO) and diffusion. The EO phenomenon is produced when water is transported from the anode to the cathode, accompanying the transport of protons. Thus, migration dehydrates the membrane area near the anode, generating an increase in resistance. Furthermore, EO tends to increase the cathode hydration, working together with the water produced by the electrochemical reaction. In general, the removal of water from the cathode occurs both by evaporation and by back-diffusion toward the anode. However, an excess of moisture at the cathode hinders its evaporation and encourages condensation within the gas diffusion layer. As a consequence, there is a decrease in the O_2 diffusion rate to the catalyst layer that reduces the PEMFCs performance [34].

According to Lu *et al.* [35], the water flux to the cathode can be expressed as the contributions of diffusion, electro-osmosis, and hydraulic permeation processes, as given in Equation (11.13):

$$j_m = -D\frac{\Delta c_{c-a}}{\delta_m} + n_d\frac{I}{F} - \frac{K}{\mu_1}\Delta p_{c-a}\frac{\rho}{M_{H_2O}} \tag{11.13}$$

where I is the current density, is the water molar density, δ_m is the membrane thickness, F is the Faraday constant, K is the hydraulic permeability, n_d is the water electro-osmotic drag coefficient, μ_1 is the viscosity of liquid water, D is the diffusion coefficient, M_{H_2O} is the water molecular weight, and Δc_{c-a} and Δp_{c-a} are the difference in water concentration and hydraulic pressure across the membrane, respectively. A schematic diagram that summarizes the different processes is shown in Figure 11.11.

Equation (11.13) can be simplified by quantifying the net water flux through the membrane by a net water transport coefficient, according to Equation (11.14):

$$j_n = \alpha\frac{I}{F} \tag{11.14}$$

Furthermore, if the water produced as a product of the electrochemical reaction, whose generation rate is given by $I/2F$, is taken into account, then the total rate of water at the cathode can be expressed as [Equation (11.15)]:

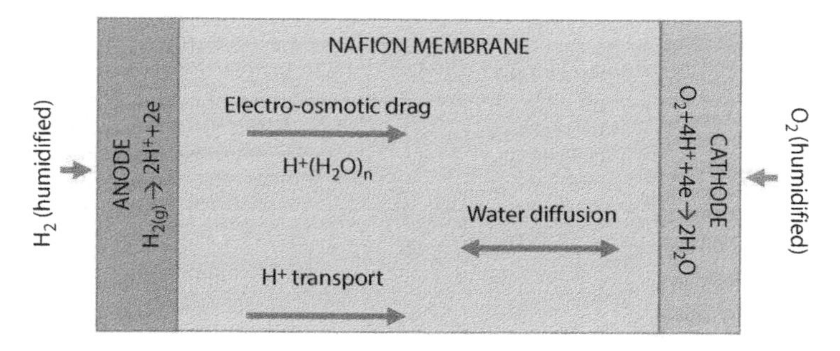

Figure 11.11 Diagram of the water transport processes within a PEMFCs (from the authors).

Table 11.6 Summary of EOD Studies (VE, vapor equilibrated; LE, liquid equilibrated; CP, chemical potential; AP, applied potential; APS, applied pressure; EC, electrochemical). Reprinted (adapted) with permission from [37]. Copyright 2011 American Chemical Society.

Technique (Driving force)	**Source**	**Conditions**	**EO Drag Coef., n_d**
Electro-osmotic drag cell (AP)	Zawodzinski *et al.* [24]	VE, 30°C	0.9
		LE, 30°C	2–2.9
Activity gradient (CP)	Fuller and Newman [38]	VE, 25°C–37.5°C	1.4
	Zawodzinski *et al.* [39]	VE, 30°C	1.0
		LE, 30°C	2.5
	Gallagher *et al.* [40]	VE, 25°C–10°C	~1
Streaming potential (APS)	Xie and Okada [41]	VE, 25°C	2.6
Electrophoretic NMR (AP)	Ise *et al.* [42]	VE, 27°C	1.5–2.5
Methanol fuel cell (EC)	Ren and Gottesfeld [43]	LE, 15°C	2.0
		LE, 130°C	5.1
Hydrogen fuel cell (EC)	Park and Caton [44]	VE, 70°C	0.5–0.82
Hydrogen pump (AP)	Weng *et al.* [45]	VE, 135°C–185°C	0.2–0.6
	Ge *et al.* [46]	VE, 30°C–50°C	0.2–0.9
		LE, 15°C–85°C	1.8–2.7
	Ye and Wang [47]	VE, 80°C	1.1
	Luo *et al.* [48]	LE, 20°C–90°C	2–3.4
		VE, 25°C	1.2–2.0

$$j_{H_2O} = \left(\alpha + \frac{1}{2}\right)\frac{I}{F} \tag{11.15}$$

However, the hydraulic permeability and diffusion processes occur less frequently with thicker membranes. For this reason, an approximation that exclusively take into account the electro-osmotic drag process is assumed in Nafion and, as consequence, is valid $\alpha \approx n_d$. Furthermore, the electro-osmotic drag coefficient can be defined as the number of H_2O molecules per H^+. The mentioned coefficient usually expresses its value as a function of membrane hydration (λ) and, generally, the values reported in bibliographies present a small dispersion. Some noteworthy examples are those of Zawodzinski *et al.* [24], who found n_d values between 2.5 and 2.9 at temperature of 30°C with fully hydrated Nafion in liquid water ($\lambda = 22$), and a value of 0.9 when $\lambda = 11$. Likewise, Onda *et al.* [36] reported that n_d coefficient depends only on temperature when the membrane has been obtained using chemical plating. These authors found an experimentally relationship between these two variables, which is shown in Equation (11.16):

$$n_d = 0.0134T + 0.03 \tag{11.16}$$

Other relevant results are summarized in Table 11.6, which was compiled by Cheah *et al.* [37] and provides a collection of the different values of electro-osmotic drag coefficients in Nafion membranes obtained under different experimental conditions.

11.7 Gas Permeation

Gas permeability through Nafion membranes is a common research topic and has been analyzed by methods such as gas chromatography, electrochemical techniques, and positron annihilation lifetime spectroscopy, among others [49, 50]. It should be noted that, in theory, the membrane is impermeable to gas transport. However, as a result of its structural characteristics, in particular the presence of high porosity with an elevated level of liquid water absorption, the dissolution of gases is allowed.

Note that the permeability is expressed as the product of the diffusion coefficient (D) and solubility (S).

$$P_m = D \times S \tag{11.17}$$

Then, Battino *et al.* [51, 52] derived the equations for the molar fraction of H_2 and O_2 solubility (x_{H_2} and x_{O_2}) in water in the temperature range between 273 K and 350 K, based on data reported in the bibliography, obtaining the following relationships:

$$ln\ X_{H_2} = -48.1611 + \frac{5528.45}{T} + 16.8893\ ln\left(\frac{T}{100}\right) \quad (11.18)$$

$$ln\ X_{O_2} = -66.73538 + \frac{8747.547}{T} 24.45264\ ln\left(\frac{T}{100}\right) \quad (11.19)$$

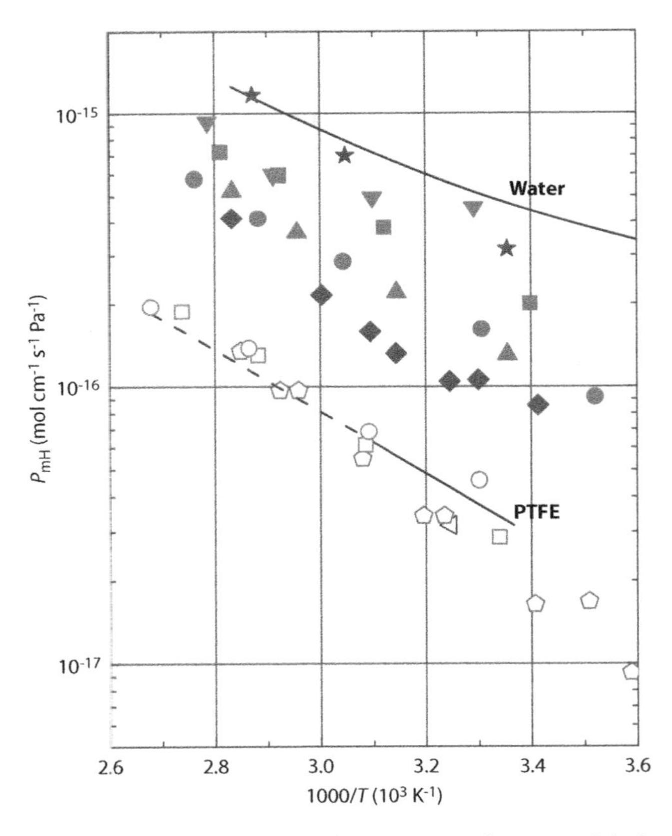

Figure 11.12 Arrhenius plot of H_2 permeability (P_{mH}) in Nafion 117 at dehydrated (empty icons) and swelled (filled icons) states obtained from literature. Reprinted from [50], Copyright 2011, with permission from Elsevier.

On the one hand, Wise and Houghton [53] proposed an exponential dependence of the diffusion coefficient with the temperature.

$$D = D_o exp\left(-\frac{E_D}{RT}\right) \tag{11.20}$$

Similarly, Ito *et al.* [50] presented a summary of the gas hydrogen and oxygen permeability in two Arrhenius plots (Figures 11.12 and 11.13) and in a table where the measurement techniques and conditions used in each case are explained (Table 11.7). In both figures, the line representing the permeability in water was calculated by substituting the values

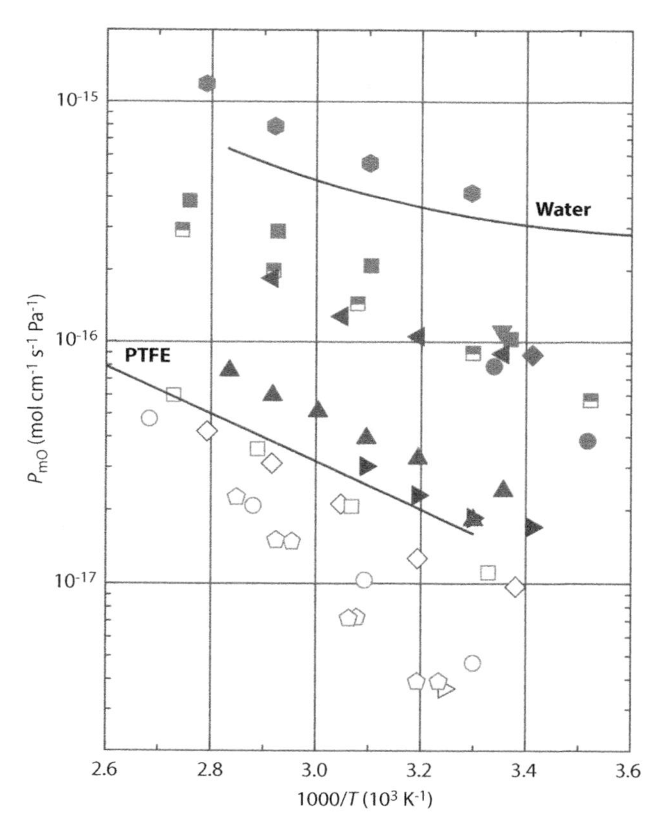

Figure 11.13 Arrhenius plot of O_2 permeability (P_{mO}) in Nafion 117 at dehydrated (empty icons) and swelled (filled icons) states obtained from different authors. Reprinted from [50], Copyright 2011, with permission from Elsevier.

Table 11.7 Most relevant measurements for diffusion coefficient (*D*), gas solubility (*S*), and permeability properties (P_m) of Nafion membranes found in the literature. Reprinted from [50], Copyright 2011, with permission from Elsevier.

Reference	Presented data	Gas[a]	Media[b]	Technique	Temperature	Pressure
Kocha *et al.* [49]	P_m	H_2	Nafion 111, Nafion 112 (Wet)	Electrochemical	20°C–80°C	1–2 atm
Ogumi *et al.* [55]	D, S	O_2	Nafion 120 (Wet)	Electrochemical	20°C–50°C	1 atm
Sakai *et al.* [56]	P_m	H_2/O_2	Nafion 117, Nafion 125 (Wet/Dry)	Volumetric	20°C–90°C	1–30 atm
Sakai *et al.* [57]	P_m, D	H_2/O_2	Nafion 117, Nafion 125 (Wet/Dry)	Time-lag	−10°C–100°C	4–14 atm
Chiou and Paul [58]	P_m, D, S	H_2/O_2	Nafion 117 (Dry)	Time-lag	35°C	1 atm
Parthasarathy *et al.* [59]	D, S	O_2	Nafion 117 (Wet)	Electrochemical	25°C–80°C	5 atm
Yoshitake *et al.* [60]	P_m	H_2/O_2	Nafion 117 (Wet)	Gas chromatography	30°C–85°C	1 atm

(Continued)

Table 11.7 Most relevant measurements for diffusion coefficient (*D*), gas solubility (*S*), and permeability properties (P_m) of Nafion membranes found in the literature. Reprinted from [50], Copyright 2011, with permission from Elsevier. (*Continued*)

Reference	Presented data	Gas[a]	Media[b]	Technique	Temperature	Pressure
Broka and Ekdunge [61]	P_m	H_2/O_2	Nafion 117 (Wet)	Gas chromatography	20°C–100°C	n/a
Lehtinen *et al.* [62]	D, S	O_2	Nafion 117 (Wet/Dry)	Electrochemical	20°C	n/a (1 atm)[c]
Haug and White [63]	D, S	O_2	Nafion 117 (Wet)	Electrochemical	25°C	n/a (1 atm)[c]
Barbir [64]	P_m	H_2	Nafion 117 (Wet)	n/a	20°C–75°C	16–220 atm
Mohamed *et al.* [65]	P_m	H_2/O_2	NRE212 (Wet/Dry)	Positron annihilation lifetime	−30°C–80°C	n/a

[a] Tested gases without H_2 and O_2 are not added to the list;
[b] Tested membranes without Nafion series are not added to the list;
[c] Pressures in parentheses are approximate values.

found in the literature with Equation (11.17), while the line for PTFE film was obtained from Pasternak *et al.* [54]. The other points shown in the graph were obtained using the corresponding equations according to the data reported in each paper. In addition, the data obtained when the membrane is hydrated or dry is distinguished.

The underlying evidence in these graphs indicates that permeability is a predominant process in wet membranes and by contrast in dry state significantly decreases reaching similar values to PTFE. Besides, the experimental results showed elevated dispersion when the membrane contains high humidity levels. For this inconvenience, a comparison between different authors has always been problematic. In this sense, the origin of this phenomenon has been subject of an intense discussion. A possible explanation for this behavior could be related with the previous membrane processing (chemical or physical). In this sense, several authors performed permeability studies in very similar conditions and finally determined that hydrogen is more permeable than oxygen gas.

11.8 Final Comments

Since its discovery in the decade of 1960s, the Nafion membrane has received an increased and special attention as a conducting electrolyte of hydrogen ion in PEMFCs, owing to its exceptional mechanical and thermal characteristics. Currently, numerous investigations are being carried out based on the properties and benefits of the Nafion membrane, and the results are mainly related with specific applications of FCs.

The constant collective effort has made possible a broad understanding of this technologically relevant material. However, there are many characteristics on which research has not achieved a consensus. Nonetheless, there is no dispute in the unique properties of Nafion, which encourages further research to improve the efficiency of the membrane and to investigate similar materials that can substitute for it where it fails to meet application requirements.

Acknowledgements

The authors would like to thank CONICET and SECYT-UNC by providing the necessary funds to conduct this study and to the National University of Cordoba for the physical space. S.P. Fernandez Bordín and J.A. Chinellato Díaz would like to thank CONICET–Argentina for their doctoral scholarship.

References

1. Wilberforce, T., Alaswad, A., Palumbo, A., Dassisti, M., Olabi, A.G., Advances in stationary and portable fuel cell applications. *Int. J. Hydrog. Energy*, 41, 16509, 2016.
2. Alaswad, A., Palumbo, A. *et al.*, Fuel cell technologies, applications, and state of the art: A reference guide, in: *Reference Module in Materials Science and Materials Engineering*, Elsevier BV, Netherlands, 2015.
3. Nunes, S.P. and Klaus-Viktor, P., *Membrane technology*, pp. 12–33, Wiley-VCH, Weinheim, New York, 2001.
4. Li, J., Pan, M., Tang, H., Understanding short-side-chain perfluorinated sulfonic acid and its application for high temperature polymer electrolyte membrane fuel cells. *RSC Adv.*, *4*, 3944, 2014.
5. Shin, D.W., Guiver, M.D., Lee, Y.M., Hydrocarbon-based polymer electrolyte membranes: importance of morphology on ion transport and membrane stability. *Chem. Rev.*, 117, 4759, 2017.
6. Grot, W.G.N., Landau, U., Yeager, E., Kortan, D., ® Membrane and its Applications, in: *Electrochemistry in Industry*, pp. 73–87, Springer, Boston, MA, 1982.
7. Banerjee, S. and Curtin, D.E., Nafion® perfluorinated membranes in fuel cells. *J. Fluor. Chem.*, 125, 1211, 2004.
8. Zhang, D., Xin, L., Xia, Y., Dai, L., Qu, K., Huang, K., Yiqun, F., Xu, Z., Advanced Nafion hybrid membranes with fast proton transport channels toward high-performance vanadium redox flow battery. *J. Membr. Sci.*, 624, 119047, 2021.
9. Li, D., Qi, R., Zhang, L.Z., Performance improvement of electrolytic air dehumidification systems with high-water-uptake polymer electrolyte membranes. *J. Appl. Polym. Sci.*, 136, 47676, 2019.
10. Chabi, S., Papadantonakis, K.M., Lewis, N.S., Freund, M.S., Membranes for artificial photosynthesis. *Energy Environ. Sci.*, 10, 1320, 2017.
11. Romero, M.R., Ahumada, F., Garay, F., Baruzzi, A.M., Amperometric biosensor for direct blood lactate detection. *Anal. Chem.*, 82, 5568, 2010.
12. Fernandez Bordín, S.P., Andrada, H.E., Carreras, A.C., Castellano, G., Schweins, R., Cuello, G.J., Mondelli, C., Galván Josa, V.M., Water channel structure of alternative perfluorosulfonic acid membranes for fuel cells. *J. Membr. Sci.*, 636, 119559, 2021.
13. Karimi, M.B., Mohammadi, F., Hooshyari, K., Recent approaches to improve Nafion performance for fuel cell applications: A review. *Int. J. Hydrog. Energy*, 44, 28919, 2019.
14. Sigwadi, R., Dhlamini, M.S., Mokrani, T., Nemavhola, F., Nonjola, P.F., Msomi, P.F., The proton conductivity and mechanical properties of Nafion®/ZrP nanocomposite membrane. *Heliyon*, 5, e02240, 2019.
15. Park, H.S., Kim, Y.J., Hong, W.H., Lee, H.K., Physical and electrochemical properties of Nafion/polypyrrole composite membrane for DMFC. *J. Membr. Sci.*, 272, 28, 2006.

16. Gierke, T.D., Munn, G.E., Wilson, F., The morphology in nafion perfluorinated membrane products, as determined by wide-and small-angle x-ray studies. *J. Polym. Sci.: Polym. Phys. Ed.*, 19, 1687, 1981.
17. Fujimura, M., Hashimoto, T., Kawai, H., Small-angle X-ray scattering study of perfluorinated ionomer membranes. 1. Origin of two scattering maxima. *Macromolecules*, 14, 1309, 1981.
18. Aldebert, P., Dreyfus, B., Gebel, G., Nakamura, N., Pineri, M., Volino, F., Rod like micellar structures in perfluorinated ionomer solutions. *J. Phys. France*, 49, 2101, 1988.
19. Haubold, H.G., Vad, T., Jungbluth, H., Hiller, P., Nano structure of Nafion: a SAXS study. *Electrochimica Acta*, 46, 1559, 2001.
20. Schmidt-Rohr, K. and Chen, Q., Parallel cylindrical water nanochannels in Nafion fuel-cell membranes. *Nat. Mater.* 7, 75, 2008.
21. Fernandez Bordín, S.P., Andrada, H.E., Carreras, A.C., Castellano, G.E., Oliveira, R.G., Galvan Josa, V., Nafion membrane channel structure studied by small-angle X-ray scattering and Monte Carlo simulations. *Polymer*, 155, 58, 2018.
22. Van der Heijden, P.C., Rubatat, L., Diat, O., Orientation of drawn Nafion at molecular and mesoscopic scales. *Macromolecules*, 37, 5327, 2004.
23. Alberti, G., Narducci, R., Sganappa, M., Effects of hydrothermal/thermal treatments on the water-uptake of Nafion membranes and relations with changes of conformation, counter-elastic force and tensile modulus of the matrix. *J. Power Sources*, 178, 575, 2008.
24. Zawodzinski, T.A., Derouin, C., Radzinski, S., Sherman, R.J., Smith, V.T., Springer, T.E., Gottesfeld, S., Water uptake by and transport through Nafion® 117 membranes. *J. Electrochem. Soc.*, 140, 1041, 1993.
25. Barbir, F., *PEM Fuel Cells: Theory and Practice*, Academic press, New York, 2012.
26. Agmon, N., The grotthuss mechanism, in: *Chemical Physics Letters*, vol. 244, p. 456, 1995.
27. Sone, Y., Ekdunge, P., Simonsson, D., Proton conductivity of Nafion 117 as measured by a four-electrode AC impedance method. *J. Electrochem. Soc.*, 143, 1254, 1996.
28. Choi, P., Jalani, N.H., Datta, R., Thermodynamics and proton transport in nafion: II. Proton diffusion mechanisms and conductivity. *J. Electrochem. Soc.*, 152, E123, 2005.
29. Silva, R.F., De Francesco, M., Pozio, A., Tangential and normal conductivities of Nafion® membranes used in polymer electrolyte fuel cells. *J. Power Sources*, 134, 18, 2004.
30. Ochi, S., Kamishima, O., Mizusaki, J., Kawamura, J., Investigation of proton diffusion in Nafion® 117 membrane by electrical conductivity and NMR. *Solid State Ionics*, 180, 580, 2009.

31. Barique, M.A., Tsuchida, E., Ohira, A., Tashiro, K., Effect of elevated temperatures on the states of water and their correlation with the proton conductivity of Nafion. *ACS Omega*, 3, 349, 2018.
32. Springer, T.E., Zawodzinski, T.A., Gottesfeld, S., Polymer electrolyte fuel cell model. *J. Electrochem. Soc.*, 138, 2334, 1991.
33. Kopitzke, R.W., Linkous, C.A., Anderson, H.R., Nelson, G.L., Conductivity and water uptake of aromatic-based proton exchange membrane electrolytes. *J, Electrochem. Soc.*, 147, 1677, 2000.
34. Bellows, R.J., Lin, M.Y., Arif, M., Thompson, A.K., Jacobson, D., Neutron imaging technique for *in situ* measurement of water transport gradients within Nafion in polymer electrolyte fuel cells. *J. Electrochem. Soc.*, 146, 1099, 1999.
35. Lu, G.Q., Liu, F.Q., Wang, C.Y., Water transport through Nafion 112 membrane in DMFCs. *Electrochem. Solid-State Lett.*, 8, A1, 2004.
36. Onda, K., Murakami, T., Hikosaka, T., Kobayashi, M., Ito, K., Performance analysis of polymer-electrolyte water electrolysis cell at a small-unit test cell and performance prediction of large stacked cell. *J. Electrochem. Soc.*, 149, A1069, 2002.
37. Cheah, M.J., Kevrekidis, I.G., Benziger, J., Effect of interfacial water transport resistance on coupled proton and water transport across Nafion. *J. Phys. Chem. B*, 115, 10239, 2011.
38. Fuller, T.F. and Newman, J., Experimental determination of the transport number of water in Nafion 117 membrane. *J. Electrochem. Soc.*, 139, 1332, 1992.
39. Zawodzinski, T.A., Davey, J., Valerio, J., Gottesfeld, S., The water content dependence of electro-osmotic drag in proton-conducting polymer electrolytes. *Electrochimica Acta*, 40, 297, 1995.
40. Gallagher, K.G., Pivovar, B.S., Fuller, T.F., Electro-osmosis and water uptake in polymer electrolytes in equilibrium with water vapor at low temperatures. *J. Electrochem. Soc.*, 156, B330, 2008.
41. Xie, G. and Okada, T., Water transport behavior in Nafion 117 membranes. *J. Electrochem. Soc.*, 142, 357, 1995.
42. Ise, M., Kreuer, K.D., Maier, J., Electroosmotic drag in polymer electrolyte membranes: an electrophoretic NMR study. *Solid State Ionics*, 125, 213, 1999.
43. Ren, X. and Gottesfeld, S., Electro-osmotic drag of water in poly (perfluorosulfonic acid) membranes. *J. Electrochem. Soc.*, 148, A87, 2001.
44. Park, Y.H. and Caton, J.A., An experimental investigation of electro-osmotic drag coefficients in a polymer electrolyte membrane fuel cell. *Int. J. Hydrog. Energy*, 33, 7513, 2008.
45. Weng, D., Wainright, J.S., Landau, U., Savinell, R.F., Electro-osmotic drag coefficient of water and methanol in polymer electrolytes at elevated temperatures. *J. Electrochem. Soc.*, 143, 1260, 1996.
46. Ge, S., Yi, B., Ming, P., Experimental determination of electro-osmotic drag coefficient in Nafion membrane for fuel cells. *J. Electrochem. Soc.*, 153, A1443, 2006.

47. Ye, X. and Wang, C.Y., Measurement of water transport properties through membrane-electrode assemblies: I. Membranes. *J. Electrochem. Soc.*, 154, B676, 2007.
48. Luo, Z., Chang, Z., Zhang, Y., Liu, Z., Li, J., Electro-osmotic drag coefficient and proton conductivity in Nafion® membrane for PEMFC. *Int. J. Hydrog. Energy*, 35, 3120, 2010.
49. Kocha, S.S., Deliang Yang, J., Yi, J.S., Characterization of gas crossover and its implications in PEM fuel cells. *AIChE J.*, 52, 1916, 2006.
50. Ito, H., Maeda, T., Nakano, A., Takenaka, H., Properties of Nafion membranes under PEM water electrolysis conditions. *Int. J. Hydrog. Energy*, 36, 10527, 2011.
51. Battino, R., Clever, H.L., Young, C.L., *Hydrogen and Deuterium*, Pergamon Press, Oxford, 1981.
52. Battino, R., Clever, H.L., Young, C.L., *IUPAC Solubility Data Series, O2 and O3*, Pergamon Press, Oxford, 1981.
53. Wise, D.L. and Houghton, G., The diffusion coefficients of ten slightly soluble gases in water at 10–60 C. *Chem. Eng. Sci.*, 21, 999, 1966.
54. Pasternak, R.A., Christensen, M.V., Heller, J., Diffusion and permeation of oxygen, nitrogen, carbon dioxide, and nitrogen dioxide through polytetrafluoroethylene. *Macromolecules*, 3, 366, 1970.
55. Ogumi, Z., Takehara, Z., Yoshizawa, S., Gas permeation in SPE method: I. Oxygen permeation through Nafion and NEOSEPTA. *J. Electrochem. Soc.*, 131, 769, 1984.
56. Sakai, T., Takenaka, H., Wakabayashi, N., Kawami, Y., Torikai, E., Gas permeation properties of solid polymer electrolyte (SPE) membranes. *J. Electrochem. Soc.*, 132, 1328, 1985.
57. Sakai, T., Takenaka, H., Torikai, E., Gas diffusion in the dried and hydrated Nafions. *J. Electrochem. Soc.*, 133, 88, 1986.
58. Chiou, J.S. and Paul, D.R., Gas permeation in a dry Nafion membrane. *Ind. Eng. Chem. Res.*, 27, 2161, 1988.
59. Parthasarathy, A., Srinivasan, S., Appleby, A.J., Martin, C.R., Temperature dependence of the electrode kinetics of oxygen reduction at the platinum/Nafion® interface—a microelectrode investigation. *J. Electrochem. Soc.*, 139, 2530, 1992.
60. Yoshitake, M., Tamura, M., Yoshida, N., Ishisaki, T., Studies of perfluorinated ion exchange membranes for polymer electrolyte fuel cells. *Denki Kagaku oyobi Kogyo Butsuri Kagaku*, 64, 727, 1996.
61. Broka, K. and Ekdunge, P., Oxygen and hydrogen permeation properties and water uptake of Nafion® 117 membrane and recast film for PEM fuel cell. *J. Appl. Electrochem.*, 27, 117, 1997.

62. Lehtinen, T., Sundholm, G., Holmberg, S., Sundholm, F., Björnbom, P., Bursell, M., Electrochemical characterization of PVDF-based proton conducting membranes for fuel cells. *Electrochimica Acta*, 43, 1881, 1998.
63. Haug, A.T. and White, R.E., Oxygen diffusion coefficient and solubility in a new proton exchange membrane. *J. Electrochem. Soc.*, 147, 980, 2000.
64. Barbir, F., PEM electrolysis for production of hydrogen from renewable energy sources. *Solar Energy*, 78, 661, 2005.
65. Mohamed, H.F., Ito, K., Kobayashi, Y., Takimoto, N., Takeoka, Y., Ohira, A., Free volume and permeabilities of O2 and H2 in Nafion membranes for polymer electrolyte fuel cells. *Polymer*, 49, 3091, 2008.

12

Solid Polymer Electrolytes for Proton Exchange Membrane Fuel Cells

Nitin Srivastava and Rajendra Kumar Singh*

Ionic Liquid and Solid-State Ionics Lab, Department of Physics, Institute of Science, Banaras Hindu University, Varanasi, India

Abstract

The topic of renewable energy is very important because of rising power consumption and environmental concerns. Among various types of renewable energy sources (RESs), fuel cell is achieving more attention because of their high efficiency, environmental friendliness, and its cost effectiveness. This chapter deals with various type of fuel cells and their working principle along with their applications. Now, among the various type of fuel cells, we have studied in detail about proton exchange membrane fuel cell (PEMFC) due to its several advantage like high protonic conductivity, good chemically and thermally stable properties, better mechanical characteristics, and low permeability to fuel cell. In addition, this chapter presents several types of solid polymer electrolyte membranes and its applications, which are used in PEMFC. Various classes of polymer membranes like sulfonated hydrocarbon polymer membrane and acid base polymer membrane have also been addressed.

***Keywords*:** Fuel cells (FCs), membrane, solid polymer electrolyte, FC efficiency, high power density

12.1 Introduction

Scarcity of energy and high dependency on fossil fuels for the main energy resources lead to serious environmental issues such as global warming, environmental pollution, and carbon dioxide (CO_2) emission.

**Corresponding author*: rksingh_17@rediffmail.com; rajendrasingh.bhu@gmail.com

Inamuddin, Omid Moradi and Mohd Imran Ahamed (eds.) Proton Exchange Membrane Fuel Cells: Electrochemical Methods and Computational Fluid Dynamics, (331–352) © 2023 Scrivener Publishing LLC

These energy crises in the recent decades have motivated the researchers toward the new, clean, and inexpensive energy storage devices mainly based on the renewable energy resources. The renewable energy resources include solar, wind, and geo-thermal energy. In this perspective, fuel cells (FCs) gained much attention because of their high-power density with zero CO_2 emission [1–3]. In addition, as compared to other batteries like rechargeable lithium and sodium, power and capacitance are independent of each other that are much desirable in the wide range of applications. Moreover, several power plants based on the fuel cell have been operated successfully in the range from 10 megawatts to a few milliwatts. Fuel cells are high-efficiency electrochemical devices that transform chemical energy directly into electricity. In FCs, hydrogen or another fuel is fed into the anode side and oxygen is supplied on the cathode side. At anode, oxidation takes place by means of catalyst (like platinum-containing compound) and the proton generated at the side of anode is migrated to other electrode, i.e., cathode through the electrolyte, whereas electrons get injected into external circuit and thus reducing the oxygen at cathode. This reaction leads to the formation of water and heat at the cathode [4]. Further, unlike in batteries, where the electrodes are consumed and also take part in electrochemical reaction irreversibly in primary battery and reversibly in secondary battery; in FCs, there is no consumption of electrode, and also, they do not participate in electrochemical reaction and it functions as a continuous power source as long as fuel is available to them [5]. A wide variety of fuels are used in FCs, the most common is hydrogen fuel. The fundamental structure of FC consists of ionic conductor electrolyte layer that is in contact with the two electrodes [6]. The basic structure of fuel cell is shown in Figure 12.1. In general, the electrochemical reaction takes place at electrode, the electrode which is positively charged, known as cathode and one that is negatively charged termed as anode. Besides that, the nature of electrolyte also affects the performance of FCs. The electrolyte must have high ionic conductivity with negligible electronic conductivity, good mechanical properties, and adequate stability during the working circumstances. FCs are categorized into different types on the basis of nature of electrolyte like polymer electrolyte membrane (PEM) fuel cells, solid oxide fuel cells (SOFCs), alkaline fuel cells (AFCs), phosphoric acid fuel cells (PAFCs), molten carbonate fuel cells (MCFCs), and PEM fuel cells [7–13]. Among them, solid PEMFC, which is also known as PEM fuel cell, has drawn more attention within the research field due to its high-power density, i.e., 250–1,000 W kg^{-1} and zero emission [1]. In PEMFCs, electrolyte is proton exchange membrane, and it has ability of transferring the proton from anode to the cathode and also acts as electronic insulator for electrons. In general, the proton exchange

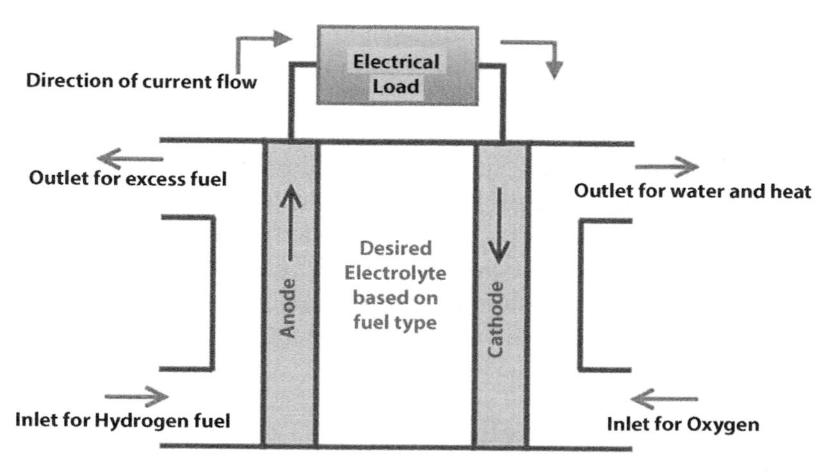

Figure 12.1 Basic structure of fuel cell.

membrane is placed between the anode and cathode, and the hydrogen gas is fed into the anode, while, at the cathode, air is supplied. The hydrogen gas is oxidized at the anode and produced proton and electron. The produced proton is migrated from anode to cathode through proton exchange membrane catalytically and combines with the oxygen at the other electrode, i.e., cathode that results in water formation and heat generation [13]. In addition, solid PEMs have superiority over the liquid electrolyte since they are easy to handle and compact and have excellent resistance and good portability due to their excellent mechanical properties [14–16]. Therefore, PEMFCs are considered to be one of the promising power tools for a vehicle. Besides that, PEM that is used as electrolyte in PEMFC should have high value of proton conductivity along with low electronic conductivity, high stability under the operating condition, good mechanical properties, low cost, and adequate water transport properties to avoid flooding and dehydration problem of PEM [4, 17]. A number of materials have been developed for PEMFC in order to meet the above requirements. The first PEM was developed by Dupont, which is based on sulfonated polytetrafluoro-ethylene commercially known as Nafion [18]. The Nafion as membranes has been considered as standard membrane for FCs due to its high thermal stability, high hydrophobic and hydrophilic properties, and high proton conductivity 70–80 mS cm^{-1} at 30°C. Besides that, it was reported that at 80°C, water content in the Nafion membrane that is responsible for proton conductivity decreases and, hence, proton conductivity decreases from 18×10^{-2} to 6×10^{-2} S cm^{-1} [17, 19–24]. Therefore, this type of membrane is operated below 100°C, which limits their uses for high temperature applications. Operating Nafion membranes at high temperature led to the dehydration and lower proton conductivity. Further, other factor of Nafion as membrane such as high-cost and low-glass transition temperature (80°C–120°C) also limits their application in industrial field [17, 25]. To overcome these limitations of Nafion membranes, enormous materials have been developed for PEMFCs that have high performance as well as efficiency and low cost than Nafion membrane. Therefore, in this chapter, we discuss various types of fuel cells, the concept of PEMFC, and their classification. In addition, the recent progress of PEM for FCs and their applications are the main focus of this chapter.

12.2 Type of Fuel Cells

There are different types of fuel cell, which are given below and schematically represented in Figure 12.2. and also summarized in Table 12.1 [18].

12.2.1 Alkaline Fuel Cells

These FCs are advanced FC technologies, having been utilized by NASA in Apollo and space shuttle projects from the mid-1960s. The electrolyte in this fuel cell is an aqueous solution of concentrated (85%) potassium hydroxide (KOH) that is performed at a high temperature of 250°C, while the operating temperature for less concentrated KOH (35%–50%) is below 120°C. This electrolyte is usually kept inside a porous matrix usually made from asbestos. In fuel cells, electrocatalysts used are metal oxide, Ni, Ag, etc. The problem associated with AFC is that (1) it is very tactful to carbon dioxide (CO_2) as it will react with KOH and results in K_2CO_3 formation and (2) carbon monoxide (CO) which is produced during reaction is hazardous to metal oxide catalyst. This reaction led to the degradation of electrolyte in fuel cells and hence eventually decreased cell efficiency. The difficulty arises in such a type of fuel cell is handling and immobilization of liquid electrolyte. In addition, it was not able to absorb CO_2 efficiently, which results in lower conductivity value and precipitation of carbonate species [26, 27]. The basic reaction of AFC is as follows:

$$Anode: H_2 + 2OH^- \rightarrow 2H_2 + 2e^- \tag{12.1}$$

$$Cathode: \frac{1}{2}O_2 + H_2O + 2e^- \rightarrow 2OH^- \tag{12.2}$$

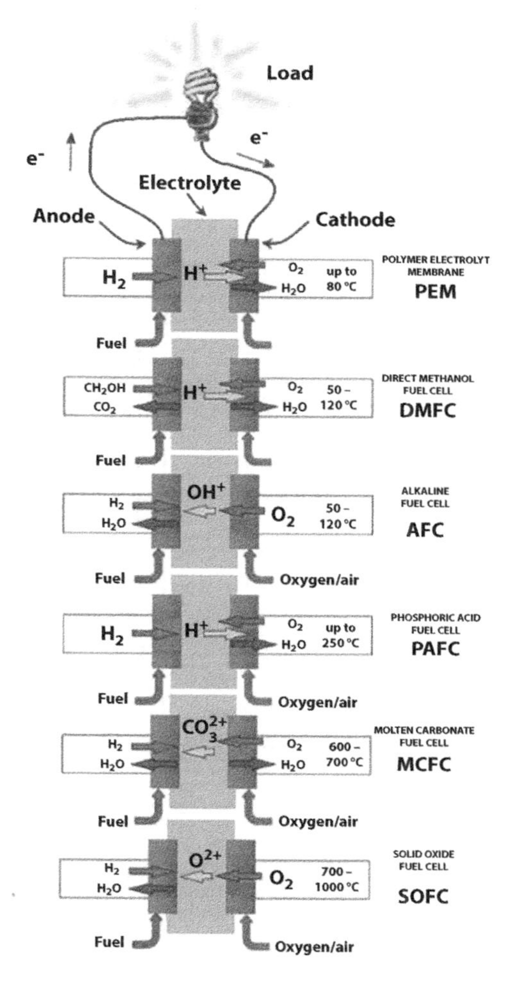

Figure 12.2 Various types of fuel cells (reprinted with the permission from ref. [4] Copyright 2014, American Chemical Society).

$$Overall: H_2 + \frac{1}{2}O_2 \rightarrow 2H_2O + electricity + heat \qquad (12.3)$$

12.2.2 Polymer Electrolyte Fuel Cells

This type of fuel cell is also known as PEMFC, because the type of electrolyte that is used for the proton conduction is solid polymer electrolyte (SPE). Perfluorinated polymer electrolytes have generally been employed

Table 12.1 Properties of various types of fuel cells.

Type of fuel cell	Anode	Electrolyte	Cathode	Operating temperature (°C)	Power density (mW/cm²)	Fuel efficiency	Application
Proton Exchange Membrane (PEM)	Pt	Perflourinated sulfonic acid	Pt	50–80	350	45–60	Vehicle and portable
Phosphoric Acid Fuel Cell (PAFC)	Pt	Phosphoric acid in SiC	Pt	160–220	200	55	Stand-alone and combined heat and power
Alkaline Fuel Cell (AFC)	Ni/Ag	KOH in Asbestos	Metal oxides	60–90	100–200	40–60	Space application
Molten Carbonate Fuel Cell (MCFC)	Ni	Alkali Carbonate in $LiAlO_3$	NiO	600–700	100	60–65	Central, stand-alone, and combined heat and power
Solid Oxide Fuel Cell	$Co-ZrO_2$ $Ni-ZrO_2$ Cermet	Y_2O_3 stabilized ZrO_2	Sr doped $LaMnO_3$	800–1,000	240	55–65	Central, stand-alone, and combined heat and power

as the membrane to separate the electrodes from oxygen and hydrogen (or methanol) gas streams in these fuel cell. The catalytic layer is a key component in this FCs. This layer is composed of the ionomer of the electrolyte membrane and nanomaterial of the catalyst on which electrochemical reaction occurs. This catalytically active layer is located next to the PEM and is supported by polytetrafluoroethylene (PTFE) treated carbon paper, which acts as a current collector and gas diffusion layer [28].

The properties related to PEMFCs are given as follows:

(a) It has high power density
(b) Rapid start-up
(c) Low temperature of operation (80°C to 120°C)

The main operating concept of this fuel cell is based on the anode-oxidation of hydrogen to protons, which is given as follows:

$$H_2 \rightarrow 2H^+ + 2e^- \quad (12.4)$$

and at the cathode side, reduction of oxygen to water:

$$4H^+ + O_2 + 4e^- \rightarrow 2H_2O \quad (12.5)$$

12.2.3 Phosphoric Acid Fuel Cells

In PAFCs, acid electrolyte is used as an electrolyte which is insensitive to CO_2 operated at a temperature 170°C–220°C. In this fuel cell, liquid phosphoric acid (H_3PO_4), which is dispersed in a silicon carbide matrix, acts as an electrolyte. The electrocatalysts in both electrodes are platinum (Pt). Here, concentrated electrolyte minimizes the water vapor pressure and absorbs CO_2 to a large extent and is less prone to absorb CO. Therefore, in this cell, water management is not difficult. Hydrogen ion (H^+ proton) acting as a charge carrier is migrated to the cathode through electrolyte. At the anode side, hydrogen breaks into electron and proton. The protons are migrated through the electrolyte toward the cathode side and electrons flow to the cathode side through an external circuit. Finally, the oxygen (O_2) is combined with an electron and proton at one electrode, i.e., cathode side and the formation of water (H_2O) takes place [29, 30]. The reactions occurring at anode and cathode are respectively as follows:

$$H_2 \rightarrow 2H^+ + 2e^- \quad (12.6)$$

$$0.5O_2 + 2H^+ + 2e^- \rightarrow H_2O \tag{12.7}$$

12.2.4 Molten Carbonate Fuel Cells

MCFCs are high temperature fuel cells (~600°C–700°C). In this fuel cell, molten carbonate salt mixture is dispersed in the ceramic matrix solid electrolyte, which acts as the electrolyte. At higher temperature, the salt in the MCFC starts to melt and gives CO_3^{2-} at the cathode side that goes to the anode side. At the anode, it combines with hydrogen and gives water, CO_2, and electrons. The produced electrons at another electrode, i.e., anode side are migrated to another electrode, i.e., cathode side via external circuit and result in current and heat. The reaction kinetics are such that the cell does not require no noble metal electrocatalysts. The cost of such types of fuel cells is less as compared to other fuel cell as a high rate of electrode reaction kinetics can easily be achieved without using platinum catalyst and thus reduces its cost. It can use gases which are derived from coal or carbon dioxide as fuel. It has efficiency higher than 60% while operating at higher temperature [11, 31]. The cell reactions of the MCFC occuring at both electrodes are as follows:

$$\text{At anode: } 2H_2 + 2CO_3^{2-} \rightarrow 2H_2O + 2CO_2 + 4e^- \tag{12.8}$$

$$\text{At cathode: } O_2 + 2CO_2 + 4e^- \rightarrow 2CO_3^{2-} \tag{12.9}$$

$$\text{The overall cell reaction: } 2H_2 + O_2 \rightarrow 2H_2O \tag{12.10}$$

12.2.5 Solid Oxide Fuel Cells

As compared to the other FCs, operating temperatures of SOFCs are generally high (~800°C–1,000°C). It consists of two porous electrodes separated by a dense oxide ion conducting electrolyte. SOFC uses a hard, ceramic compound of metal (like calcium or zirconium) oxides (chemically, O_2) as electrolyte. In this fuel cell, oxygen is fed into the cathode, and it reacts with the electrons that are injected into the external circuit, which results in oxide ion. The oxide ion formed at the cathode migrated to anode side through the oxide ion conducting electrolyte and combines with hydrogen/CO and H_2O/CO_2, respectively, and liberates electrons. It has several advantages like high efficiency, long-term stability, fuel flexibility, low emission, and relatively low cost [7, 8, 32]. The following reactions take place at both electrodes:

$$\text{At anode: } H_2 + O^{2-} \rightarrow H_2O + 2e^- \tag{12.11}$$

$$CO + O^{2-} \rightarrow CO_2 + 2e^- \tag{12.12}$$

$$\text{At cathode: } O_2 + 4e^- \rightarrow 2O^{2-} \tag{12.13}$$

$$\text{Overall: } H_2 + CO + O_2 \rightarrow H_2O + CO_2 \tag{12.14}$$

Among the mentioned various types of FCs, PEMFCs are the main focus of this chapter.

12.3 Basic Properties of PEMFC

Among the various types of fuel cells, PEMFC is the most promising clean energy technology. PEMFC consists of catalyst layers, gas diffusion, bipolar plates, and a PEM. The schematic view of PEMFC is shown in Figure 12.3. In this fuel cell, hydrogen and oxygen are fed into the anode and cathode, respectively. The two electrodes are separated by proton exchange/PEM. On the anode, hydrogen gas flows through the gas diffusion layer and splits into protons and electrons when it reaches the catalyst layer. These two protons are migrated through PEM to the catalyst layer on cathode side and electrons injected into the external circuit. On the other hand, at the cathode side, through the gas diffusion layer, oxygen reaches the catalyst layer and finally reacts with protons and electrons, which gives heat and water. The cell reaction is given as follows:

$$\text{At anode: } H_2 \rightarrow 2H^+ + 2e^- \tag{12.15}$$

$$\text{At cathode: } \frac{1}{2}O_2 + 2H^+ + 2e^- \rightarrow H_2O \tag{12.16}$$

The PEMFC cell reaction is as follows:

$$\frac{1}{2}O_2 + H_2 \rightarrow H_2O + Heat + electrical\ energy \tag{12.17}$$

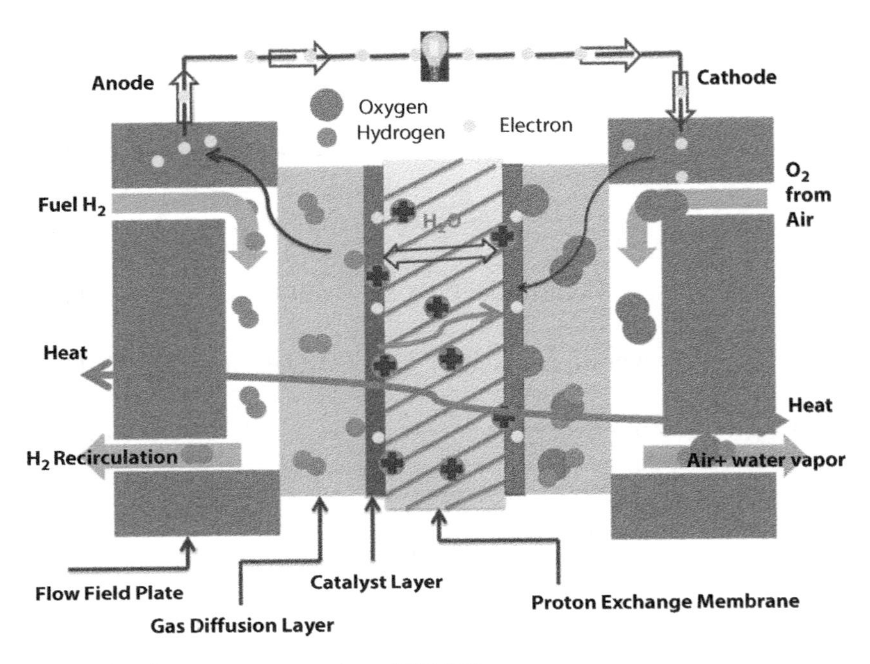

Figure 12.3 Schematic representation of PEMFC.

Generally, H_2/O_2 fuel cell and direct methanol FC (DMFC) use the SPE membrane. However, DMFC suffers from some drawbacks such as high methanol/water permeability and low operating temperature (<100°C). In that context, PEMFC has gained more attention because its power density is very high. The key component of a PEMFC is a dense proton exchange membrane, which is responsible for proton migration from anode to cathode. Therefore, it is important to know the properties for solid PEM for achieving high efficiency for PEMFC which are as follows:

(a) Low electron conductivity
(b) High proton conductivity
(c) Low fuel and oxidant permeability
(d) Good thermal and hydrolytic stability
(e) Significant dimensional and morphological stability
(f) Good thermal and mechanical stability
(g) Electrochemically stable under working temperature.
(h) Substantial morphological and dimensional stability
(i) Suppressed water transport through diffusion and electroosmosis.

12.4 Classification of Solid Polymer Electrolyte Membranes for PEMFC

Solid PEMs are classified on the basis of membrane material used for FCs and, under this categorization, they fall into five groups.

12.4.1 Perfluorosulfonic Membrane

The perfluorosulfonic acid membranes are generally represented by Nafion. Nafion belongs to the family of cation-exchange polymers and is an ionomer. It consists of hydrophobic part poly-(tetra-fluoroethylene) as the main backbone and a hydrophilic part pendant side chain that terminated with sulfonic group. The hydrophobic parts have structural integrity properties while hydrophilic parts provide adequate proton conductivity on fully hydration. When the pendant sulfonic acid group is incorporated into the backbone of polymer the Nafion ionic characteristic improves. Nafion works as a superacid catalyst and hence has high proton conductivity. On hydration of Nafion, water molecules get accumulated around the sulfonic group attached to the hydrophobic part of Nafion. Proton transport mechanism in Nafion is usually described by Grotthuss mechanism also known as structural diffusion (mainly associated with proton hopping) and the vehicle mechanism. λ is usually used to describe the number of H_2O content per sulfonate group and usually given by the following equation [22]:

$$\lambda = \frac{\textit{Number of water contents}}{\textit{Number of ionic headgroups}}$$

For $\lambda < 2$, Nafion acts as an insulator due to low water content, whereas for $\lambda \sim 2$, threshold of proton conductivity occurs. As the water content increases (for $2 < \lambda < 5$), Nafion gets splitted into hydrophobic and water filled hydrophilic cluster domains and water clusters are formed, which are large in size. Most of water molecules are present at the border of cluster, which result in the accumulation of water molecules in sulfonic solvated group shells. This accumulation of water molecules inside the percolation leads to high activation energy and impedes "Grotthuss"-type proton conductivity also known as structural diffusion and results in vehicle mechanism. For $5 < \lambda < 7$, the water cluster is expanded and develops a continuous path for proton conduction. At this point, the Grotthuss mechanism coexists with vehicle mechanisms. Finally, for $\lambda > 7$, the proton conductivity increases monotonically through the structural transport mechanism because of a well-developed percolated network and increase of free water content [17, 22, 33]. This highly dependency of Nafion on the content of water makes it undesirable at temperatures lesser than 0°C and greater than 100°C because of significant decrease in conductivity value due to freezing and boiling properties of water respectively. Besides this limitation, Nafion also suffers from some problems such as poor chemical stability at high temperature, high cost, and insufficient resistance. Several approaches are adopted for the modification of perfluorosulfonic membrane or Nafion. Choon *et al.* [34] modified the Nafion membrane by plasma etching and palladium sputtering. They found that the membrane of the surface gets rough and methanol permeability also decreases. Lin *et al.* [35] have studied the proton conductivity and PEMFC performance of Nafion membranes by applying the electric field during solution casting and found better PEMFC performance than Nafion 117. Other methods such as doping of bifunctional inorganic filler like silica,

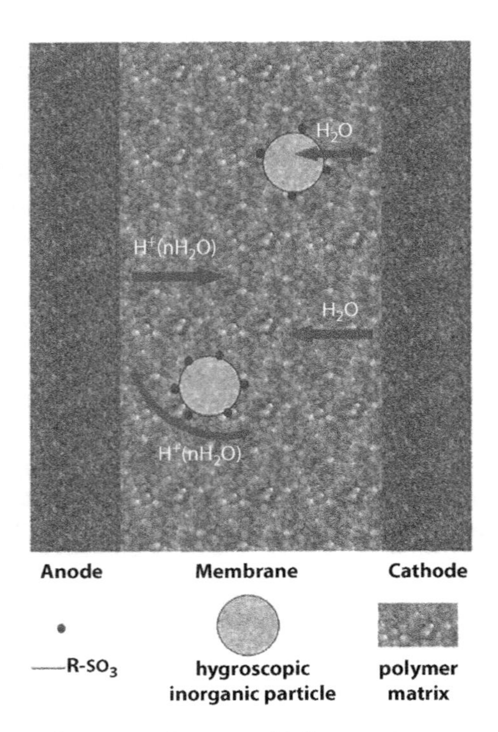

Figure 12.4 Proton conduction mechanism in Nafion membrane as a result of its hygroscopic as well as proton conductive characteristics (reprinted with the permission from ref. [17], Copyright 2012, American Chemical Society).

silicon, and polysiloxane into the Nafion matrix have been also studied. Silicon-modified Nafion membrane shows better properties such as lower methanol permeability and better water management. The presence of hygroscopic and proton conducting characteristics of bifunctional inorganic filler improves water management, which is shown in Figure 12.4 [36–46]. Further, Nafion composite membranes have the potential to be used in high temperature applications.

12.4.2 Partially Fluorinated Polymers

Partially fluorinated acid membrane refers to the grafted with radiation membrane and membrane with blending using commercial fluoropolymer as the key backbone. It is less expensive. The main steps involved in the fabrication of radiation induced grafting membrane are as follows: (1) pre-irradiation, (2) grafting, and (3) sulfonation, as shown in Figure 12.5 [47]. The common membranes lying under this category are poly(vinyl fluoride) (PVF), PTFE (Teflon), etc. Because of good chemical stability and mechanical stability (except PTFE and PVF), perfluorinated membranes are widely used in PEMFCs. While the PVDF-based partially fluorinated acid membrane has excellent mechanically, thermally, and chemically stability properties. Nasef *et al.* [48] studied the formation of proton exchange membrane and found that proton (H^+) conductivity is ~114 mS cm^{-1} with 65% grafting level [48]. Kim and co-workers [49–52] reported the P(VDF-co-CTFE)-g-PSSS and found the proton conductivity 0.074 S cm^{-1}. In addition, the cross-linking of P(VDF-co-CTFE)-g-PSSA membrane PSSA (73 wt.%) reduces conductivity value from 74.0 to 68.0 mS cm^{-1} at room temperature [49–52]. The characteristics of the grafted membranes are heavily influenced by the substrate structure. The stability of membranes grafted with styrene sulfonic acid is limited, but those grafted with α substituted styrene sulfonic acid have better stability.

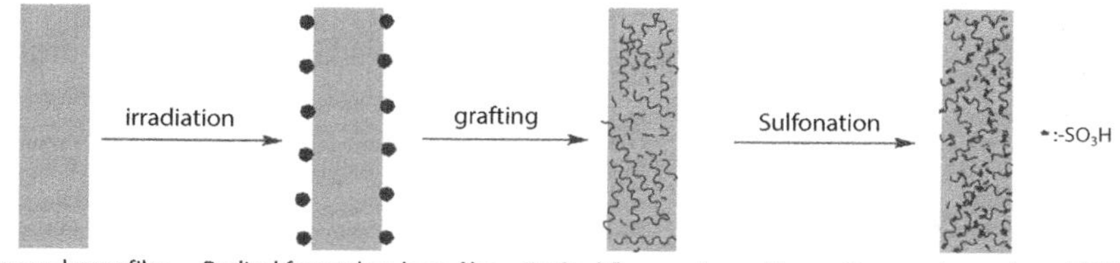

Figure 12.5 Radiation Grafting method for the synthesis of PEM membranes (reprinted with the permission from ref. [17], Copyright 2012, American Chemical Society).

Polymer blending is also an effective technique for modifying the properties of fluoropolymer-based membranes. Blending of polymer like hydrocarbons type and partially fluorinated type is a very easy technique for fabrication of PEMs. The blending of P(S-co-SSA)-b-PMMA and PVDF polymer shows better proton conductivity and a well-ordered structure than membranes having random order morphology.

12.4.3 Non-Fluorinated Hydrocarbon Membrane

Non-fluorinated hydrocarbons with aliphatic or aromatic polymers are utilized in the fabrication of PEMs for PEMFCs. In these polymers, a benzene ring is attached to the backbone of the polymer. These hydrocarbon polymers for polymer backbone are used as PEMs and are one of the best ways to achieve high performance PEMFC. There are many advantages of these hydrocarbon polymers over other polymeric membranes given as follows:

(a) Hydrocarbon polymers are more cost effective than perfluorinated ionomers.
(b) By conventional method, these polymers are easy to recycle.
(c) The polymers, which contain polar groups, have high water uptake over a wide range of temperatures.

12.4.4 Nonfluorinated Acid Membranes With Aromatic Backbone

Presently, one of the important ways to increase the performance and stability of proton conductivity of PEM is by incorporating the aromatic hydrocarbon directly into the hydrocarbon polymer backbone. Such polymer membranes are better than perfluorinated membranes as higher water uptake by the polar group associated with the hydrocarbon polymer, wide range of temperature operations, and easy recyclability by conventional method.

12.4.5 Acid Base Blend

Acid-base complex–based membranes are those membranes, which are having high value of conductivity at higher temperature without suffering from the dehydration effect. In these membranes, addition of a component of acid into alkaline polymer base will take place for the promotion of proton conduction. The phosphoric acid–doped polybenzimidazole (PBI/H_3PO_4) membranes are the best system for high temperature PEMFC. Acid base polymer blend is one of the most promising materials for the development of low cost with good performance fuel cell. In this membrane, the interaction between the acid and base polymer like ionic cross-linking and hydrogen bonding bridges reduces the swelling of membrane without deformation in flexibility. Hence, this type of membrane has the following properties: good mechanical flexibility, high proton conductivity, good thermal stability, and low water uptake. Dan Wu *et al.* [53] prepared a hybrid acid-base polymer membrane by blending method using different polymers through a sol-gel process. They found that this membrane composition has high proton conductivity of 72.3 mS cm^{-1} and water uptake of 30.9%, which improved the performance of fuel cell. The conductivity of this type of membrane is sensitive to the doping concentration and temperature. At the doping concentration of 450% and temperature of 165°C, the conductivity of PBI membrane was found to be 0.046 mS cm^{-1}. Kerres *et al.* [54] showed excellent thermal stability of membrane using PBI and sulfonated poly(etheretherketone) sPEEK and as the basic and acidic compound, respectively. Vargas *et al.* [55] found highest electrical conductivity (~10^{-1} S cm^{-1}) for polyvinyl alcohol (PVAL) and hyphophosphorous acid. Paul *et al.* [56] found that fuel cell operated with doping of PBI/H_3PO_4 membrane at temperature of 190°C and at atmospheric pressure gives a power density value 550 mWcm^{-2} and a current density value 1,200 mA cm^{-2}.

12.5 Applications

Fuel cells are used in stationary applications like back-up power supplies, power generation for remote locations, and distributed generation for buildings as shown in Figure 12.6. PEMFCs are used in portable and stationary applications. PEMFC provides a continuous supply of electrical energy at high efficiency level and power density so this fuel cell is very suitable for transportation application [57]. The most potential use of the PEMFC technology is fuel cell vehicles. In PEMFCs, the principal components such as bipolar plates and membrane electrode assembly (MEA) that consist of gas diffusion layer, cathode, anode, proton conductive electrolyte, and the electrocatalyst layer, which facilitates heat and water management. Each component plays a unique role in power system applications. The bipolar plate with the gas diffuser enables water and heat management

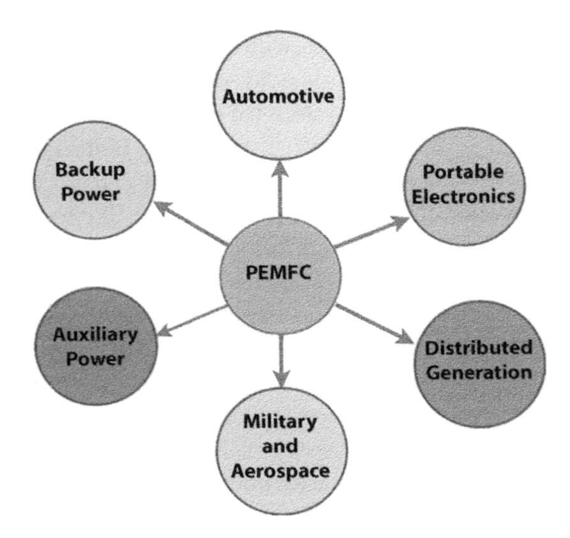

Figure 12.6 Schematic view of applications of PEMFCs.

and facilitates electric current transfer. Due to unique feature of each component, PEMFCs are widely used in transportation as discussed as follows.

12.5.1 Application in Transportation

In transportation, the issues such as gas emission from vehicles and environmental pollution demand energy storage devices that must be much cleaner without affecting the performance of machines or vehicles. In that context, PEMFCs provide a viable alternative due to zero emission and high-power density. Various applications of PEMFCs in transportation are discussed as follows:

a. Land transportation: In land transportation, PEMFC technology has been widely used in mainly in making bicycle, car, light traction vehicles, light-duty fuel cell electric vehicles (L-FCEVs), heavy-duty fuel cell electric vehicles (H-FCEVs), etc. Among them, L-FCEVs and H-FCEVs are equipped with hydrogen-based fuel cells. For such types of EVs, researchers developed a new designed FC setup consisting of a thermoelectric generator equipped with an ammonia hydrogen internal combustion engine. This setup provides the overall good performance of a fuel cell, and also, it is cost effective and durable. Besides that, in many countries such as the United States, fuel cell powered cars are widely used.

b. Air transportation: In air transportation, PEMFCs are also widely used. The air transportation activities such as aeroplanes, aerial jet propeller, and jet are running on PEMFCs. This fuel cell system is based on Uncrewed Aerial Vehicles (UAVs), which is developed by Intelligent Energy UK in 2018. Nowadays, battery technology is replaced by fuel cell technology due to their unique features mainly in drone technology like in security, search and rescue, surveillance research, and development system.

c. Water transportation: In marine transportation such as boat, ships, and canoe, fuel cell technology is widely used as they offer low noise and emission, high efficiency, and specific power. In addition, merchant ships and warships are run on the hydrogen-based fuel cell system due to stable supply and low emissions.

12.6 Conclusions

Energy shortages and pollution have become a major issue for humanity today. To solve these concerns, several efforts have been made in order to find the alternative of fossil fuels with other energy sources, like clean energy fuel. Because of its unique features such as high power density, zero emission, and clean technology, fuel cells are poised to usher in a massive revolution in the world of energy. Due to their characteristics properties, continuous research in the development of PEMFC is in progress; therefore, in this chapter, we summarize the characteristic properties of PEMFC and their potential application in power system technology.

References

1. Ma, S., Lin, M., Lin, T.E., Lan, T., Liao, X., Maréchal, F., Van herle, J., Yang, Y., Dong, C., Wang, L., Fuel cell-battery hybrid systems for mobility and off-grid applications: A review. *Renew. Sust. Energ. Rev.*, 135, 110119, 2021.
2. Wang, Y., Kwok, H.Y.H., Zhang, Y., Pan, W., Zhang, H., Lu, X., Leung, D.Y.C., A flexible paper-based hydrogen fuel cell for small power applications. *Int. J. Hydrogen Energy*, 44, 29680, 2019.

3. Wang, G., Yu, Y., Liu, H., Gong, C., Wen, S., Wang, X., Tu, Z., Progress on design and development of polymer electrolyte membrane fuel cell systems for vehicle applications: A review. *Fuel Process. Technol.*, 179, 203, 2018.
4. Kraytsberg, A. and Ein-Eli, Y., Review of advanced materials for proton exchange membrane fuel cells. *Energ. Fuel.*, 28, 7303, 2014.
5. Manoharan, Y., Hosseini, S.E., Butler, B., Alzhahrani, H., Senior, B.T.F., Ashuri, T., Krohn, J., Hydrogen fuel cell vehicles; Current status and future prospect. *Appl. Sci.*, 9, 2296, 2019.
6. Akinyele, D., Olabode, E., Amole, A., Review of fuel cell technologies and applications for sustainable microgrid systems. *Inventions*, 5, 1, 2020.
7. Minh, N.Q., Solid oxide fuel cell technology - Features and applications. *Solid State Ion.*, 174, 271, 2004.
8. Singhal, S.C., Advances in solid oxide fuel cell technology. *Solid State Ion.*, 135, 305, 2000.
9. Gülzow, E., Alkaline fuel cells: A critical view. *J. Power Sources*, 61, 99, 1996.
10. Sammes, N., Bove, R., Stahl, K., Phosphoric acid fuel cells: Fundamentals and applications. *Curr. Opin. Solid State Mater. Sci.*, 8, 372, 2004.
11. Dicks, A.L., Molten carbonate fuel cells. *Curr. Opin. Solid State Mater. Sci.*, 8, 379, 2004.
12. Song, Y., Zhang, C., Ling, C.Y., Han, M., Yong, R.Y., Sun, D., Chen, J., Review on current research of materials, fabrication and application for bipolar plate in proton exchange membrane fuel cell. *Int. J. Hydrogen Energy*, 45, 29832, 2020.
13. Elwan, H.A., Mamlouk, M., Scott, K., A review of proton exchange membranes based on protic ionic liquid/polymer blends for polymer electrolyte membrane fuel cells. *J. Power Sources*, 484, 229197, 2021.
14. Srivastava, N., Singh, S.K., Gupta, H., Meghnani, D., Mishra, R., Tiwari, R.K., Patel, A., Tiwari, A., Singh, R.K., Electrochemical performance of Li-rich NMC cathode material using ionic liquid based blend polymer electrolyte for rechargeable Li-ion batteries. *J. Alloys Compd.*, 843, 155615, 2020.
15. Meghnani, D., Gupta, H., Singh, S.K., Srivastava, N., Mishra, R., Tiwari, R.K., Patel, A., Tiwari, A., Singh, R.K., Enhanced cyclic stability of $LiNi_{0.815}Co_{0.15}Al_{0.035}O_2$ cathodes by surface modification with $BiPO_4$ for applications in rechargeable lithium polymer batteries. *ChemElectroChem*, 8, 2867, 2021.
16. Singh, S.K., Dutta, D., Singh, R.K., Enhanced structural and cycling stability of Li_2CuO_2-coated $LiNi_{0.33}Mn_{0.33}Co_{0.33}O_2$ cathode with flexible ionic liquid-based gel polymer electrolyte for lithium polymer batteries. *Electrochim. Acta*, 343, 136122, 2020.
17. Zhang, H. and Shen, P.K., Recent development of polymer electrolyte membranes for fuel cells. *Chem. Rev.*, 112, 2780, 2012.
18. Peighambardoust, S.J., Rowshanzamir, S., Amjadi, M., Review of the proton exchange membranes for fuel cell applications. *Int. J. Hydrogen Energy*, 35, 9349, 2010.

19. Kreuer, K.D., Proton Conductivity: Materials and applications. *Chem. Mater.*, 8, 610, 1996.
20. Gebel, G., Structural evolution of water swollen perfluorosulfonated ionomers from dry membrane to solution. *Polymer*, 41, 5829, 2000.
21. Haubold, H., Vad, T., Jungbluth, H., Hiller, P., Nano structure of NAFION : A SAXS study. *Electrochim. Acta*, 46, 1559, 2001.
22. Eikerling, M., Kornyshev, A.A., Stimming, U., Electrophysical properties of polymer electrolyte membranes: A random network model. *J. Phys. Chem. B*, 101, 10807, 1997.
23. Gierke, T.D., Munn, G.E., Wilson, F.C., The morphology in nafion* perfluorinated membrane products, as determined by wide- and small- angle x-ray studies. *J. Polym. Sci.*, 19, 1687, 1981.
24. Halim, J., Brhm, F.N., Stamm, M., Scherer, G.G., Characterization of perfluorosulfonic acid membranes by conductivity measurements small-angle x-ray scattering. *Electrochim. Acta*, 39, 1303, 1994.
25. Kreuer, K.D., Ion conducting membranes for fuel cells and other electrochemical devices. *Chem. Mater.*, 26, 361, 2014.
26. Gülzow, E., Alkaline fuel cells. *Fuel Cells*, 4, 251, 2004.
27. Kalogirou, S.A., *Solar Energy Engineering: Processes and Systems*, pp. 397–429, Elsevier, USA, 2014.
28. Brandon, N.P. and Parkes, M.A., Fuel Cells: Materials, in: *Reference Module in Materials Science and Materials Engineering*, vol. 1, 2016.
29. Sudhakar, Y.N., Selvakumar, M., Bhat, D.K., *Biopolymer Electrolytes: Fundamentals and Applications in Energy Storage*, pp. 151–66, Elsevier, India, 2018.
30. Neergat., M. and Shukla, A.K., A high-performance phosphoric acid fuel cell. *J. Power Sources*, 102, 317, 2001.
31. Dincer, I. and Rosen, M.A., *Exergy: Energy, Environment and Sustainable Development*, pp. 479–514, Elsevier, UK, 2021.
32. Ormerod, R.M., Solid oxide fuel cells. *Chem. Soc. Rev.*, 32, 17, 2003.
33. Smitha, B., Sridhar, S., Khan, A.A., Solid polymer electrolyte membranes for fuel cell applications — a review. *J. Membr. Sci.*, 259, 10, 2005.
34. Choi, W.C., Kim, J.D., Woo, S.I., Modi ® cation of proton conducting membrane for reducing methanol crossover in a direct-methanol fuel cell. *J. Power Sources*, 96, 411, 2001.
35. Lin, H., Yu, T.L., Han, F., A method for improving ionic conductivity of Nafion membranes and its application to PEMFC. *J. Polym. Res.*, 13, 379, 2006.
36. Tominaga, Y., Hong, I., Asai, S., Sumita, M., Proton conduction in Nafion composite membranes filled with mesoporous silica. *J. Power Sources*, 171, 530, 2007.
37. Lin, Y., Yen, C., Ma, C.M., Liao, S., Lee, C., Hsiao, Y., Lin, H., High proton-conducting Nafion ® /– SO_3H functionalized mesoporous silica composite membranes. *J. Power Sources*, 171, 388, 2007.

38. Su, L., Li, L., Li, H., Tang, J., Zhang, Y., Yu, W., Zhou, C., Preparation of polysiloxane modified perfluorosulfonic acid composite membranes assisted by supercritical carbon dioxide for direct methanol fuel cell. *J. Power Sources*, 194, 220, 2009.
39. Woo, J., Chang, Y., Jun, Y., Dong, S., Yong, H., Jung, W., Effects of organofunctionalization and sulfonation of MCM-41 on the proton selectivities of MCM-41/Nafion composite membranes for DMFC. *Microporous Mesoporous Mater.*, 114, 238–49, 2008.
40. Wei, S., Zhang, X., Liu, Z., Hong, L., Hwa, S., Composite Nafion® membrane embedded with hybrid nanofillers for promoting direct methanol fuel cell performance. *J. Membr. Sci.*, 32, 139, 2008.
41. Wang, K., Mcdermid, S., Li, J., Kremliakova, N., Kozak, P., Song, C., Tang, Y., Zhang, J., Zhang, J., Preparation and performance of nano silica/Nafion composite membrane for proton exchange membrane fuel cells. *J. Power Sources*, 184, 99, 2008.
42. Kang, J., Ghil, L., Kim, Y., Kim, Y., Rhee, H., Preparation of Nafion® nanocomposite membrane modified by phosphoric acid-functionalized 3-APTES. *Colloids Surf. A Physicochem. Eng. Asp.*, 314, 207, 2008.
43. Nam, S., Kim, S., Kang, Y., Wook, J., Lee, K., Preparation of Nafion/sulfonated poly (phenylsilsesquioxane) nanocomposite as high temperature proton exchange membranes. *J. Membr. Sci.*, 322, 466, 2008.
44. Lavorgna, M., Mascia, L., Mensitieri, G., Gilbert, M., Scherillo, G., Palomba, B., Hybridization of Nafion membranes by the infusion of functionalized siloxane precursors. *J. Membr. Sci.*, 294, 159, 2007.
45. Yen, C., Lee, C., Lin, Y., Lin, H., Hsiao, Y., Liao, S., Chuang, C., Ma, C., Sol – gel derived sulfonated-silica / Nafion® composite membrane for direct methanol fuel cell. *J. Power Sources*, 173, 36, 2007.
46. Ladewig, B.P., Knott, R.B., Martin, D.J., Nafion-MPMDMS nanocomposite membranes with low methanol permeability. *Electrochem. Commun.*, 9, 781, 2007.
47. Scherer, G.G. and Fuel cell, I., *Advances in Polymer Science*, Springer, Berlin, Heidelberg, 2008.
48. Mahmoud, M., Saidi, H., Zaman, K., Dahlan, M., Single-step radiation induced grafting for preparation of proton exchange membranes for fuel cell. *J. Membr. Sci.*, 339, 115, 2009.
49. Kim, Y. and Cho, J., Lithium-Reactive $Co_3(PO_4)_2$ nanoparticle coating on high-capacity $LiNi_{0.8}Co_{0.16}Al_{0.04}O_2$ cathode material for lithium rechargeable batteries. *J. Electrochem. Soc.*, 154, A495, 2007.
50. Choi, J.K., Kim, Y.W., Koh, J.H., Kim, J.H., Proton conducting membranes based on poly (vinyl chloride) graft copolymer electrolytes. *Polym. Adv. Technol.*, 19, 915, 2008.
51. Patel, R., Tae, J., Seok, W., Hak, J., Ryul, B., Composite polymer electrolyte membranes comprising P (VDF- co -CTFE) - g -PSSA graft copolymer and zeolite for fuel cell applications. *Polym. Adv. Technol.*, 20, 1146, 2009.

52. Roh, D.K., Park, J.T., Koh., J.H., Proton-conducting composite membranes from graft copolymer electrolytes and phosphotungstic acid for fuel cells. *Ionics*, 15, 439, 2009.
53. Wu, D., Xu, T., Wu, L., Wu, Y., Hybrid acid – base polymer membranes prepared for application in fuel cells. *J. Power Sources*, 186, 286, 2009.
54. Kerres., J., Ullrich, A., Meier, F., Haring, T., Synthesis and characterization of novel acid – base polymer blends for application in membrane fuel cells. *Solid State Ion.*, 125, 243, 1999.
55. Vargas, M.A., Vargas, R.A., Mellander, B.E., New proton conducting membranes based on PVAL/H3PO2/H2O. *Electrochimica Acta.*, 44, 4227–4232, 1999.
56. Steiner, P., Sandor, R., Polybenzimidazole prepreg: Improved elevated temperature properties with autoclave processability. *High Perform. Polym.*, (UK), 3, 139–50, 1991.

13

Computational Fluid Dynamics Simulation of Transport Phenomena in Proton Exchange Membrane Fuel Cells

Maryam Mirzaie[1]* and Mohamadreza Esmaeilpour[2]

[1]*Department of Chemical Engineering, Faculty of Engineering, Vali-E-Asr University of Rafsanjan, Rafsanjan, Iran*
[2]*Department of Chemical Engineering, Faculty of Engineering, Shahid Bahonar University of Kerman, Kerman, Iran*

Abstract

In the recent years, by increasing the consumption of fossil fuels, the carbon dioxide level increases, which affects human health and climate change. One of the important technologies to solve these problems is fuel cell vehicles (FCVs), and, due to low operating temperature and high efficiency, the proton exchange membrane fuel cells (PEMFCs) have been more considered among various types of FCVs. Computational fluid dynamics (CFD) and numerical modeling have a great potential to study the detailed physical phenomena in the fuel cells such as mass, heat and energy transport, electrode kinetics, and potential fields and can reveal which factors in a real system are most important from a sensitivity point of view

We reviewed several existing works relevant to CFD simulation and mathematical modeling of PEMFCs, which were performed by different software and codes like ANSYS Fluent, COMSOL, lattice Boltzmann, MATLAB codes, and OpenFOAM. The key factors that affect performance and failure modes of PEMFCs were discussed based on these simulations. One of these factors is water management that can be improved by using appropriate flow field configuration. The various configurations were reviewed. In addition, the effect of humidification of inlet reactants on the performance of PEMFCs was discussed.

Keywords: Fuel cells, proton exchange membrane fuel cells, transport phenomena, CFD simulation, mathematical modeling

**Corresponding author*: m.mirzaie@vru.ac.ir

Inamuddin, Omid Moradi and Mohd Imran Ahamed (eds.) Proton Exchange Membrane Fuel Cells: Electrochemical Methods and Computational Fluid Dynamics, (353–394) © 2023 Scrivener Publishing LLC

13.1 Introduction

The successful conversion of chemical energy into electricity was the first things that fuel cells demonstrated several years ago. Fuel cells are generally referred to as a device that can convert the chemical energy of a fuel, which is normally hydrogen, into electrical energy. Fuel cells are widely produced in different types and fall into various categories [1].

Different types of fuel cells that have been developed and studied so far can be summarized in three categories: alkaline fuel cell (AFC), polymeric electrolyte membrane fuel cell and phosphoric acid fuel cell (PAFC) [2]. Fuel cells generally have many advantages like no moving parts, no pollution generation, high efficiency in various operating conditions, and easy use that lead to the widespread use of fuel cells in cases such as transportation as an energy source.

Polymer electrolyte fuel cells are types of fuel cells commonly known as proton exchange membrane fuel cells (PEMFCs). This kind of fuel cell has a lot of advantages like low weight, compactness, the potential of working at high current density, long time, working at low temperature, and possibility of use in discontinuous processes [3]. In addition, low power levels (less than 1 kW) to medium power levels (up to 50 kW) in these cells, along with the ability to set up quickly and rapid response, have made the use of these cells attractive.

These characteristics of polymer cells have led to the widespread use of this cell model in various fields. Use in the form of portable cells, micro power to large scale, and power plants can be the features of these cells. For these reasons, many companies active in the field of fuel cells, automobiles, and electricity are interested in using PEMFCs.

The main and important part of these fuel cells is their dense proton exchange membrane, which is responsible for moving protons from the anode to the cathode. In order to produce protons, in these fuel cells, the protons are catalytically oxidized at the anode. In these cells, the membrane is located in the space between the anode and the cathode. The produced protons migrate through the anode to the cathode. As a result of this displacement of protons, water and heat are produced at the cathode due to the reaction of protons with oxygen. Despite all the advantages and positive features of these cells, they also have disadvantages, which indicate that there is still a long way to go before these cells can replace traditional methods of energy production [3].

The main parts of PEMFCs are polymer membrane, flow channels, and catalyst layers (CLs), gas diffusion layers (GDLs), and current collectors. The main operating mechanism of PEMFCs is based on the electrical reaction between fuel (H_2) provided by flow field plate on anode side and existed oxidant (O_2) in the air (on cathode side) in the presence of catalyst (that is platinum). The hydrogen gas is decomposed into positive hydrogen ions and negative electrons and reacts with oxygen to produce water [4, 5]. A schematic diagram of PEMFCs is showed in Figure 13.1 [6].

Only positive ions allow transfer across the polymer membrane, and electron transfer along an external circuit and electrical current is produced. An important advantage of PEMFCs is low operating temperature, i.e., 40°–90°C, and the fuel cell performance depends on flow field design and interface between reactants. The reactant gas distribution and controlling heat and water management are affected by the flow field behavior. The gas diffusivity is lower on the cathode side with respect to anode side; therefore, the optimization of flow field on the cathode side is more important [7, 8].

For appropriate performance of PEMFCs, the flow field must have the characteristics like uniform distributions of reactant gases on CL surface, temperature, produced water and current density, minimum pressure drop, removing the produced water to the outside of cell, and proper electron exchange with CL surface [9]. If produced water does not exit from

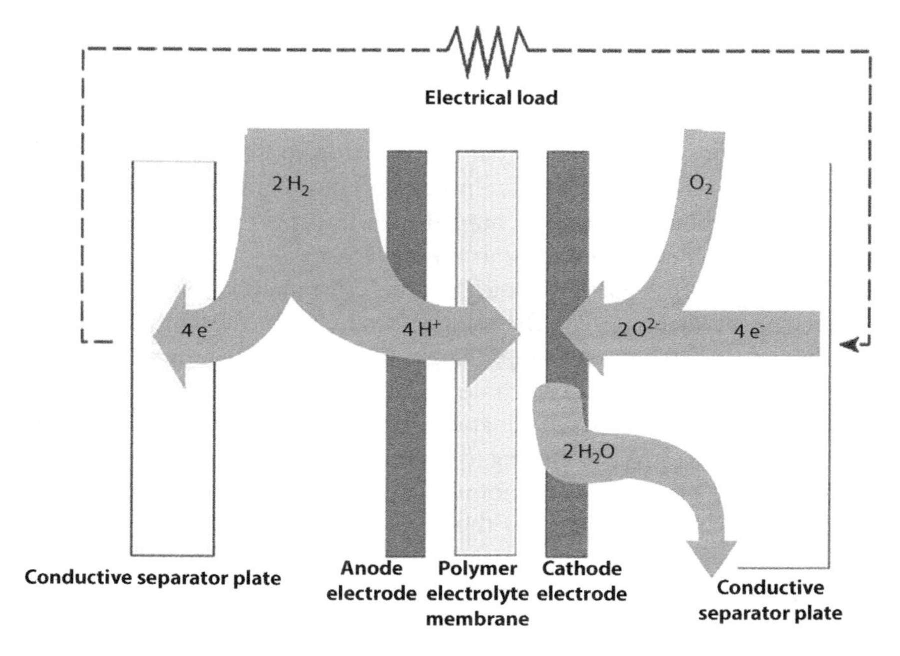

Figure 13.1 A schematic diagram of a PEMFC. Modified after Wilberforce *et al.* (2017) [6].

the cell, then flooding phenomenon happens. In the flooding, the oxygen diffusivity decreases, and this leads to high temperature and then membrane is dried. In these conditions, due to increment of the internal cell resistance, the PEMFC efficiency decreases. According to these facts, an appropriate configuration of the flow field leads to a uniform gas distribution in PEMFC. Therefore, the design of the flow field must be such that these conditions have been met [10].

Computational flow dynamics (CFD) is a powerful tool for prediction of the flows of oxygen and hydrogen inside the fuel cell, prediction of pressure drop, and studying the effects of inlet gas relative humidity (RH) and cell geometry on the cell flow field. Therefore, many researchers applied this method to analysis and optimization of PEMFC performance by prediction of velocity profile, pressure gradients, and transfer of species inside the cell [9, 11, 12].

This chapter reviewed the application of CFD simulation for transport phenomena in PEMFCs. The CFD simulations were reviewed from the studies by ANSYS Fluent and COMSOL software, lattice Boltzmann method (LBM), MATLAB codes, and OpenFOAM. The effects of important parameters on the performance of PEMFCs were discussed.

13.2 PEMFC Simulation and Mathematical Modeling

The detailed behavior and performance of PEMFCs under different operating conditions are well understood by using mathematical modeling and CFD simulation of the cells. In detailed modeling, the electrochemistry and thermodynamics are coupled with transport phenomena (mass, momentum, and energy) in porous media [13–15].

There are three categories of models for fuel cells: analytical, semi-empirical, or mechanistic that based on the solution strategy; the mechanistic models are divided into two main groups: single-domain or multi-domain. In the multi-domain method, the separate equations are solved in each cell but the single-domain consists of equations governing the entire fuel cell, with source and sink terms accounting for species consumption and generation within the cell [4, 16, 17]. In another classification, two main class models were used by researchers for the prediction of fuel cell behaviors: the static model and the dynamic linear model. In the first model, the curve of polarization at fixed operating conditions is predicted, while by use of the second method, the detailed behavior of cell in each operating condition is investigated accurately [18, 19].

In the mechanistic model, all detailed of chemistry and physics are used to study and optimization of fuel cell performance. By applying this type of models, the internal reaction mechanism is known precisely, but this method due to numerous unknown parameters is very complicated. In the semi-empirical method, the experimental data are used to determine the parameters and many assumptions are considered, which this causes the model results are not well fitted by actual data. An appropriate method is applying the semi-empirical models, and, based on what information is needed, researchers apply one of these models. When the phenomena at pore level must be known or three phases must be considered, calculations are complicated and time consuming [20]. Table 13.1 provides common assumptions considered in PEMFC modeling.

Table 13.1 The common considered assumptions for PEMFC modeling. Modified after Siegel *et al.* (2008) [4].

Thermal considerations	- When analyzing two-phase flow, distribution of temperature is very important (saturation pressure curve) - In order to determine the local all hot-spots, a two-equation system for solid and fluid phase can be used. - When the difference temperature of two-phase is high, thermal diffusion should be considered.
Two-phase flow	- The general approach to determine the phase velocity is mixture multi-phase model, but this approach needs the higher number of dependent variables and phases are coupled. - In the mixture model, the phase transport is calculated by using mixture properties and the level of water saturation is obtained by extracted liquid concentration from solution. - The capillary is the only driving force in the moisture diffusion model, and one additional equation must be used. - The effect of gas pressure on the liquid not accounted in the moisture diffusion model. - It is assumed that the water generates only in the liquid or vapor phase. - The level of water saturation iteratively updates by the porosity correction model, while correcting the porosity. - For liquid water flux, the modified Darcy-type equation is used.

(*Continued*)

Table 13.1 The common considered assumptions for PEMFC modeling. Modified after Siegel *et al.* (2008) [4]. (*Continued*)

Pressure considerations	- In applying sandwich model, the pressure can be considered constant (isobar). - The pressure and velocity behavior must be considered along the channel simulation and the interactions between liquid and gas are ignored.
Electrical considerations	- Almost ignored the voltage losses and the potential of electrode considered constant. - For specific dynamic studies, the double layer charging must be interesting.
Physical parameters	- The physical properties are considered isotropic and homogeneous. - The heat transfer coefficient is considered constant - For porous media and amount of water saturation, physical parameters must corrected
Gas properties	- In the simulation of a fuel cell, commonly, the ideal gas law is used. - The gas phase in the channel is considered as a well-mixed phase. - Density is considered constant. - The gas phase is assumed to be saturated in some simulations. - The gas properties often are considered constant.
Perfect walls	- Bends and imperfection in the channel of gas are ignored. - Two-dimensional are used for channel of gas stream while with apply a volume averaged method, the results are corrected for third dimension. - Buoyancy is considered as an external force term.
Thermal considerations	- The heat transfer by convection mechanism in the gas and heat exchange not exist between the flow of gas and bipolar plate.

(*Continued*)

Table 13.1 The common considered assumptions for PEMFC modeling. Modified after Siegel *et al.* (2008) [4]. (*Continued*)

Porous media considerations	- In the simulations, porous media are defined as reactive boundary layer. Uniform distribution of catalyst particles (Pt particles) is mostly used. - In order to describe a catalyst layer, the porous electrode model or a micro or macroscopic agglomerate structure is applied. When the agglomerate model is used, the shape and size are considered identical for catalyst particles. - The constant current density is assumed over CL and electrochemical reaction takes place at the interface between the electrode.
Gas diffusion	- The effective diffusivity is calculated by adding Knudsen diffusion term into Stefan–Maxwell diffusion equation.
Porous media flow	- Flow through porous media is characterized by Darcy model. - Often, the Brinkman extend (friction from macroscopic shear) and Forchheimer term (fluid inertial energy) are neglected. - For liquid water, the driving force is only capillary pressure.
Thermal considerations	- Because of high thermal conductivity of the bipolar plates, the temperature is constant. - The adiabatic boundary condition is used for all walls.

13.2.1 Governing Equations

Several strong coupled equations must be solved for detailed modeling of mass and heat transport phenomena in a PEMFC, which can be classified to five types, namely, mass, momentum, species, and energy conservation equations; fluxes constitutive equations; reaction kinetic relations; equilibrium equations; and auxiliary relations for variables calculation.

13.2.1.1 Continuity Equation

For steady flow, the continuity equation is as follows [21, 22]:

$$\nabla.\left(\varepsilon\rho\vec{u}\right)=0 \tag{13.1}$$

where $\vec{u}$ is the velocity vector and ρ and ε are density and porosity, respectively.

13.2.1.2 Momentum Equation

The momentum equation for steady and laminar flow is given in Equation (13.2) [21]:

$$\nabla.\left(\varepsilon\rho\vec{u}\vec{u}\right)=-\varepsilon\nabla p+\nabla.\left(\varepsilon\mu\nabla\vec{u}\right)+S_u \tag{13.2}$$

For the porous media, the above equation is modified through Darcy's law as follows:

$$\nabla.\left(\varepsilon\rho\vec{u}\vec{u}\right)=-\varepsilon\nabla p+\nabla.\left(\varepsilon\mu\nabla\vec{u}\right)+S_P \tag{13.3}$$

In the above equations, p and μ are pressure and viscosity, respectively, and S_u represents source of momentum. S_p in Equation (13.3) is momentum source for porous media. Modified equation for PEMFC porous media is as follows:

$$\varepsilon\vec{u}=-\frac{K_P}{\mu}\nabla P \tag{13.4}$$

where K_p is permeability and defined as follows [23]:

$$K_P=\frac{d_P^2}{72\tau^2}\frac{\varepsilon^2}{(1-\varepsilon)^2} \tag{13.5}$$

d_p is particle mean diameter and τ is tortuosity factor.

13.2.1.3 Mass Transfer Equation

The multiphase mass transport is one of the most challenging issues in PEMFC modeling, accounting of the evaporation/condensation of water. Based on Fick's law, the equation for jth component is as follows [24]:

$$\nabla.\left(\varepsilon\vec{u}C_j\right)=\nabla.\left(D_K^{eff}\nabla C_j\right)+\nabla.\left(-\frac{n_d}{F}I\right)+S_j \tag{13.6}$$

where C_j is concentration and n_d is the memberane electroosmotic drag coefficient. S_j reperesents the source term in CLs zones and for different components are given in Equations (13.7) to (13.9) [22, 25]:

$$S_{H_2} = -\frac{j_a}{2FC_{total,a}} \tag{13.7}$$

$$S_{O_2} = -\frac{j_c}{4FC_{total,c}} \tag{13.8}$$

$$S_{H_2O} = -\frac{j_c}{2FC_{total,c}} \tag{13.9}$$

where j_a and j_c are current densities and defined in Equations (13.12) and (13.13).

Appropriate relations for effective transport coefficients are needed to simulate the PEMFCs. D_K^{eff} is calculated by the equation as follows:

$$D_K^{eff} = \varepsilon^{0.5}(1-s)^{r_s} D_K^0 \left(\frac{p_0}{p}\right)\left(\frac{T}{T_0}\right)^{1.5} \tag{13.10}$$

where r_s is pore blockage saturation exponent and D_K^0 is kth component mass diffusivity at the reference conditions, s is water saturation and calculated as follows:

$$s = \frac{Vol_{liq}}{Vol_{Mix}} \tag{13.11}$$

In the above equations j_a and j_c are current densities and are calculated by butler volume function as follows [26]:

$$j_a = (1-s)^r \varsigma_{an} j_{a,ref} \left(\frac{c_{H_2}}{c_{H_2,ref}}\right)^{\gamma_a} \left(e^{\frac{\alpha_a F}{RT}\eta_a} - e^{\frac{-\alpha_c F}{RT}\eta_a}\right) \tag{13.12}$$

$$j_c = (1-s)^r \varsigma_{cat} j_{c,ref} \left(\frac{c_{O_2}}{c_{O_2,ref}} \right)^{\gamma_a} \left(e^{\frac{\alpha_c F}{RT}\eta_c} - e^{\frac{-\alpha_c F}{RT}\eta_c} \right) \quad (13.13)$$

where r and j_{ref} represent the pore blockage and the reference exchange current density, respectively. The specific active area is ς. c and c_{ref} denote the local and reference concentration of reactants, respectively. Finally, γ and α are the concentration dependence and coefficient of transfer.

13.2.1.4 *Energy Transfer Equation*

The conservation equation for energy is given as follows [21]:

$$\nabla.\left(\varepsilon.\rho.c_p\vec{u}\mathrm{T}\right) = \nabla.(k_{eff}\nabla T) + S_h \quad (13.14)$$

k_{eff} is the effective conductivity and S_h is the source term that can be expressed as follows:

$$S_h = R_{ohm} I^2 - R_{an,cat}\,\eta_{act} + r_w h_1 + h_{reaction} \quad (13.15)$$

R_{ohm} and $R_{an,cat}$ are ohmetic resistance and over potential on anode and cathode sides, respectively. $r_w h_1$ is water vaporization or condensation and finally $h_{reaction}$ is electrochemical reaction enthalpy.

13.2.1.5 *Equation of Charge Conservation*

The electronic charge and mass transfer are the most important phenomena, responsible for the reactions in a PEMFC. The electron transfer from GDLs and CLs (current collectors) is obtained by solving the potential equation, which is as follows [26]:

$$\nabla.\,(\sigma_s \nabla \Phi_s) + j_s = 0 \quad (13.16)$$

where for anode $j_s = -j_a < 0$ and for cathode $j_s = +j_c > 0$. In the above equation σ, Φ, and j are the electrical conductivity, electric potential, and transfer of electrical current, respectively.

13.2.1.6 *Formation and Transfer of Liquid Water*

As mentioned in previous sections, the PEMFC operates at low temperature and when the current density is high, the produced water vapor may condensate and this leads to the reduction of diffusion in GDL and then the cell performance decreases. The water production equation is as follows [14]:

$$\frac{\partial(\varepsilon\rho_1 s)}{\partial t}+\nabla.(\rho_1\vec{v}_1 s)=r_w \tag{13.17}$$

where s denotes the saturation of water and r_w is the condensation rate, which is determined as follows:.

$$r_w=\begin{cases}(1-s)c_r\dfrac{P_{WV}-P_{sat}}{RT}M_{H2O} & if(P_{WV}>P_{sat})\\ sc_r\dfrac{P_{WV}-P_{sat}}{RT}M_{H2O} & if(P_{WV}<P_{sat})\end{cases} \tag{13.18}$$

where c_r is the constant of water condensation rate and M denotes the molecular weight. P_{sat} is water saturation pressure and calculated from the following equation:

$$log_{10} P_{sat} = -2.1794 + 0.02953\Delta T - 9.1837 \times 10^{-5}\,\Delta T + 1.4454 \times 10^{-5}\,\Delta T$$
$$\Delta T = T - 273.17 \tag{13.19}$$

13.3 The Solution Procedures

In Figure 13.2, the solution procedure for PEMFC model is shown. The governing equations can be solved by different software. Researchers were used ANSYS Fluent, COMSOL, and other commercial codes for simulation of PEMF and, in the next sections, a review of their studies is presented.

13.3.1 CFD Simulations

Flow field design has an important role on PEMFC performance because of its effect on pressure, current density, and temperature distributions and management of produced water [27–29]. Atyabi *et al.* (2018) proposed a three-dimensional steady-state and non-isothermal model for

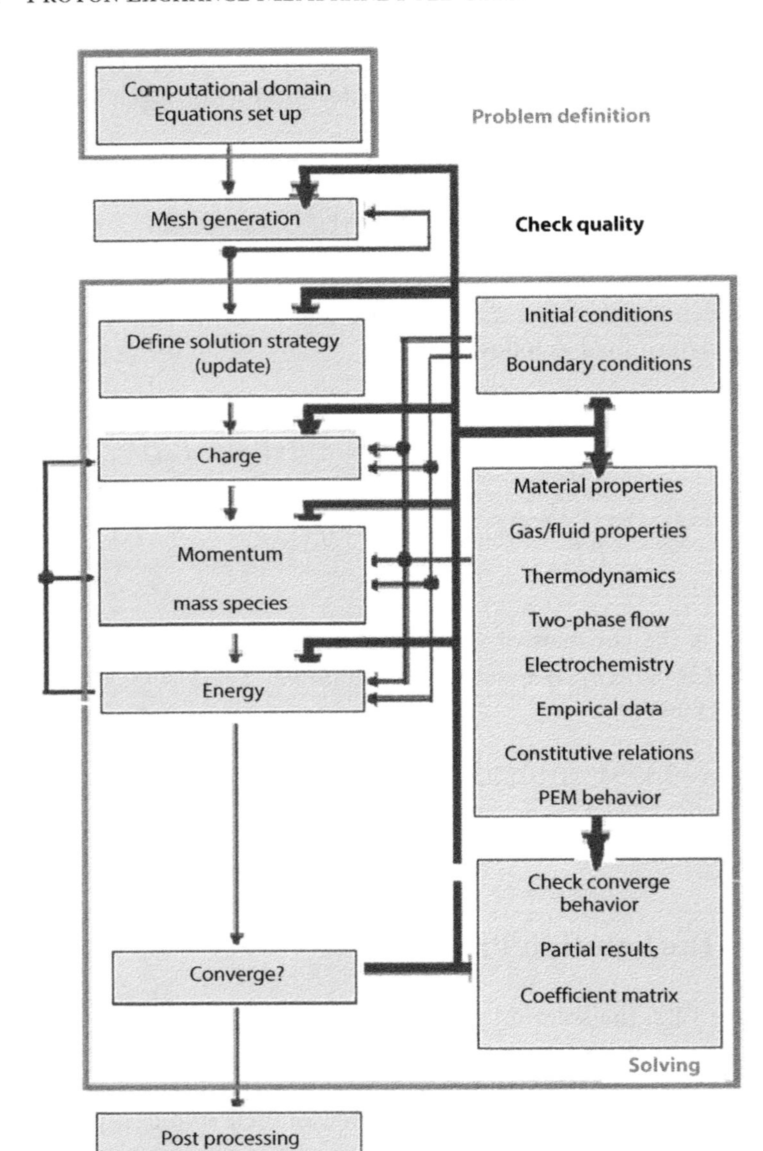

Figure 13.2 PEMFC modeling solution procedure.

PEMFC with parallel sinusoidal flow field and compared their results with straight-parallel flow field [21].

They used single-domain method to solve governing equations. Water formation and phase change in GDL and CL were considered in their simulations, and the equal heat transfer in two phases was assumed. They showed that, at the interface of GDL and CL, the maximum velocity and pressure drop were higher than that of straight-parallel flow field, and the performance of PEMFC was higher compared to sinusoidal flow field.

Since the uniform mass fraction distribution of species affected the cell performance, they calculated the uniformity index of Ψ parameter by Equation (13.20).

$$U_P + \frac{\int_A |\Psi - \Psi_{ave}|\, dA}{\int_A dA} \tag{13.20}$$

This index (Ψ) is a criterion to determine the uniformity of distributions and as close to zero indicates that the mass fraction distribution is more uniform. Their CFD results implied that the oxygen mass fraction was higher at cathode side by using sinusoidal flow field configuration and uniformity index of oxygen was 0.01286 for this configuration, which the index for parallel flow field was 0.0896.

When the oxygen distribution was uniform, the cell performance due to optimal usage of active area was improved. The uniform produced water and generated heat (and then uniform temperature distribution) were obtained by uniformity of O_2 distribution. Therefore, thermal stresses decreased. A single channel PEMFC was simulated by Bednarek *et al.* (2017) by applying a CFD code. They studied the limitations and complications of CFD modeling of PEMFCs and finally proposed a method to obtain solution convergence. They investigated the effects of bipolar plates, GDLs, CL, and proton membrane on the polarization curve. They reported their simulation results as velocity and temperature distributions, pressure losses, and reactant gas and produced water concentrations [30]. They assumed a counter-flow for the reactants gases and non-linear equation of state did not consider because of low pressure of gas flows (ideal gases assumption was valid) and used Maxwell-Stefan equation for gas diffusivity. They applied two-phase mixture model and assumed that water phase change took place in the gas diffuser and CLs.

For the validation of their proposed model, the balance of water vapor and components distributions were compared with analytical solutions based on the inlet components and reaction stoichiometry and good agreement observed between them. In addition, they showed that the lack of reliable input data was one of the issues for obtaining accurate results and the uncertainty in the values of the electrochemical constants (i.e., the open circuit voltage and exchange current density) affected the results and then these parameters could be considered as the fitting parameters to obtain agreement between experimental data and CFD results.

One of the key factors that affect fuel cells performance is the pressure drop, because of interaction with other parameters such as water management in the cell. Therefore, Wilberforce *et al.* (2017) studied the effect of different common designs of plates and flow rates on pressure drop. Their results showed that by modification of flow plates design, the pressure drop reduces, which leads to higher performance of fuel cell even though, other parameters maybe also contribute to this improvement of cell performance [6]. As shown in Figure 13.3, using a combination of both the modified serpentine and the parallel flow channels led to that the pressure drop was 50 times lower than that of the traditional serpentine design. This figure indicates that the lowest pressure drop corresponded to the modified parallel flow channel. The simulation results revealed that for all cases at air

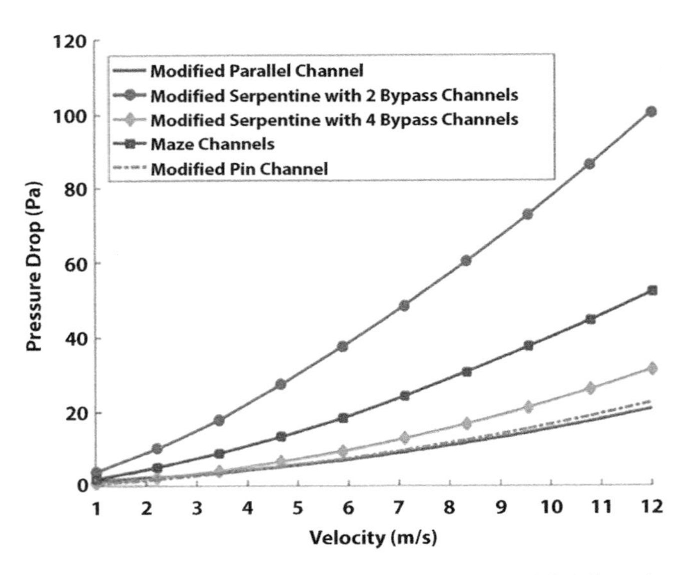

Figure 13.3 A velocity diagram against pressure drop for all modified flow plate designs. Reprinted with permission of Wilberforce *et al.* (2017) [6].

inlet and the H_2 outlet, the current density and potential had the highest values.

The oxygen concentration distribution on the cathode side was determinative of the current density and potential distributions. The oxygen diffusion rate reduces by reduction of O_2 concentration across the GDL; therefore, because of the electro-chemical reaction reduction, less current is produced.

In the recent years, the use of high temperature PEMFC (temperature higher than 100°C) was taken into consideration. This kind of PEMFCs have some advantages with respect to conventional PEMFCs, which operate at low temperature (80°C) like faster electrochemical kinetics, simpler water management, higher carbon monoxide (CO) tolerance, and easier cell cooling and waste heat recovery.

Jiao and Li (2010) simulated a three-dimensional and non-isothermal PEMFC with phosphoric acid doped polybenzimidazole (PBI) membranes at high temperature by computational fluid dynamics (CFD). The parameters such as temperature, phosphoric acid doping level, and surrounding RH effects on the membrane proton conductivity were studied and the conductivity of proton was calculated by using a semi-empirical correlation, which was based on the Arrhenius' law [16].

The performance of the fuel cell under different temperature conditions was investigated. At all temperature conditions, no obvious drop in concentration was observed due to the high stoichiometric ratio and the prevention of liquid water formation. However, it can be seen that, with increasing temperature, the peak of power density increases dramatically so that, with increasing temperature from 110°C to 150°C, and from 150°C to 190°C, the peak of power densities increases from 0.213 to 0.278 and 0.34W cm^{-2}, respectively. This reveals that the efficiency of the PEMFCs is very affected by operating temperature and its changes should be carefully monitored, because the RH is strongly affected by temperature and this can affect the performance of the fuel cell. Figure 13.4 shows the RH contours at the middle layer of the cathode plate at different temperatures. This figure indicates that more water is produced at higher temperatures, but, nevertheless, the RH is higher at lower operating temperatures, because as the temperature increases, the vapor saturation pressure also increases. It can also be generally observed that at any temperature, the RH increases in the direction of fluid flow due to the accumulation of produced water; in addition, due to the easier removal of water under the flow channel, the RH under land is higher than the RH under the flow channel.

In addition, they studied the influence of different levels of phosphoric acid of PBI membrane. It was observed that, with increasing doping level,

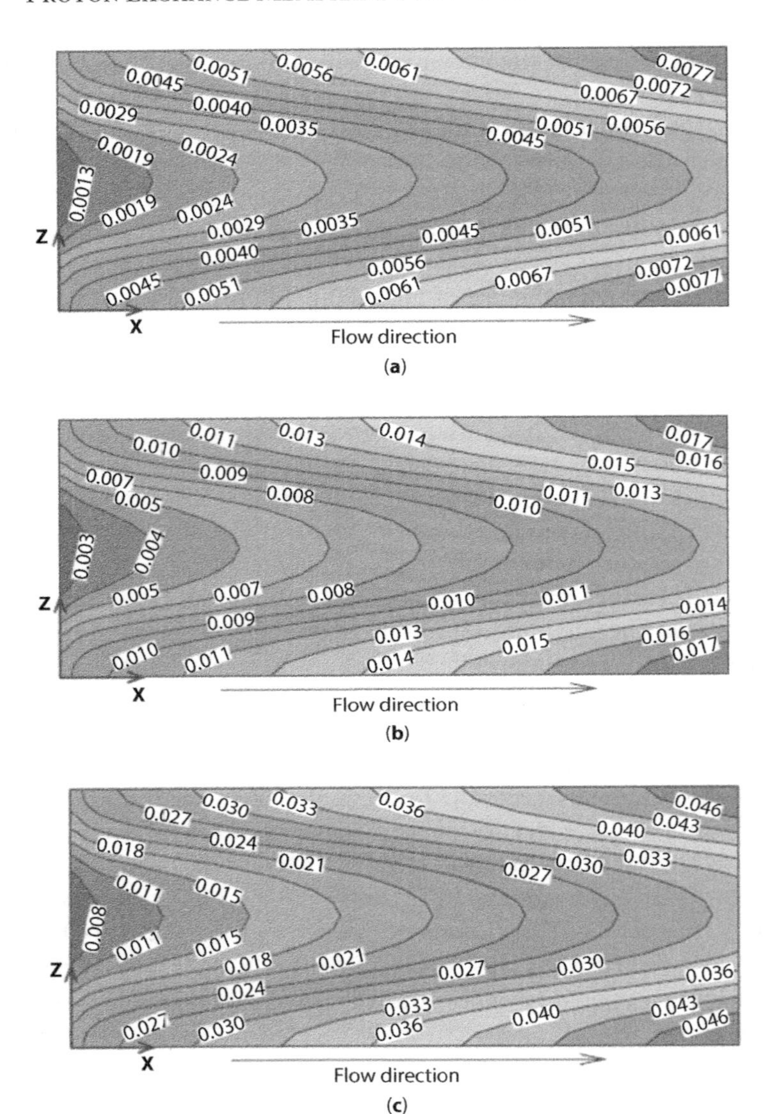

Figure 13.4 Contours of relative humidity at the middle plane of cathode CL (V = 0.6 V as operating voltage, atmospheric pressure on anode and cathode side, the stoichiometry ratio of reactants was 2 for reference current density of 1.5 A/cm^2, and the phosphoric acid doping level for the PBI membrane is 6); the operating temperature was (a) 190°C, (b) T = 150°C and (c) T = 110°C. Reprinted with permission of Jiao and Li (2010) [16].

the peak of power density increases, so that, with increasing concentration from 3 to 9, the peak of power density increases from 0.203 to 0.435 W cm^{-2}. The reason for this is that, with increasing doping level, the proton conductivity of the membrane increases, which results in an increase in current density and large ohmtic losses. Moreover, they studied the effect of RH on the cell performance and concluded that, by increasing the inlet RH, the peak of power density increases, which can be attributed to an increase in membrane proton conductivity. However, the results showed that humidifying the inlet flow at room temperature cannot significantly affect the cell performance.

Chen *et al.* (2021), in another study, simulated a PEMFC with orientated-type flow channels and investigated the effects of baffle location and height on the cell performance using COMSOL software. In this simulation, it was assumed that the fuel cell operated at steady-state conditions and, because of the low velocity of the fluid flow, the flow inside the cell was laminar; in addition, the porosity of porous surfaces such as GDL and cathode layer were considered homogenous and isotropic, and all gas phases were considered ideal [31].

Figure 13.5 shows the polarization curves of the PEMFC at three different modes to investigate the effect of baffle height. From this figure, it can be seen that by keeping the volume of the baffles at a constant value and,

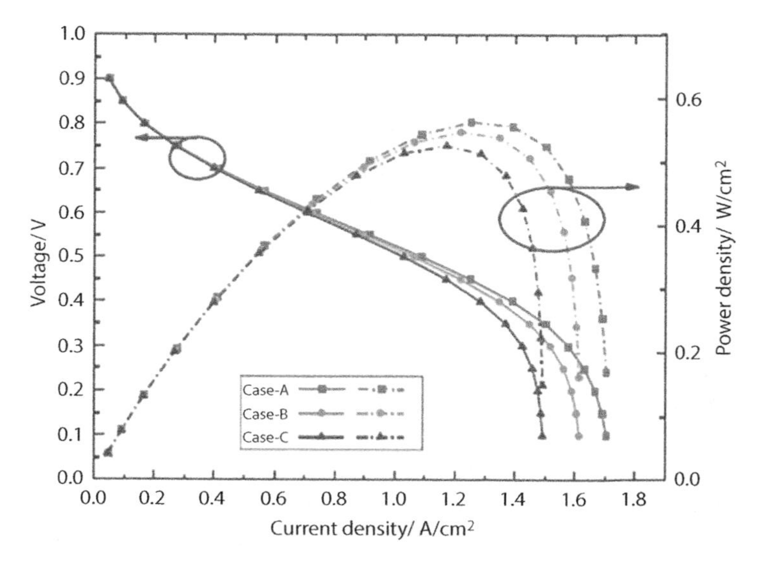

Figure 13.5 Polarization curves under baffle height effects. Reprinted with permission of Chen *et al.* (2021) [31].

by increasing their height, the efficiency of the cell can be improved (from case C to case A). It can also be stated that, by reducing the height of the baffles, their width increases, so it can be concluded that the height of the baffles has a greater effect on improving the efficiency of the PEMFC with orientated-type flow channels compared to the width of the baffles.

In total, due to the effect of the baffles, the local flux transfer of the reactants from the channel to the GDL is improved [32]. The results showed that, in fuel cells with different baffle heights, the local flux was also different and it could be said that local mass transfer fluxes improves with increasing baffle height. The reason for this can be explained that, as the height of the baffles increases, the local flow space between the baffles and the surface of the GDL decreases, so the concentration gradient of the reactants between the channel and the GDL increases, resulting in an increase in the mass flux. They concluded that, as the height of the baffles decreases, the liquid water saturation under the baffle increases. It can be understood that, as the height of the baffle decreases, the fluid region under the baffle enlarging, and as this regions enlarged, the fluid velocity decreases, and by decreasing fluid velocity and inertia, less water can remove. However, removing more water from the channel improves the transfer of water from the cathode layer to the GDL.

Based on the literature, the studies related to scale-up PEMFC with active area of 100 cm^2 are very few. The output power increases by scale-up the PEMFC in each cell and the current density decreases compared to conventional PEMFCs that have 25 cm^2 area. When a PEMFC is scaled up, one key that must be considered accurately is that the design must be such that the water and species distribution be appropriate. PEMFCs with 25 cm^2 are not suitable for commercial purposes. Staking of PEMFCs is costly and needs space. Then, before staking, to obtain appropriate efficiency, fuel cells must be scaled up and flow field must be optimized. Another parameter that affects the cell performance is the inlet humidification of reactant and oxidant.

Selvaraj *et al.* (2019) studied the effects of flow field and the humidification of reactants and oxidant on the performance of scale-up PEMFC using CFD code. They used five different flow path designs with scaled-up active area of 100 cm^2 and varying the inlet relative humidification from 10% to 100%. Figure 13.6 shows the five different flow path designs, which were used in their studied [33].

Their CFD results showed that the maximum power per unit area was obtained by using straight zigzag flow field configuration (0.3711 W/cm^2), and species distribution was more uniform in this design and concentration of water in the channels and membrane was optimal. In the serpentine

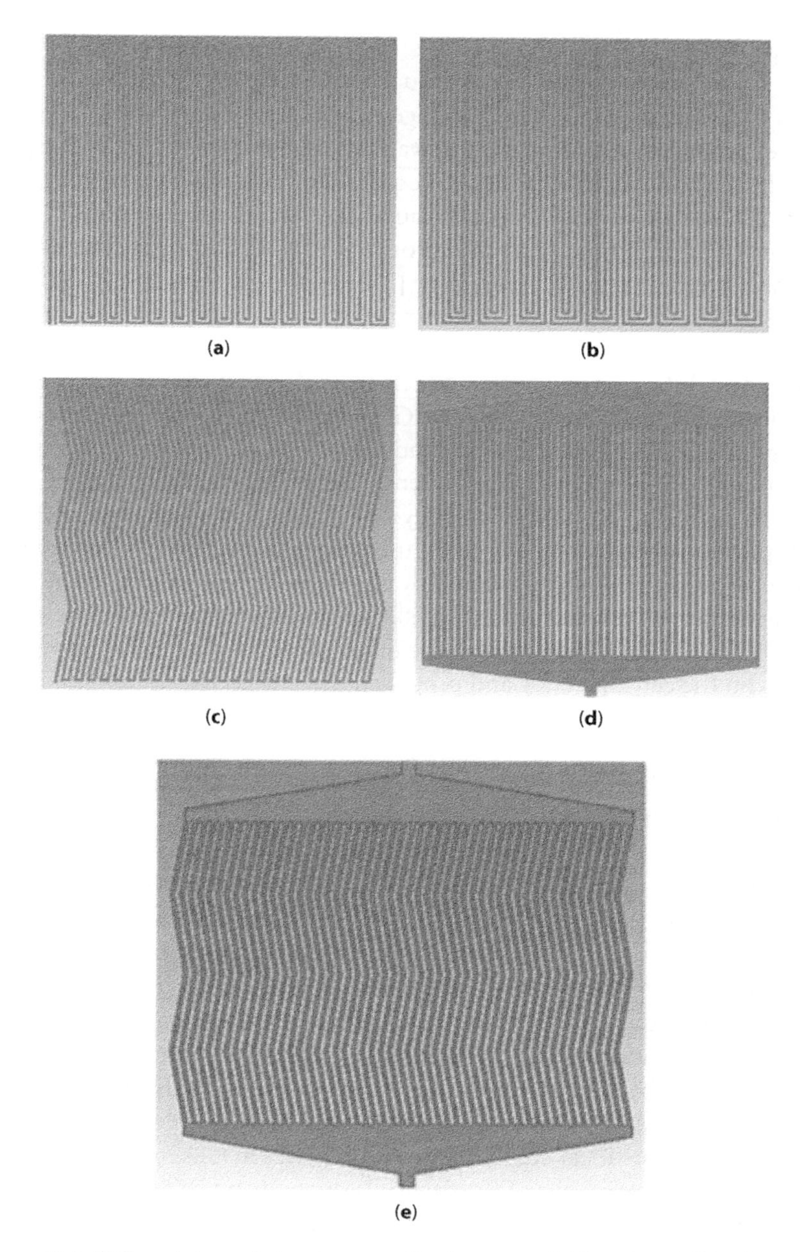

Figure 13.6 Different channel structures used in the study of Selvaraj *et al.* (2019): (a) 2-serpentine, (b) 3-serpentine, (c) serpentine zigzag, (d) straight parallel, and (e) straight zigzag. Reprinted with permission of Selvaraj *et al.* (2019) [33].

zigzag configuration, because of less uniform species and water distributions, lower power density was produced.

As mentioned above, water management is a key factor to enhance fuel cell performance. Accumulation of water in the channel increases the pressure drop, and then, more power is required for pumping. Existence of baffles in oriented-type flow channels causes the flow channel regions blocked and pressure drop increased and in order to avoid from this problem, the leeward lengths could be enlarging [31]. In the cases that PEMFC have baffles, the arrangement and dimensions of baffles affect cell performance. The transfer of reactant to proton membrane increases in orientated-type flow channels having porous blocks. Chen *et al.* (2020) simulated oriented-type flow channels in a PEMFC by CFD to obtain the configuration that could decrease the pressure drop (and then decrease pumping power) and appropriate distribution of components. They used 12 flow channel structures in their simulations [32]. They concluded that applying the baffles with porous blockage and large volume streamline baffles improved the cell performance. In porous region, water is injected more due to the effect of inertial force, and this injection is higher in the region near the baffles in GDL. In addition, they studied the effects of the location and porosity of baffles on the cell efficiency and proposed that, when the number of baffles with porous blockage was more at downstream, the removal of produced water was better.

They implied that species transport increase when the streamline baffles was used and the power loss decreases during the fluid flowing process of a PEMFC. Figures 13.7 and 13.8 show the current density distribution and polarization curve for different structures. As observed from Figure 13.8, the output performance (polarization curve) of case A (conventional straight flow channel) was lower than that of other cases, because by using the baffle the species transport along CLs increased and by installing porous blocks, this transportation was improved more.

Figure 13.9 indicates the case that has larger baffle volumes; the reactant fluxes increase by the increase the gradient of species concentration from the channel to gas diffuser layer and CL, and, then, in the region of porous blocked baffle, the local current density increases at CLs.

Since the performance of the fuel cells depends on cell geometry, many researchers, to decrease the time and cost, studied this parameter by using CFD tools or mathematical modeling and attempted to optimize this factor under different operating conditions. For experimental tests, they built the best predicted performing cathode flow channel [34–36]. For example, the effect of the channel depth on the PEMFC performance was investigated by Khanzaee and Ghazikhani (2012). Their results showed that, with

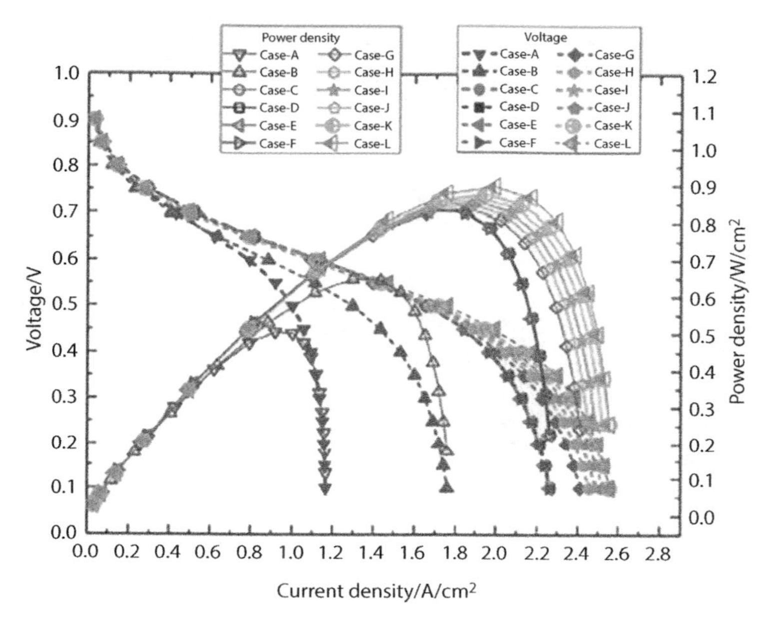

Figure 13.7 Polarization curves of different cases. Reprinted with permission of Chen *et al.* (2020) [32].

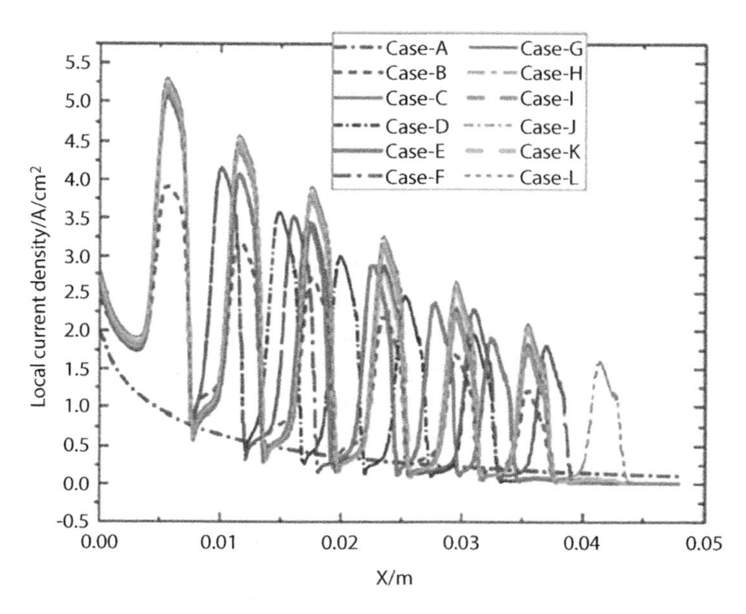

Figure 13.8 Local current density distribution in CLS of cathode sides ($V_{cell} = 0.4\ V$). Reprinted with permission of Chen *et al.* (2020) [32].

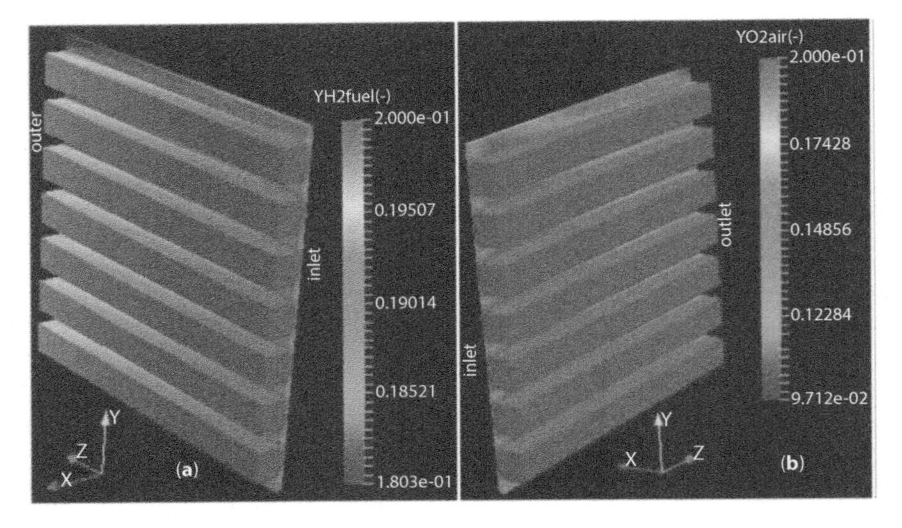

Figure 13.9 Mass fraction distribution of different species: (a) hydrogen and (b) oxygen. Reprinted with permission of Kone *et al.* (2018) [39].

increasing the channel depth on either side of the cell, the cell performance increases [37].

Carcadea *et al.* (2018) studied the effects of the width and depth of cathode flow channel using CFD and obtained the optimal width and depth of channel. Then, they built the best performing design for the cathode flow channel and investigated the sensitivity of the fuel cell to the RH and temperature [38]. Their results showed that the cell performance was not very affected by the channel dimensions at high and medium operating voltages, and the cell performance was more affected by channel size at lower operating voltages and, by reduction of the channel depth and width, the cell performance improved. By decreasing the dimensions, the velocity increases and produced water was more removed from the channel that this led to less flooded pathways between the channel and the CL and more oxygen was supplied at active site of catalyst and due to more oxygen consumption, the cell performance improved.

13.3.2 OpenFOAM

Open Source Field Operation and Manipulation (OpenFOAM) is one of the CFD simulation tools that can be used as an open source unlike software like ANSYS Fluent. In a study by Kone *et al.* (2018), OpenFOAM was used to simulate a proton exchange fuel cell [39]. The model used in this study was a non-isothermal single-phase model; it was assumed that the operation is steady-state. In addition, because of the low air velocity, the flow was considered to be laminar and incompressible; moreover, it was assumed that the membrane was fully humidified and against reactant gas was impenetrable. It was also assumed that the gases, both individually and in mixture, are ideal and that all components of the fuel cell are isotropic and homogeneous.

In this simulation, several governing equations were solved. The mass conservation equation is expressed as follows [39]:

$$\nabla.\left(\rho\vec{U}\right)=0 \tag{13.21}$$

where ρ is the density and $\vec{U}$ is the velocity vector. The following equation is also written for the conservation of the momentum [39]:

$$\nabla.\left(\rho\vec{U}\vec{U}\right)=-\nabla p+\nabla.\left(\mu\nabla\vec{U}\right)+S_m \tag{13.22}$$

where p, μ, and S_m represent pressure, viscosity, and momentum source term, respectively. The amount of this source term in channels is 0 and, in porous space, can be calculated from Equation (13.23).

$$S_m=\frac{-\mu\vec{U}}{k} \tag{13.23}$$

In addition, conservation equations for chemical species can be written as follows:

$$\nabla.\left(\rho\vec{U}y_i\right)=(\nabla.\rho D\nabla y_i) \tag{13.24}$$

In the above equation, D is the effective diffusivity of chemical type A. The mass fraction of inert species on each electrode can also be obtained by calculating the mass fraction of other species and subtracting it by one. In addition, the energy conservation equation is written as [39]:

$$\nabla.\left(\rho\vec{U}T\right)=\nabla.(k\nabla T)+S_E \tag{13.25}$$

where S_E is the source term of energy, which it is due to released heat in chemical reactions, and it can be calculated by the equation as follows:

$$S_E = \frac{1}{\delta_{MEA}}\left(\eta - \frac{T\Delta S}{n_i F}\right) \tag{13.26}$$

In total, cell voltage is expressed by the following equation:

$$V_{cell} = E_{Nernst} - \eta \tag{13.27}$$

where E is Nernst potential and η is known as over potential.

In this simulation, to determine the boundary conditions, Dirichlet boundary condition (B.C.) was used at the input boundaries, while Neumann B.C. was used at the exit boundaries. Interfaces between membranes and electrodes were considered to be insulation against mass flows and chemical species, and non-slip, non-flux boundary conditions were applied at solid surfaces. In terms of temperature, the boundary condition or zero-gradient temperature was applied on the outer surfaces of this cell.

The mass fraction distributions of hydrogen and oxygen are shown in Figure 13.9. As can be seen from this figure, the component mass fraction decreases from the inlet of the fuel cell to its outlet. The reason for this reduction in the mass fraction can be explained by the fact that the electrodes are depleted from the reactants due to the chemical reaction. In addition, as shown in Figure 13.10, the Nernst potential decreases due to the anode being depleted of fuel, due to the interference of the effects of hydrogen consumption and water production from the fuel cell inlet to its outlet.

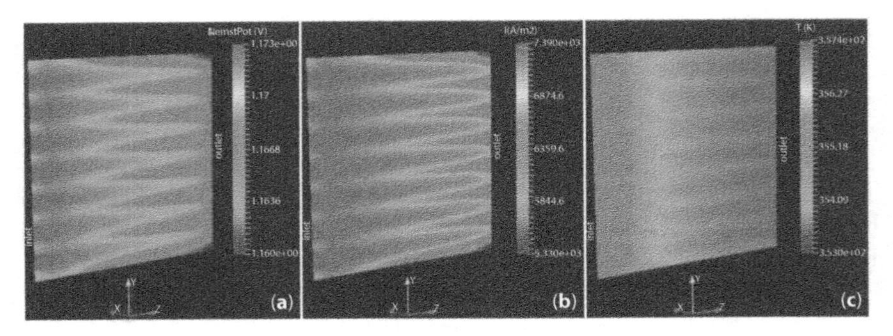

Figure 13.10 Distribution of different parameters on membrane cathode interface: (a) Nernst potential, (b) local current density, and (c) temperature. Reprinted with permission of Kone *et al.* (2018) [39].

As can be seen in this figure for Nernst-potential, for local current density, this decreasing trend observed also visible from the inlet to the outlet of the cell.

Contrary to the two decreasing trends observed for the above cases, as shown in Figure 13.10, the temperature increases from the inlet of fuel cell to its outlet. The reason for this phenomenon can be attributed to the reduction of production sources such as Joule heating and the non-uniform electrochemical reactions heat under ribs. The decrease in the heat of reaction below ribs can be considered as a decrease in mass transfer due to the lower current density in this region. In addition, as shown in Figure 13.11, in which the cell voltage is plotted in terms of current density, the results obtained by this simulation by OpenFOAM were in agreement with the experimental data obtained by Yuan *et al.* (2010) [40].

In another study by Niu *et al.* (2018), using OpenFOAM, two-phase flow in the GDL in a PEMFC was investigated using volume of fluid (VOF) model and the results were compared with experimental data. In this CFD simulation with the help of VOF model, in the governing equations, the phase fraction of liquid water (γ) can be expressed as the most important variable, which needs to be solved [41]. In a cell that is completely filled with water, the value of (γ) is equal to one and in a cell that is completely filled with air (γ) is equal to zero. Therefore, when the numerical value of

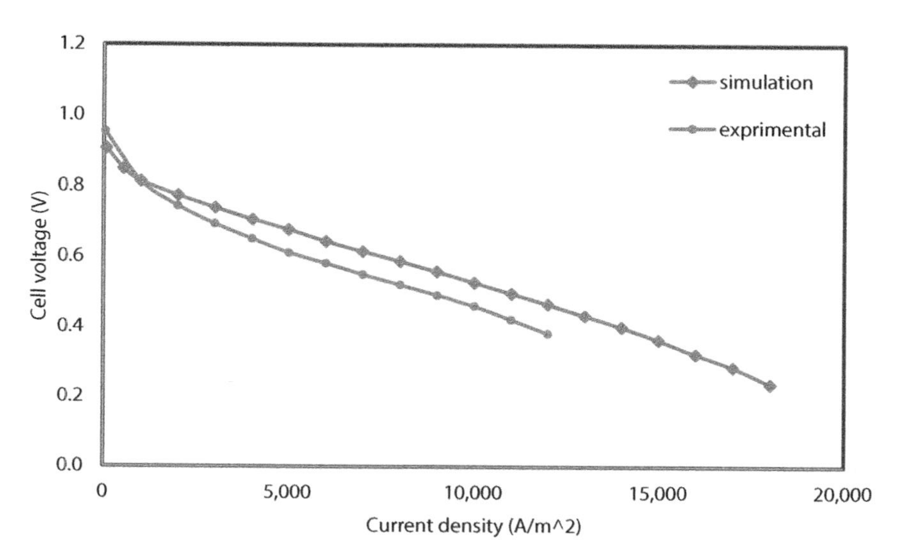

Figure 13.11 Comparison of simulation results and experimental data. Modified after Kone *et al.* (2018) [39].

(γ) is a number between zero and one, it indicates that there is liquid water and air in the cell at the same time.

In the VOF model, ρ and μ were considered as the average volumetric density and dynamic viscosity for the mixture of liquid water and air. These two parameters can be expressed based on the phase fraction of liquid water as follows [41]:

$$\rho = \rho_l \gamma + \rho_g (1 - \gamma) \tag{13.28}$$

$$\mu = \mu_l \gamma + \mu_g (1 - \gamma) \tag{13.29}$$

In the above relations, index l is for the liquid phase and index g is for the gas phase.

By assuming steady and incompressible flow, the continuity equation is written as follows [41]:

$$\nabla.\vec{U} = 0 \tag{13.30}$$

In addition, in this simulation, the equations of phase and momentum conservation are given, respectively, as follows [41]:

$$\frac{\partial \gamma}{\partial t} + \nabla.(\vec{U}.\gamma) + \nabla.[\vec{U_r}\gamma(1-\gamma)] = 0 \tag{13.31}$$

$$\frac{\partial(\rho\vec{U})}{\partial t} + \nabla.(\rho\vec{U}\vec{U}) - \nabla.(\mu\nabla\vec{U}) - (\nabla\vec{U}).\nabla\mu = -\nabla p_d - \vec{g}\vec{x}\nabla\rho + \sigma k \nabla\gamma \tag{13.32}$$

In the above equations, the velocity vector is defined as follows:

$$\vec{U} = \vec{U}_l \gamma + \vec{U}_g (1 - \gamma) \tag{13.33}$$

where σ and k are surface tension coefficient and mean curvature of phase interface, respectively. $\vec{U}_r$ is defined as the relative velocity of liquid and gas and p_d is modified pressure, which can be expressed by the following equations [41]:

$$\vec{U}_r = \vec{U}_l - \vec{U}_g \tag{13.34}$$

$$p_d = p - \rho \vec{g}.\vec{x} \tag{13.35}$$

where $\vec{g}$ is gravity vector and $\vec{x}$ is position vector.

In this CFD simulation, in determining the boundary conditions and models, in addition to VOF model, continuous surface force (CSF) model was used to calculate the effect of surface stresses at the interface between liquid and gas. In this case, GDL diameter was 1.5 mm and its thickness was considered to be 190 μm. Between the inlet and outlet (upper and lower surface of the GDL), the pressure difference condition was set as the boundary condition and on the walls and side surfaces, the no-slip condition was applied. It was also considered that at the beginning of this simulation there was no liquid water in the gas penetration layer. In the solution process, Semi-Implicit Method for Pressure-Linked Equation scheme was used to couple the pressure and velocity.

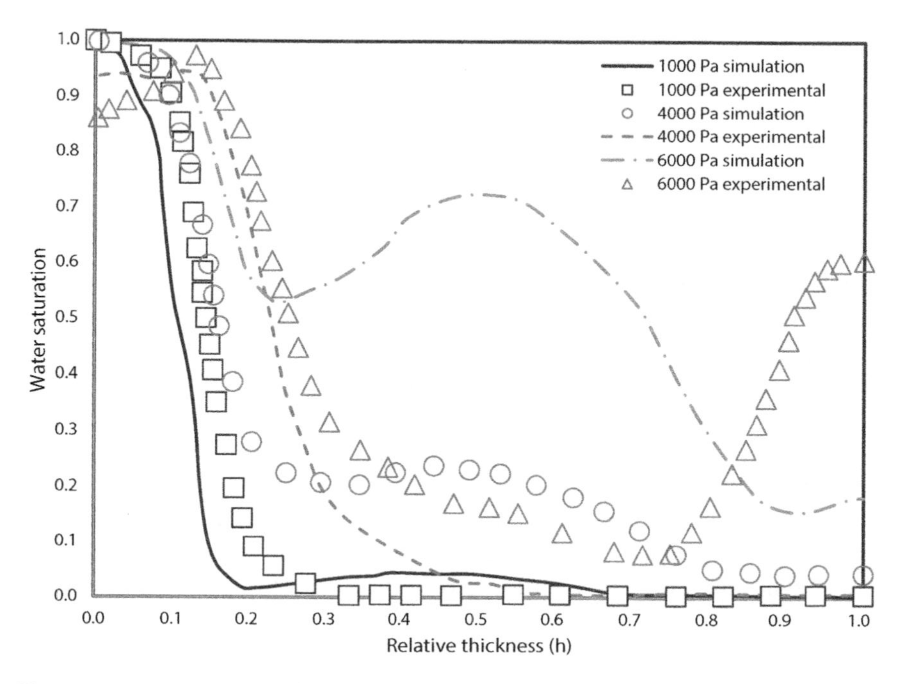

Figure 13.12 Comparison of simulation results and measured data of water saturation through plane direction at different pressures. Modified after Niu *et al.* (2018) [41].

In this simulation, local water saturation was obtained by averaging water phase fraction in porous spaces, and the results were compared with the actual values obtained from X-ray experiments. As shown in Figure 13.12, water saturation is obtained in terms of relative thickness and water profiles are plotted in two experimental and simulation modes at different pressures.

When the pressure difference between the two surfaces of the GDL was equal to 1,000 Pa, close to the inlet, water saturation decreased rapidly at once, and then, its amount remained low to the outlet (the other surface of GDL). At this pressure, the simulation results and experimental data were in good agreement. As the pressure difference increased from 1,000 to 4,000 Pa, the water saturation decreased sharply again near the inlet. Although at this pressure the profile obtained from the simulation generally had the same trend as the experimental data, but in the middle of profile obtained from the simulation, there was an increase in the value of water saturation, which led a difference between the experimental data and the simulation results.

As the pressure increased to 6000 Pa, there was a rapid drop in the water saturation close to the inlet, but in the middle of graph, the values obtained from the simulation increased, while the experimental data with a steep slope continued to decrease. The trend of experimental data changed in relative thickness = 0.7 and increased up to the outlet surface, but the simulation results at this point were completely opposite to the experimental data and with a decreasing trend reached less than the experimental data. Although the trend of changes at the 4,000-Pa pressure difference was the same for the experimental data and simulation results, the deviation at the 6,000-Pa pressure difference was quite significant. The reason for this discrepancy could be attributed to the difference between the plane porosities [42] or the different amounts of fiber or their contact angle [43].

Figure 13.13 shows the mean values of water saturation at different capillary pressures. Although at pressure of 6,000 Pa, the local values of water saturation obtained from the simulation were different from the experimental data, at this pressure, the experimental data of water saturation were in agreement with the simulation values. This adaptation occurred when non-uniform porosities were considered in the simulation. If uniform porosities were used, then a large difference between the experimental data and the simulation results could be observed. In addition, if the Leverett J-function was used, then it can be seen that, at low capillary pressures, for example, −1,000 and −2,000 Pa, the experimental data and CFD results were correlated, but at high capillary pressures (4,000 and 6,000 Pa), the difference between the results obtained by this function and experimental

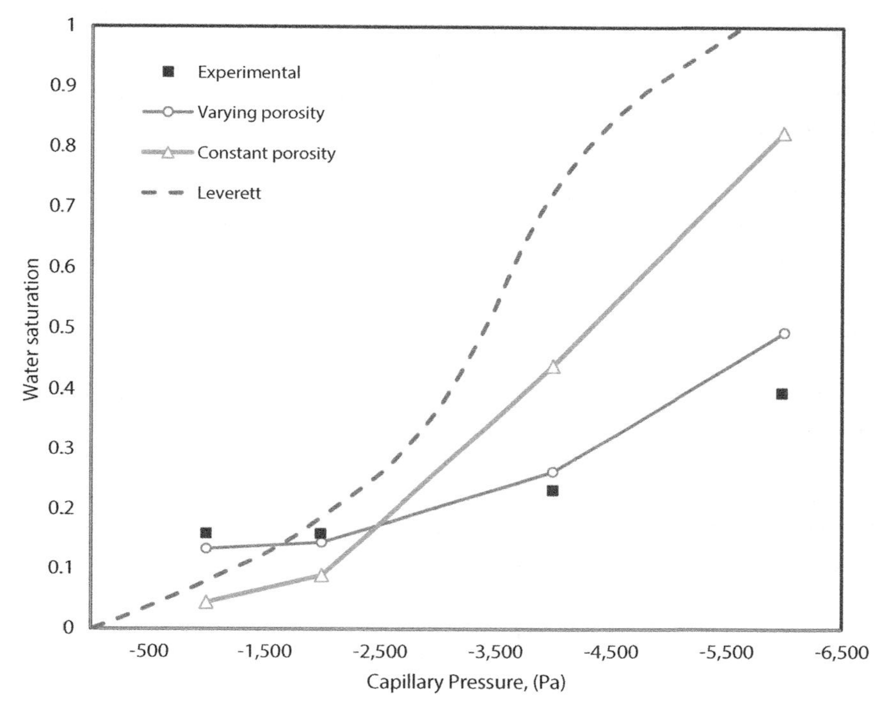

Figure 13.13 Constant and varying porosity gas diffusion layer numerical results compared to experimental data and standard Leverett capillary pressure function model. Modified after Niu *et al.* (2018) [41].

data was significant. The reason for this difference could be considered due to the special structure of the GDL that was different from sphere pack, which was used by Leverett J-function.

In addition, when the water phase fraction was 0.5 and the flow became stable, at a pressure of 1,000 Pa, only a thin layer of water was observed at the inlet and almost the rest of the space was empty of liquid water. As the pressure increased to 6,000 Pa, water penetrated the entire GDL and filled the bottom. In the lateral parts of the GDL, which had less fiber, due to low resistance and relatively large porosity, water was absorbed, while in the middle parts, due to smaller porosity, it was free of water.

13.3.3 Lattice Boltzmann

Due to its numerical stability and excellent adaptability, LBM has been used as a promising and reliable method for simulating fluid flow in recent years, which can be used efficiently in simulations of interfacial dynamic and complex geometries. In the LBM, fluid-like particles were simulated based on the Boltzmann equation at the mesoscopic level by using a conventional space grid.

Chen *et al.* (2013) modeled a PEMFC using the LBM and the finite volume method. In this modeling, two-dimensional geometry was used as the computational domain. This geometry included a rectangular domain as the gas channel, a porous space for the GDL and a microstructure as the cathode layer. The carbon fibers in the GDL represent a layer with a porosity coefficient of 0.7 [44].

The cathode layer in the PEMFC has a complex structure including three phases of platinum, carbon, and ionomer. The carbon phase allows electrons to conduct and platinum acts as a catalyst for the electrochemical reaction. The ionomer also provides a specific pathway for protons to conduct and the voids allow reactive gases and products to penetrate. In this simulation, the local current density produced by the reaction can be calculated using Tafel equation [15]:

$$I_{local} = I_{ref}\left(\frac{C_o}{C_{o,ref}}\right)^{r_c} Exp\left(\frac{\alpha F}{RT}\eta\right) \tag{13.36}$$

Where I_{ref} represents the reference exchange current, α is the coefficient of transfer, R is the gases constant coefficient, and η is also the over potential of local surface. In addition, r_c represents the dependence of cathode concentration and $C_{o,\,ref}$ represents the reference concentration of oxygen.

The microstructure of the cathode layer was similar to the work done by Wang *et al.* (2006). It was assumed that proton transfer is the limiting factor of the oxygen reduction reaction, and therefore, electron transfer is neglected. In the mixed phase, the proton conductivity can be obtained according to the following equation [45]:

$$\sigma = \sigma_0\left(\frac{\varepsilon_e}{1-\varepsilon_g}\right)^{1.5} \tag{13.37}$$

where σ_0 is the intrinsic conductivity of electrolyte, and ε_e and ε_g are the volume fraction of the electrolyte and the pore, respectively.

The computational domain of this fuel cell consists of two sub-domains: one of them includes the gas channel and the other one includes porous spaces (GDL and cathode layer). The limited volume method was used to simulate fluid flow and mass transfer in the gas channel, while the LBM was used to simulate fluid flow and mass transfer in the GDL, as well as to simulate the electrochemical reaction and proton conduction in the cathode layer and, eventually, these two methods were coupled with each other.

In applying the boundary conditions, at the inlet of gas channel, air was considered as a mixture of 79% nitrogen and 21% oxygen with uniform velocity. Moreover, in the outlet of the gas channel, according to the length of channel, a fully developed condition was used. At all walls and solid surfaces of this fuel cell, the no-slip condition was used and on all surfaces except reactive surfaces, there was no proton flux. Since a part of the inlet and outlet of the gas channel were in a domain that solved by the LBM, the boundary conditions must be defined differently in this method.

The boundary condition defined by Zou and He was used for the flow of a fluid with known velocity [46]. In addition, at the outlet of gas channel, to express the condition of fully developed, specified distribution functions at the boundary nodes were considered to be equal to the corresponding neighbor nods. Moreover, in the lattice Boltzmann region, on solid surfaces, to define the no-slip as a boundary condition, the scheme of bounce-back was applied.

In the case of mass transfer, at the entrance of gas channel in which the gas concentration was known, there was only one unknown distribution function for D2Q5 square lattice, which it was possible to obtain according to the other specified distribution functions and the same method was applied to define the fully developed condition, for determining the concentration in the outlet. In this simulation, on the active surface of the cathode layer, where there was both proton transfer and mass flux, the boundary condition presented by Kang *et al.* (2007) was applied [47]. As predicted, there was a significant difference between the velocity vectors in the gas channel and the GDL. Reactant gases moved mainly in the gas channel, and the velocity magnitude in the gas channel was significantly greater than that in the GDL, which could be attributed to the mechanism of penetration in the GDL where in this layer, velocity vectors were strongly influenced by carbon fibers.

At different values of the over-potential applied to the upper surface of computational domain, the distributions of oxygen and water vapor concentrations are shown in Figures 13.14 and 13.15, respectively. It is clear that oxygen concentration decreased in the direction of fluid flow due to oxygen consuming at cathode layer; in addition, by increasing the applied over-potential, oxygen concentration decreased rapidly, which could be ascribed to the increase in the rate of electrochemical reaction.

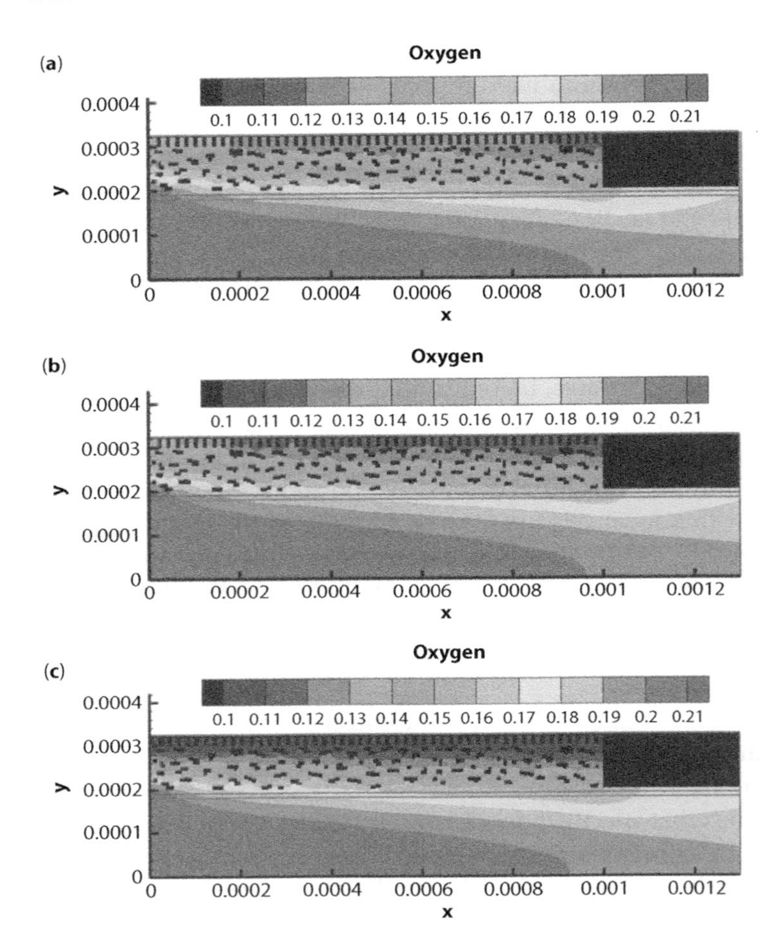

Figure 13.14 The effect of different over-potentials on the distribution of oxygen mole fraction: (a) 0.2 V, (b) 0.6 V, and (c) 0.8 V. Reprinted with permission of Chen *et al.* (2013) [44].

Contrary to the decreasing trend of oxygen concentration, water vapor was continuously produced due to electrochemical reactions and as a result, the concentration of water vapor increased at the flow direction. It can also be seen that, as the over-potential increased, the concentration of water vapor increased too. In addition, as the applied over-potential increased, the rate of electrochemical reactions increased and a greater potential drop occurred as shown in Figure 13.16. In addition, more current density was produced as a result of the increase in applied over-potential.

In another study by Molaeimanesh and Akbari (2015), the pore size of a protein exchange membrane fuel cell was investigated by the LBM.

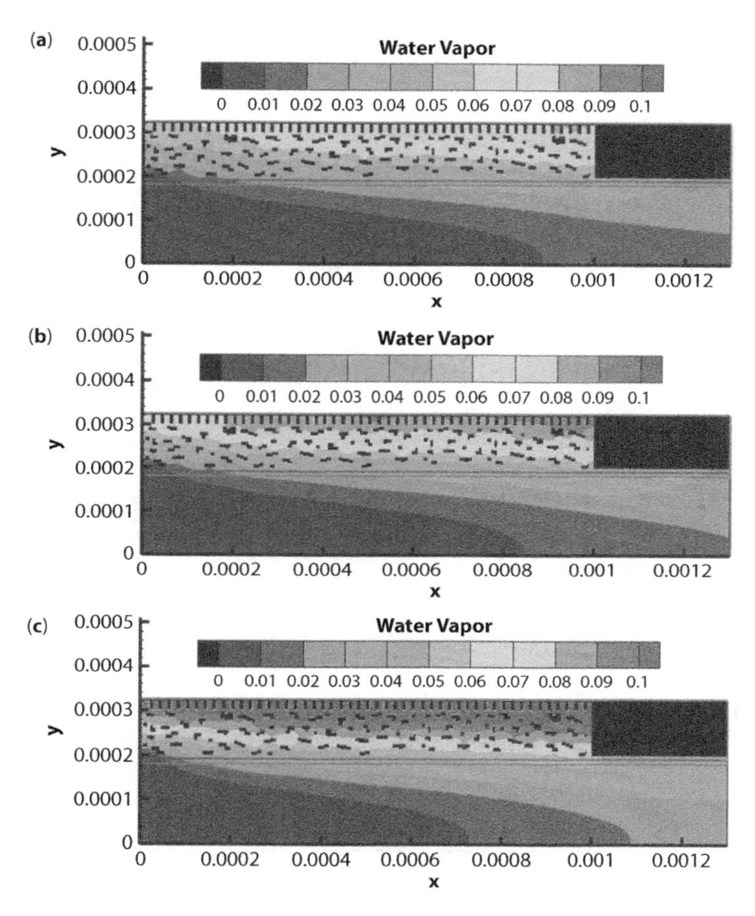

Figure 13.15 The effect of different over-potentials on the distribution of water vapor mole fraction: (a) 0.2 V, (b) 0.6 V, and (c) 0.8 V. Reprinted with permission of Chen *et al.* (2013) [44].

They found that molar fractions of oxygen and water vapor decreased and increased as they moved from inlet to outlet, respectively. This could be attributed to the electrochemical reaction on the cathode layer. In addition, due to an increase in the reaction rate with increasing over-potential, as the result of other studies, a decrease in the molar fraction of oxygen and an increase in the molar fraction of water vapor could be observed by increasing the applied over-potential [48]. As can be predicted from Bulter-Volmer equation, as the results shown in Figure 13.17, a higher current density can also be obtained. The results given in this figure were in

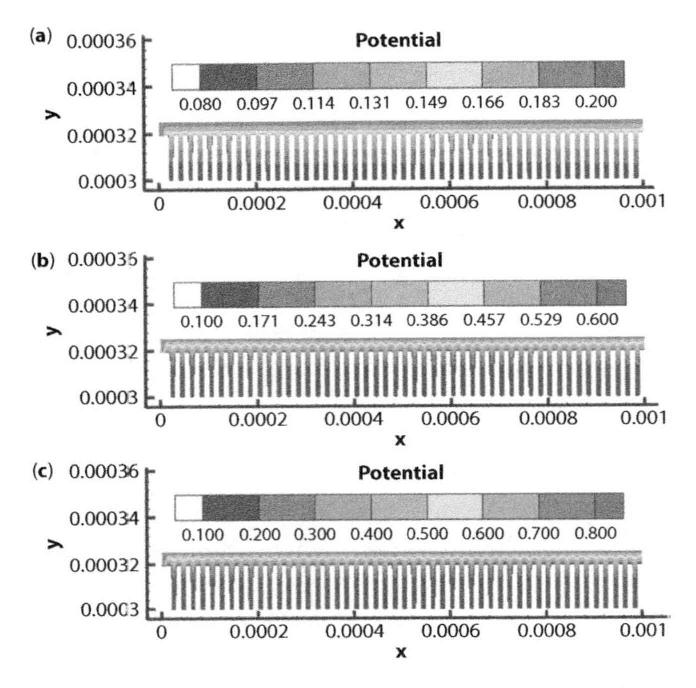

Figure 13.16 The effect of different over-potentials on the distribution of potential: (a) 0.2 V, (b) 0.6 V, and (c) 0.8 V. Reprinted with permission of Chen *et al.* (2013) [44].

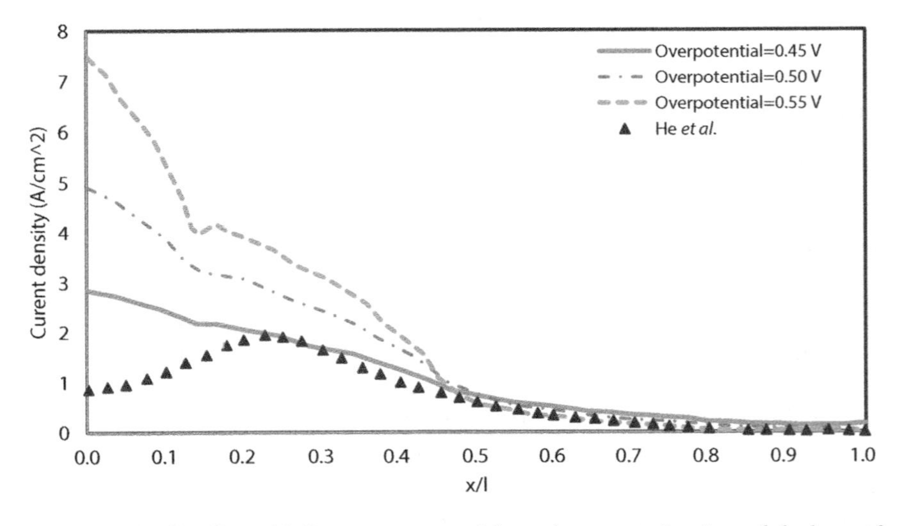

Figure 13.17 The effect of different over-potentials on the current density and the layer of catalyst. Modified after Molaeimanesh and Akbari (2015) [48].

accordance with the results of a CFD study conducted by He *et al.* (2000) [49].

In investigating the effect of difference between the inlet pressure and outlet pressure on the molar fraction of oxygen, this molar fraction was examined at four different pressure differences. The results showed that the molar fraction of oxygen was significantly affected by the pressure difference between the inlet and outlet, which could be attributed to an increase in air flow rate due to an increase in pressure difference. In addition, according to Figure 13.18, it can be seen that at higher pressure differences, more current density can be achieved, which can be stated that, with increasing air flow rate due to increasing pressure difference, oxygen penetration in the GDL was improved.

By examining the effect of land width, it was observed that, with decreasing the land width, the amount of inlet air increased due to the widening of the channel width and as the results shown in Figure 13.19, by reducing the width of the land and increasing the width of the channel (reducing their width ratio), a larger current density was obtained.

Molaeimanesh and Akbari (2014) in another study investigated the dynamics of water droplets during removal from a PEMFC using the LBM [49]. A two-dimensional geometry was considered as the computational domain that was a part of a cross-flow fuel cell, which includes half of the

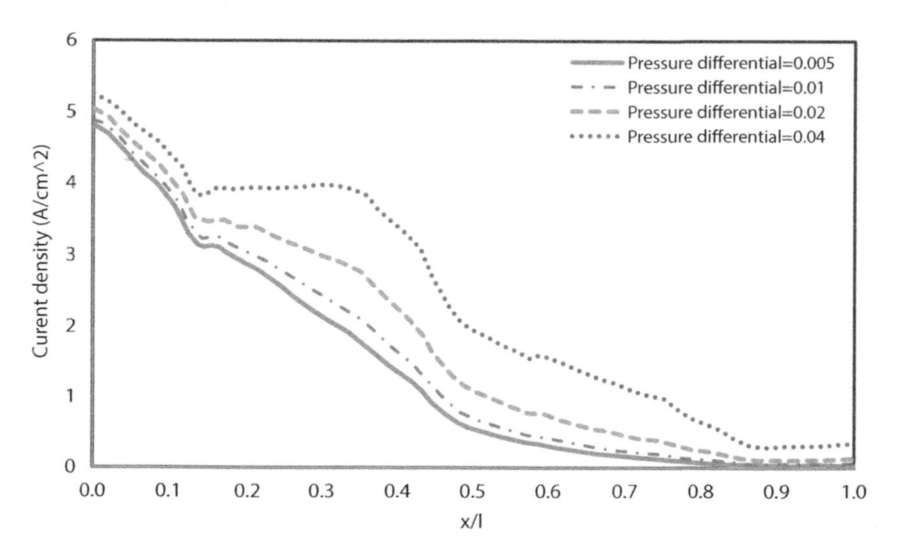

Figure 13.18 The effect of difference between inlet pressure and outlet pressure on the current density and the layer of catalyst. Modified after Molaeimanesh and Akbari (2015) [48].

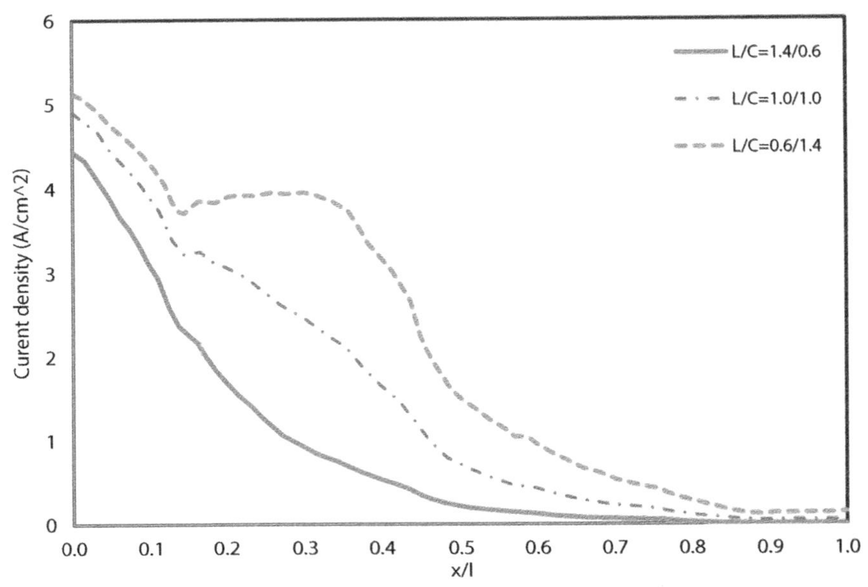

Figure 13.19 The effects of different ratios of land width to channel width (L/C) on the current density and the layer of catalyst. Modified after Molaeimanesh and Akbari (2015) [48].

gas channel inlet, half of the gas channel outlet, land, and the GDL, while the structure of GDL was adapted from Chen *et al.* (2012) [20].

A semicircle with a radius of 100 μm was considered as a water droplet at two cases, one attached to land and other attached to microporous layer as the most common places for water droplets to form. When the droplet was stuck to microporous layer, by assuming uniform wettability of the GDL, solid surfaces of carbon fibers had the same contact angle.

In the process of removing water from the hydrophilic GDL where the contact angle is 80°, it was observed that a small droplet separated from the initial droplet and adheres to the carbon fibers. Then, due to its hydrophilic properties, the droplet spreads on the surface of carbon fiber, so that the fluid flow can move it. Over time, the initial droplet adheres to series of carbon fibers and cannot move anymore. If the GDL has hydrophobic characteristics (contact angle 150°), then the droplet attached to microporous layer breaks into smaller droplets, but due to the hydrophobic characteristics of the GDL, it does not adhere to carbon fibers and moves with the fluid flow and comes out of the GDL. In addition, according to Figure 13.20, it can be concluded that, by increasing the contact angle, the process of removing water from the GDL improves.

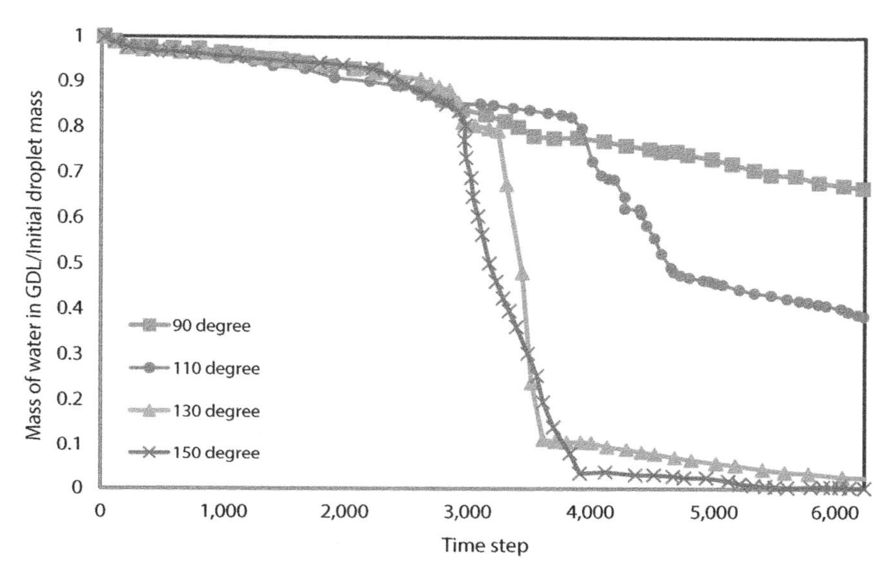

Figure 13.20 The effect of different contact angles on water removing from gas diffusion layer with uniform wettability. Modified after Molaeimanesh and Akbari (2014) [48].

Moreover, it can be stated that, when the water droplets attached to the land, depending on the GDL hydrophilic or hydrophobic characteristics, like water droplets attached to the microporous layer, the droplet can adhere to the carbon fibers in the GDL and remain, or get out of it, during water removal process.

13.4 Conclusions

In recent years, with the advancement of technology and the development of very modern computers, mathematical modeling and simulation of processes and equipment have received much attention due to low cost and low time consuming for optimization or scale up of the processes. In this chapter, several CFD simulations and mathematical modelings of PEMFCs were reviewed. Researchers applied different software and codes to simulate or modeling of PEMFCs such as ANSYS, COMSOL, lattice Boltzmann, MATLAB codes, and OpenFOAM. In a detailed PEMFC model, the electrochemistry and thermodynamics are coupled with mass, momentum, and energy transport in porous media.

The results of published studies reveal that the CFD is a powerful tool for prediction of pressure drop and the flows of oxygen and hydrogen inside PEMFCs. The impacts of operating conditions like the inlet gas RH, cell geometry, and physical parameters on PEMFC performance are qualitatively and quantitatively better predicted using CFD tools.

References

1. Grove, W.R., XXIV. On voltaic series and the combination of gases by platinum. *Lond. Edinb. Dublin Philos. Mag. J. Sci.*, 14, 127, 1839.
2. Steele, B.C. and Heinzel, A., Materials for fuel-cell technologies, in: *Materials for sustainable energy: a collection of peer-reviewed research and review articles from nature publishing group*, vol. 224, 2011.
3. Wee, J.H., Applications of proton exchange membrane fuel cell systems. *Renew. Sust. Energy Rev.*, 11, 1720, 2007.
4. Siegel, C., Review of computational heat and mass transfer modeling in polymer-electrolyte-membrane (PEM) fuel cells. *Energy.*, 33, 1331–1352, 2008.
5. Peavey, M.A., *Fuel from water: energy independence with hydrogen*, 11th, Louisville, KY:Merit Inc, Louisville, KY, USA, 2003.
6. Wilberforce, T., El-, Hassan, Z., Khatib, F.N., Al Makky, A., Mooney, J., Development of Bi-polar plate design of PEM fuel cell using CFD techniques. *Int. J. Hydrogen Energy.*, 23, 25663, 2017.
7. Carton, J.G. and Olabi, A.G., Design of experiment study of the parameters that affect performance of three flow plate configurations of a proton exchange membrane fuel cell. *Energy*, 35, 2796, 2010.
8. Carton, J.G. and Olabi, A.G., Representative model and flow characteristics of open pore cellular foam and potential use in proton exchange membrane fuel cells. *Int. J. Hydrog. Energy.*, 40, 5726, 2015.
9. Maher, A.R. and Al-Baghdadi, S., A CFD study of hydrothermal stresses distribution in PEM fuel cell during regular cell operation. *Renew Energy.*, 34, 674—82, 2009.
10. Sunden, B. and Faghri, M., *Transport phenomena in fuel cells—series: developments in heat transfer*, WIT Press, Southampton, 2005.
11. Hashemi, F., Rowshanzamir, S., Rezakazemi, M., CFD simulation of PEM fuel cell performance: effect of straight and serpentine flow fields. *Math. Comput. Model.*, 55, 1540, 2012.
12. Haghayegh, M., Eikani, M.H., Rowshanzamir, S., Modeling and simulation of a proton exchange membrane fuel cell using computational fluid dynamics. *Int. J. Hydrogen Energy.*, 42, 21944, 2017.
13. Bao, C. and Bessler, W.G., A computationally efficient steady-state electrode-level and 1D+ 1D cell-level fuel cell model. *J. Power Sources.*, 210, 67–80. 2012.

14. Mazumder, S. and Cole, J.V., Rigorous 3-D mathematical modeling of PEM fuel cells. *J. Electrochem. Soc*, 150, A1503, 2003.
15. Kim, S.H. and Pitsch, H., Reconstruction and effective transport properties of the catalyst layer in PEM fuel cells. *J. Electrochem. Soc*, 156, 673, 2009.
16. Jiao, K. and Li, X., A Three-Dimensional Non-isothermal Model of High Temperature Proton Exchange Membrane Fuel Cells with Phosphoric Acid Doped Polybenzimidazole Membranes. *Fuel Cells*, 10, 351, 2010.
17. Meng, H. and Ruan, B., Numerical studies of cold-start phenomena in PEM fuel cells: A review. *Int. J. Energy Res.*, 35, 2, 2011.
18. Yao, K.Z., Karan, K., McAuley, K.B., Oosthuizen, P., Peppley, B., Xie, T., A review of mathematical models for hydrogen and direct methanol polymer electrolyte membrane fuel cells. *Fuel Cells*, 4, 329, 2004.
19. Bıyıkoglu, A., Review of proton exchange membrane fuel cell models. *Int. J. Hydrog. Energy*, 30, 1181–1212, 11812005.
20. Chen, L., Luan, H.B., He, Y.L., Tao, W.Q., Pore-scale flow and mass transport in gas diffusion layer of proton exchange membrane fuel cell with interdigitated flow fields. *Int. J. Therm. Sci.*, 51, 132, 2012.
21. Atyabi, A. and Afshari, E., A numerical multiphase CFD simulation for PEMFC with parallel sinusoidal flow fields. *J. Therm. Anal. Calorim.*, 135, 1823, 2019.
22. Toghyan, S., MoradiNafchi, F., Afshari, E., Hasanpour, K., Baniasadi, E., Atyabi, S.A., Thermal and electrochemical performance analysis of a proton exchange membrane fuel cell under assembly pressure on gas diffusion layer. *Int. J. Hydrogen Energy.*, 43, 4534, 2018.
23. Mazumder, S. and Cole, J.V., Rigorous 3-D mathematical modeling of PEM fuel cells. *J. Electrochem. Soc*, 150, A1510, 2003.
24. Afshari, E. and Jazayeri, S.A., Effects of the cell thermal behavior and water phase change on a proton exchange membrane fuel cell performance. *Energy Convers. Manage.*, 51, 655, 2010.
25. Hao, L. and Cheng, P., Lattice-Boltzmann simulations of anisotropic permeabilities in carbon paper gas diffusion layers. *J. Power Sources.*, 186, 104, 2009.
26. Mann, R.F., Amphlett, J.C., Peppley, B.A., Thurgood, C.P., Application of Butler–Volmer equations in the modeling of activation polarization for PEM fuel cells. *J. Power Sources.*, 161, 775, 2006.
27. Schmittinger, W. and Vahidi, A., A review of the main parameters influencing long-term performance and durability of PEM fuel cells. *J. Power Sources*, 15, 180, 1, 2008.
28. Secanell, M., Wishart, J., Dobson, P., Computational design and optimization of fuel cells and fuel cell systems: a review. *J. Power Sources.*, 196, 3690, 2011.
29. Bao, C. and Bessler, W.G., Two-dimensional modeling of a polymer electrolyte membrane fuel cell with long flow channel. Part II. Physics-based electrochemical impedance analysis. *J. Power Sources.*, 278, 675, 2015.

30. Bednarek, T. and Tsotridis, G., Issues associated with modeling of proton exchange membrane fuel cell by computational fluid dynamics. *J. Power Sources.*, 343, 550, 2017.
31. Chen, H., Guo, H., Ye, F., Ma, C.H.F., A numerical study of baffle height and location effects on mass transfer of proton exchange membrane fuel cells with orientated-type flow channels. *Int. J. Hydrog. Energy*, 4, 7528–7545 2021.
32. Chen, H. *et al.*, A numerical study of orientated-type flow channels with porous-blocked baffles of proton exchange membrane fuel cells. *Int. J. Hydrog. Energy*, 2020.
33. Selvaraj, A.S. and Rajagopal, T.K.R., Effect of flow fields and humidification of reactant and oxidant on the performance of scaled-up PEM-FC using CFD code. *Int. J. Energy Res.*, 1, 7254–7274, 2019.
34. Ferng, Y.M., Su, A., Lu, S.M., Experiment and simulation investigations for effects of flow channel patterns on the PEMFC performance. *Int. J. Energy Res.*, 32, 12, 2008.
35. Inoue, G., Matsukuma, Y., Minemoto, M., Effect of gas channel depth on current density distribution of polymer electrolyte fuel cell by numerical analysis including gas flow through gas diffusion layer. *J. Power Sources.*, 157, 36, 2006.
36. Sun, W., Peppley, B.A., Karan, K., An improved two-dimensional agglomerate cathode model to study the influence of catalyst layer structural parameters. *Electrochim. Acta*, 50, 3359–3374, 2005.
37. Khanzaee, A. and Ghazikhani, M., Numerical simulation and experimental comparison of channel geometry on performance of a PEM fuel cell. *Arab J. Sci. Eng.*, 37, 2297, 2012.
38. Carcadea, E., Varlam, M., Ingham, D.B. *et al.*, The effects of cathode flow channel size and operating conditions on PEM fuel performance: A CFD modelling study and experimental demonstration. *Int. J. Energy Res.*, 1, 2789–2804, 2018.
39. Kone, J.P., Zhang, X., Yan, Y., Hu, G., Ahmadi, G., CFD modeling and simulation of PEM fuel cell using OpenFOAM. *Energy Procedia.*, 145, 6469, 2018.
40. Yuan, W., Tang, Y., Pan, M., Li, Z., Tang, B., Model prediction of effects of operating parameters on protonexchange membrane fuel cell performance. *Renew. Energy.*, 35, 656, 2010.
41. Niu, Z., Wang, Y., Jiao, K., Wu, J., Two-phase flow dynamics in the gas diffusion layer of proton exchange membrane fuel cells: volume of fluid modeling and comparison with experiment. *J. Electrochem. Soc*, 165, F613, 2018.
42. Zenyuk, I.V., Parkinson, D.Y., Hwang, G., Weber, A.Z., Probing water distribution in compressed fuel-cell gas-diffusion layers using X-ray computed tomography. *Electrochem. Commun.*, 53, 27, 2015.
43. Ito, H., Abe, K., Ishida, M., Nakano, A., Maeda, T., Munakata, T., Nakajima, H., Kitahara, T., Effect of through-plane distribution of polytetrafluoroethylene

in carbon paper on in-plane gas permeability. *J. Power Sources.*, 248, 822, 2014.

44. Chen, L., Feng, Y.L., Song, C.X., Chen, L., He, Y.L., Tao, W.Q., Multi-scale modeling of proton exchange membrane fuel cell by coupling finite volume method and lattice-Boltzmann method. *Int. J. Heat Mass Transfer.*, 63, 268, 2013.
45. Wang, G., Mukherjee, P.P., Wang, C.Y., Direct numerical simulation (DNS) modeling of PEFC electrodes: Part I. Regular microstructure. *Electrochim. Acta*, 51, 3139, 2006.
46. Zou, Q. and He, X., On pressure and velocity boundary conditions for the lattice-Boltzmann BGK model. *Phys. Fluids.*, 9, 1591, 1997.
47. Kang, Q., Lichtner, P.C., Zhang, D., An improved lattice-Boltzmann model for multicomponent reactive transport in porous media at the pore scale. *Water Resour. Res.*, 43, 2007.
48. Molaeimanesh, G. and Akbari, M.H., Water droplet dynamic behavior during removal from a proton exchange membrane fuel cell gas diffusion layer by Lattice-Boltzmann method. *Korean J. Chem. Eng.*, 31, 598, 2014.
48. He, W., Yi, J.S., Nguyen, T., Two-phase flow model of the cathode of PEM fuel cells using interdigitated flow fields. *AIChE J.*, 46, 20532000.
49. Molaeimanesh, G.R. and Akbari, M.H., A pore-scale model for the cathode electrode of a proton exchange membrane fuel cell by lattice-Boltzmann method. *Korean J. Chem. Eng.*, 32, 397, 2015.

Index

Also of Interest

Check out these other forthcoming and published titles from Scrivener Publishing

Books by the same editor from Wiley-Scrivener

MATERIALS FOR HYDROGEN PRODUCTION, CONVERSION, AND STORAGE, edited by Inamuddin, Tariq Altalhi, Sayed Mohammed Adnan, and Mohammed A. Amin, ISBN: 9781119829348. Edited by one of the most well-respected and prolific engineers in the world and his team, this book provides a comprehensive overview of hydrogen production, conversion, and storage, offering the scientific literature a comprehensive coverage of this important fuel.

FUNDAMENTALS OF SOLAR CELL DESIGN, edited by Inamuddin, Mohd Imran Ahamed, Rajender Boddula, and Mashallah Rezakazemi, ISBN: 9781119724704. Edited by one of the most well-respected and prolific engineers in the world and his team, this book provides a comprehensive overview of solar cells and explores the history of evolution and present scenarios of solar cell design, classification, properties, various semiconductor materials, thin films, wafer-scale, transparent solar cells, and other fundamentals of solar cell design.

Biofuel Cells, Edited by Inamuddin, Mohd Imran Ahamed, Rajender Boddula, and Mashallah Rezakazemi, ISBN: 9781119724698. This book covers the most recent developments and offers a detailed overview of fundamentals, principles, mechanisms, properties, optimizing parameters, analytical characterization tools, various types of biofuel cells, edited by one of the most well-respected and prolific engineers in the world and his team.

Biodiesel Technology and Applications, Edited by Inamuddin, Mohd Imran Ahamed, Rajender Boddula, and Mashallah Rezakazemi, ISBN: 9781119724643. This outstanding new volume provides a comprehensive overview on biodiesel technologies, covering a broad range of topics and practical applications, edited by one of the most well-respected and prolific engineers in the world and his team.

Applied Water Science Volume 1: Fundamentals and Applications, Edited by Inamuddin, Mohd Imran Ahamed, Rajender Boddula and Tauseef Ahmad Rangreez, ISBN: 9781119724766. Edited by one of the most well-respected and prolific engineers in the world and his team, this is the first volume in a two-volume set that is the most thorough, up-to-date, and comprehensive volume on applied water science available today.

Applied Water Science Volume 2: Remediation Technologies, Edited by Inamuddin, Mohd Imran Ahamed, Rajender Boddula and Tauseef Ahmad Rangreez, ISBN: 9781119724735. The second volume in a new two-volume set on applied water science, this book provides understanding, occurrence, identification, toxic effects and control of water pollutants in aquatic environment using green chemistry protocols.

Potassium-Ion Batteries: Materials and Applications, Edited by Inamuddin, Rajender Boddula, and Abdullah M. Asiri, ISBN: 9781119661399. Edited by one of the most well-respected and prolific engineers in the world and his team, this is the most thorough, up-to-date, and comprehensive volume on potassium-ion batteries available today.

Rechargeable Batteries: History, Progress, and Applications, edited by Rajender Boddula, Inamuddin, Ramyakrishna Pothu, and Abdullah M. Asiri, ISBN: 9781119661191. Edited by one of the most well-respected and prolific engineers in the world and his team, this is the most thorough, up-to-date, and comprehensive volume on rechargeable batteries available today.

Printed and bound by CPI Group (UK) Ltd, Croydon, CR0 4YY

27/02/2023

03195034-0003